Reliability and Safety Analyses under Fuzziness

Studies in Fuzziness

Editor-in-chief
Prof. Janusz Kacprzyk
Systems Research Institute
Polish Academy of Siences
ul. Newelska 6
01-447 Warsaw, POLAND
E-mail: kacprzyk@ibspan.waw.pl

Vol. 1. J. Kacprzyk and M. Fedrizzi (Eds.)
Fuzzy Regression Analysis, 1992
ISBN 3-7908-0591-2 (ISBN 83-85262-07-5)

Vol. 2 A. M. Delgado et al. (Eds.)
Fuzzy Optimization, 1994
ISBN 3-7908-0749-4

Vol. 3. A. Geyer-Schulz
Fuzzy Rule-Based Expert Systems and Genetic Machine Learning, 1995
ISBN 3-7908-0830-X

Takehisa Onisawa · Janusz Kacprzyk
Editors

Reliability and Safety Analyses under Fuzziness

With 106 Figures

Springer-Verlag Berlin Heidelberg GmbH

Professor Takehisa Onisawa
Institute of Engineering Mechanics
University of Tsukuba
1-1-1, Tennodai
Tsukuba 305, Japan

Professor Janusz Kacprzyk
Systems Research Institute
Polish Academy of Sciences
ul. Newelska 6
01-447 Warsaw, Poland

CIP-Titelaufnahme der Deutschen Bibliothek
Reliability and safety analyses under fuzziness / Takehisa Onisawa; Janusz Kacprzyk, ed.

(Studies in fuzziness; Vol. 4)

DOI 10.1007/978-3-7908-1898-7

NE: Onisawa, Takehisa [Hrsg.]; GT

Originally published by Physica-Verlag Heidelberg in 1995
MyCopy version of the original edition 1995

88/2202-543210 - Printed on acid-free paper
www.springer.com/mycopy

FOREWORD

In recent years fuzzy sets theory has been playing a remarkable role in many diverse areas of science and technology, notably in control and knowledge-based systems. Reliability and safety analyses of systems is also one of the most important fields in which fuzzy sets theory may find applications. The collection of contributions to this volume represents numerous important developments in reliability and safety analyses under fuzziness. It is very exciting to learn what has been successfully achieved in this area by applying fuzzy theory.

Fuzzy theory has certainly brought about a new scope in the world of probabilities. Obviously probability theory has been and still is considered to be a dominant tool to deal with reliability and safety issues. Hence fuzzy theory researchers are faced with a lot of criticism from the probabilists in this area.

However, as a matter of fact, fuzzy theory, for example possibility theory, is compatible with probability theory. What we have shown so far is that probability theory needs not be replaced by fuzzy theory, but rather that the former works much better in applications if it is combined with the latter. In fact, it is said that there are two essential uncertainties in the field of reliability and safety analyses: one is a probabilistic uncertainty which is more relevant for mechanical systems and the natural environment, and the other is fuzziness (imprecision) caused by the existence of human beings in systems. The classical probability theory alone is therefore not sufficient to deal with uncertainties in humanistic system.

In such a context this collection of works will put a milestone in the arguments of probability theory and fuzzy theory. This volume covers fault analysis, life time analysis, reliability, quality control, safety analysis and risk analysis. The works of such a wide range were collected by two editors, Professor Takehisa Onisawa and Professor Janusz Kacprzyk. The editors have played a major role in the development of these important branches of fuzzy theory. Thus I have no doubt of the success of this volume.

Tokyo, June, 1994 Michio Sugeno

Table of Contents

1

INTRODUCTORY SECTIONS

FOUNDATIONS OF RELIABILITY AND SAFETY

Janusz Karpiński, Włodzimierz Lewin and Jerzy Rudnicki
Polish Academy of Sciences, Systems Research Institute
Newelska 6, 01-447 Warszawa, Poland

Abstract: Basic notions and ideas used in reliability and safety analysis are presented. The outline of fundamental methods of analysis is given.

Keywords: reliability, safety, statistical methods.

1. PROLOGUE: THE INDUSTRIAL NEED

All man-made technical items are imperfect. The source of imperfections can be as follows:

a) some important facts has been neglected while designing, concerning either the operation of an item or the environment in which an item operates;

b) some defects have been introduced into an item when producing it;

c) the item has been made from the materials containing some weaknesses.

These imperfections can lead to a failure of the item. A natural item deterioration as ,example, metal fatigue or corrosion, can be another cause of failure. Sometimes, failure is caused by mistakes of the operator or service personnel. It is obvious that in practice item failure is mostly caused simultaneously by several factors mentioned above.

Failures are undesirable, or even dangerous, from two points of view. First, they have *economic consequences*, not only because of necessary repairs after component failure but, first of all, because of other failure effects such as a loss of production. Secondly, failures are dangerous from *safety of human beings* point of view in such systems as nuclear power plants, aircraft or medical instrumentation. Safety issues have became increasingly important in the last decade, especially in connection with public discussions concerning nuclear power and environmental pollution.

The above facts constitute an incentive and need for industry to perform systematic research for the identification and possible elimination of failure causes, quantification of failure occurrences and for the reduction of failure consequences. These studies should be performed parallel to the design process in order to attain a final product in the most economic way. These needs created *reliability theory* which has became an important tool in hands of scientists and *reliability engineers*.

Reliability theory and engineering have grown out of the demands of modern technology, particularly after the World War II experiences with complex military systems. Early problems attempted were machine maintenance and system replacements. Now the reliability approach is used in majority of real technical problems. If the purpose of the study is to determine safety parameters, it is necessary to consider, in addition to equipment failure, the possibility of damage by (or to) the system. If this phase of the safety study suggests that there could be system failures, then a *risk study* is done to determine the consequence of the failure in terms of possible damage to property or people.

2. INTRODUCTION

Reliability of an item is its ability to perform a required function under stated conditions for a stated period of time. Respectively, the failure of an item is the termination of the ability of an item to perform a required function. It means that we can analyze reliability of an item and its failure only in connection with the task it should execute.

Execution of the task consists of a correct compliance of specified function in specified conditions and in a determined time interval. If t denotes work time of the item, ϕ - the function performed by the item and κ - working conditions, then the task to be fulfill can be described by the ordered 3-tuple (t,Φ,κ). Knowing the function ϕ one can fix such a set of requirements ω_Φ for substantial attributes of the item in a given application the compliance of which is a necessary and sufficient condition for correctly fulfilling the function ϕ by the considered item. If we assume that the criterion of a correct performance of the function ϕ has only two values, i.e. the function ϕ is fully performed or it is not completely executed ("all or nothing"), then each physical state of an item conforming the requirements ω_ϕ is called *a good performance state* and an item being in such a state is called *a functioning item*. On the other hand, each physical state of an item characterized by the non-conforming of the requirements is called *a failed state* and an item being in this state is *a failed item*. Such "all or nothing" items are called *two-state items in the sense of reliability*. In the present paper such items are considered.

From the above considerations it follows that for two state items the set of requirements ω_ϕ is a fixed set of physical good performance states of the item under the given function ϕ and the task of the item to be performed can be described by the 3-tuple (t,ω_Φ,κ). Usually it is assumed that the pair (ω_Φ,κ) is fixed and item reliability is considered as a function of t of its time to failure.

If $A(s,\omega_\Phi,\kappa)$ denotes an event which means that the item is in a good performance state at time s under the given set of requirements ω_ϕ and the given working conditions κ, then the item will perform its task described by the 3-tuple (t,ω_Φ,κ) if the event $A(s,\omega_\Phi,\kappa)$ occurs for all s from the interval $[0,t]$. The item reliability measure with respect to the 3-tuple (t,ω_Φ,κ) is the probability of performing of this task:

$$R = R(t,\omega_\Phi,\kappa) = \Pr\{A(s,\omega_\Phi,\kappa);\quad 0 \le s \le t)\} \ . \qquad (1)$$

If by R_0 we denote a so called *initial item reliability*, i.e. probability that at the moment of initiating the task (s=0) the item is functioning, where

$$R_0 = \Pr\{A(0,\omega_\Phi,\kappa)\} \ , \qquad (2)$$

then the previous formula can be written in the form:

$$R = R_0 R(t) \ , \qquad (3)$$

where

$$R(t) = \Pr\{A(s,\omega_\Phi,\kappa);0 \le s \le t|A(0,\omega_\Phi,\kappa)\} \quad (4)$$

is the probability that the item being in a good performance state at the beginning of the task will perform this task. The function $R(t)$ is called *a reliability function* or *a survival function*.

The initial reliability $R_0 = R_0(\omega_\Phi,\kappa)$ and the reliability function $R(t) = R(t,\omega_\Phi,\kappa)$ take on values in the interval [0,1]. Moreover, the function $R(t)$ is nondecreasing, i.e.

$$R(t_1) \ge R(t_2) , \quad \text{for} \quad t_1 \le t_2, \quad (5)$$

and satisfies the equalities

$$R(0) = 1 , \qquad R(\infty) = 0 . \quad (6)$$

It is obvious that at any time the functioning item can change its properties and consequently can become the failed one. Passing the two state item from the good performance state to the failed state is called *a failure*, whereas restoring the item to good performance state is called *a repair* of the item. A repair is *full* if item reliability after the repair is the same as reliability before starting the task. A full repair is also called *a renovation*.

An item which is discarded the first time it ceases to perform satisfactory is called *a nonrepairable item*. However, many items are designed to function in a long time period. It can be achieved by repairing an item after its failure. *A repairable item* is an item which, after failing to perform one or more of its functions satisfactory, can be restored to satisfactory performance by any method, other than replacement of the entire item. *Repair* means operation of restoring the item to the satisfactory state, whereas restoring an item to the good performance state is called *renovation*.

Taking into consideration a quality of repairs the following classification of repairs has been generally accepted (Apostolakis and Chu, 1980; Ascher, 1968; Ascher and Feingold, 1984; Beichelt and Franken, 1983; Foglio and Garriba, 1980, Garriba and others, 1976):

1) *a full repair* or *ideal repair*, named also a *renewal*, when all parameters of the item are restored to such state that it can be treated as a "good-as-new", (for example, replacement is a full repair);
2) *an incomplete repair*, when after the repair the item has got worse parameters than a new one (for example, if only failed components have been replaced and the others still work).

One can mention the following incomplete repairs:

a) *a minimal repair* - an item after the repair has the same values of parameters as at the moment immediately before the failure (for example, some adjustments);
b) *a repair with worsening* - an item after the repair has worse parameters than at the moment immediately before the failure (for example, if during the repairing some wear out processes have taken place in non-failed components or the repair has been done incorrectly).

From the point of view of repair time the following types of repairs can be distinguished:

1) *repairable items with negligible repair time* - when repair time is very short and can be omitted in item reliability analysis;
2) *repairable items with non-zero repair time* - when repair time cannot be omitted in item reliability analysis.

One should bear in mind that acceptance of the repair time as zero is subjective and depends on aims of item reliability analysis and demands made for the item. For example, the repair time equal to 10 minutes be neglected in many cases, whereas in other cases it is essential and cannot be omitted.

Items with zero repair time are a subject of investigation of renewal theory or, more generally, the theory of point processes (Barlow and Proschan, 1965; Beichelt and Franken, 1983; Cox, 1962).

The present paper contains a very general outlook of method related to reliability problems and presents a rather traditional and well known approach. The authors hope that it will be useful in understanding the main ideas and methods of reliability theory and practice.

3. RELIABILITY CHARACTERISTICS OF TWO-STATE ITEMS

3.1. Nonrepairable items

The time interval between the beginning of the task and its failure is called *the time to failure* or *life time*. Denote by τ the item time to failure. It is a random variable of positive values ($\tau > 0$) and distribution depending on physical properties of the item, working conditions κ and a fixed set of requirements ω_Φ.

We will denote the distribution function of the item time to failure τ by $F(t)$. We will also assume that there exists the density $f(t) = dF(t)/dt$ of the item time to failure.

Assume that working conditions κ and the set of requirements ω_Φ of the considered item are given. If at the beginning of execution of the fixed task the item is in a good performance state, its reliability is univocally determined by the reliability function $R(t)$, where

$$R(t) = \Pr\{\tau>t;\ \omega_\Phi\} = 1 - F(t) \ , \qquad (7)$$

since the event $\tau > t$ means that if the item is in a good performance state at the beginning of the work then it functions at any time instant s from the interval $[0,t]$. In many cases while analyzing the nonrepairable items reliability, it is convenient to use other characteristics. It is the case when there are difficulties with determining the reliability function $R(t)$ or distribution function $F(t)$, whereas the other measures contain sufficient information making it possible to perform the needed reliability analyses. Furthermore, even if there is no problems with determining the reliability function, sometimes it is more convenient to use other reliability characteristics.

Differentiability of the distribution function $F(t)$ implies that the reliability function $R(t)$ is differentiable and

$$\frac{dR(t)}{dt} = - f(t) \ , \qquad (8)$$

where $f(t)$ is a *failure density function*. The expression $f(t)\Delta t + o(\Delta t)$ is equal to the probability that the item will fail in the time interval $(t,t+\Delta t)$.

An important reliability measure of nonrepairable item is *the mean time to failure* (**MTTF**): $T = E[\tau]$. Under a given pair (ω_Φ, κ) it is specified by the formula

$$T = E[\tau] = \int_{\infty}^{\infty} tf(t)dt = \int_{0}^{\infty} R(t)dt = \int_{0}^{\infty} [1-F(t)]dt. \qquad (9)$$

A failure intensity function - called also a *hazard rate function* or *failure rate function* - is the next important reliability characteristic of nonrepairable items. This function is usually denoted by $\lambda(t)$ and is a conditional density of the item time to failure at the instant t under the condition that the item has not failed till t. It can be described by the formula:

$$\lambda(t) = \lim_{\Delta t \longrightarrow 0} \frac{\Pr\{t<\tau\le t+\Delta t \mid \tau > t\}}{\Delta t} =$$

$$= \lim_{\Delta t \longrightarrow 0} \frac{\Pr\{t<\tau\le t+\Delta t\}}{\Pr\{\tau>t\}\Delta t} = \frac{f(t)}{R(t)} \qquad (10)$$

Notice that $\lambda(t)$ can be given as

$$\lambda(t) = \frac{d\,ln R(t)}{dt}. \qquad (11)$$

3.2. Repairable items

The two-state repairable item is usually modeled as a binary stochastic process $\{Z(t): t\in[0,+\infty)\}$ defined on a probabilistic space $(\Omega, \mathfrak{M}, \Pr)$, with values in $\{0,1\}$:

$$Z(t)=\begin{cases} 1 \text{ if at instant } t \text{ the item is in a good state,} \\ 0 \text{ if at instant } t \text{ the item is failed.} \end{cases}$$

Let S_m denote the time instant of the m-*th* succeeding item failure, whereas S_m^* - the time instant of termination of the *m-th* succeeding repair, and denote by $U_1, U_2, \ldots, U_m, \ldots$ the succeeding intervals of functioning (staying in the good performance state), while by $D_1, D_2, \ldots, D_m, \ldots$ - the succeeding intervals of the repairs (staying in the failed state):

$$U_m = S_m - S^*_{m-1} \quad , \quad D_m = S^*_m - S_m, \quad m \geq 1.$$

From the above assumptions it follows that $U_1, U_2, \ldots$ and $D_1, D_2, \ldots$ are random variables with finite values, i.e. the times of staying in the good performance state as well as the times of staying in failed state are finite and strictly positive.

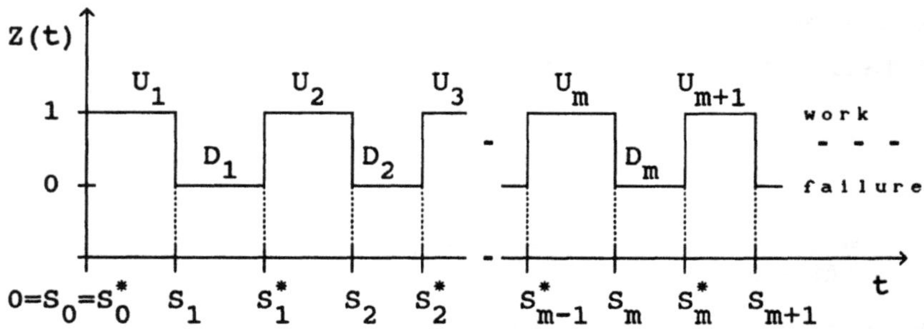

3.2.1. Rezation of the pro {(t)} for a repairable system

{Z(t)} is called *a performance process*.

Below several more important reliability characteristics for repairable items are listed (Barlow and Proschan, 1965; Beichelt and Franken, 1983; Henley and Kumamoto, 1981).

1. *Reliability function* R(t):

$$R(t) = \Pr\{\ Z(s) = 1 \ \textit{for all}\ s \in [0,t]\ \} = \Pr\{\ U_1 > t\ \} \tag{12}$$

- probability that an item will not fail in the period [0,t]. This characteristic is also called *reliability at* t or *survival probability*.

2. *Unreliability function* F(t):

$$F(t) = \Pr\{ Z(s) = 0 \textit{ for some } s \in [0,t] \} =$$

$$= \Pr\{ U_1 \le t \} = 1 - R(t) \qquad (13)$$

- probability of item failure until the time instant t. It is also called *unreliability at* t, *failure probability or cumulative distribution function.*

3. *Mean Time to First Failure* (MTFF)

$$\text{MTFF} = T = E[U_1] = \int_0^\infty R(t)dt = \int_0^\infty t dF(t) \qquad (14)$$

4. *Instant ability* $K_g(t)$

$$K_g(t) = \Pr\{ Z(t) = 1 \} \qquad (15)$$

- probability that at the time instant t an item will be in the good performance state . In case of nonrepairable items $K_g(t) = R(t)$. In general, however, $K_g(t) \ge R(t)$

Similarly, the instant unavailability is defined as

$$K_n(t) = \Pr\{ Z(t) = 0 \} = 1 - K_g(t) \qquad (16)$$

5. *Mean number of failures* W(t) *in a time interval* [0,t]

$$W(t) = E[N_o(t)] = \sum_{n=1}^{\infty} F_{S_n}(t), \qquad (17)$$

where $F_{S_n}(t) = \Pr\{ S_n \le t \}$.

6. *Mean number of repairs* V(t) *in a time interval* [0,t]:

$$V(t) = E[N_1(t)] = \sum_{n=1}^{\infty} F_{S_n^*}(t), \qquad (18)$$

where $F_{S_n^*}(t) = Pr\{ S_n^* \leq t \}$.

7. *Failure frequency* w(t) *at* t *(unconditional failure rate)*

$$w(t) = \frac{d}{dt}W(t) = \lim_{h\to 0} (W(t+h)-W(t))/h \qquad (19)$$

- mean number of item failures in time unit.

8. *Repair frequency* v(t) *at* t *(unconditional repair rate)*

$$v(t) = \frac{d}{dt}V(t) = \lim_{h\to 0} (V(t+h)-V(t))/h \qquad (20)$$

- mean number of repairs in time unit.

9. *Conditional failure intensity* $\lambda^*(t)$

$$\lambda^*(t) = \frac{w(t)}{K_g(t)} \qquad (21)$$

- instantaneous probability that the item will fail in time unit after t under condition that it is in the good performance state at the time instant t.

For a small dt the value $\lambda^*(t)dt$ is equal to the probability of failure in the time interval [t,t+dt) under condition that an item is in the good performance state at t. For nonrepairable items $\lambda^*(t) = \frac{f(t)}{R(t)} = \lambda(t)$ is equal to failure intensity. In general, $\lambda^*(t) \neq \lambda(t)$.

For the majority of reliability characteristics described above it is possible to define *stationary (steady-state) characteristics* for $t\to\infty$. Usually, they are defined as limit values of corresponding nonstationary characteristics.

4. TWO-STATE COMPLEX SYSTEMS

4.1. Definitions

In the theory of reliability, *simple* and *complex* items are distinguished. *A simple item* is an item considered - in some cases - as a whole, whereas *a complex item* is an arranged collection of simple items. Usually, a simple item is called a *component* and a complex item is called a *system* or, sometimes, a *complex system*. Notice that a simple item sometimes can be a very complicated technical construction. For instance, a TV is rather complicated device but in reliability analysis it can be treated - depending on the purpose of analysis - as a component or as a system. If we are interested in reliability analysis of a TV set without penetrating its structure, then it is useful to assume that the TV set is a simple item. Otherwise, if we want to analyze mutual connections between the TV set's failure and failures of its separate parts, then it should be treated as a system.

In reliability analysis a general, but simultaneously a very simple definition of a system is used: **a system is a collection of two or more parts (components) which is designed to perform one or more functions. Mutual connections between system components defining dependence of system failure upon failures of sets of components is called a** ***reliability structure*** **of the system.**

Consider a system with $n \geq 1$ components $\mathbb{C} = \{c_1, c_2, \dots, c_n\}$. Define binary indicators: X_i, $i = 1, \dots, n$ for the components, and φ for the system. An indicator is equal 1 if an item (component or system) is in a good performance state and 0 if it is failed.

A binary vector $\underline{X} = (X_1,X_2,...,X_n)$ describes the state of all components and is called *a component state vector* (c.s.v).

Denote by $\mathbf{S}$ the set of all c.s.v.'s:

$$\mathbf{S} = \{0,1\}^n = \{\underline{X}: \underline{X} = (X_1,...,X_n), X_i \in \{0,1\}, i=1,...,n\}.$$

It is assumed that the state of the system is univocally determined by the states of its components, i.e. by the vector $\underline{X}$. It means that φ is a function specified on S with values in {0,1}. Therefore, $\varphi(\underline{X}) = \varphi(X_1,X_2,...,X_n)$ is the state of a system corresponding with the states $X_1,X_2,...,X_n$ of its components. Function φ is called *a system structure function*. The pair $<\mathbb{C},\varphi>$ is called *a binary* or *two-state system*.

A component $i \in \mathbb{C}$ is *irrelevant* if function φ doesn't depend on X_i, otherwise it is called *relevant*.

A binary system $<\mathbb{C},\varphi>$ is *monotonic* (precisely, *non-decreasing*) if φ is a non-decreasing function with respect to each variable.

A binary system $< \mathbb{C},\varphi >$ is *non-trivial* if φ is not constant.

Notice that if a system is monotonic and non-trivial then $\varphi(\underline{0})=0$ and $\varphi(\underline{1})=1$.

4.2. Basic structures of two-state systems

In real cases structures of technical systems are a superposition of structures defined bellow ($\mathbb{C} = \{1,2...n\}$).

1. Series structure

The structure of a system $<\mathbb{C},\varphi>$ is series if a failure of any component causes a system failure, i.e. a system is in the good performance state if and only if all its components are in the good performance states (see fig. 4.2.1.):

$$\varphi(\underline{X}) = \min(X_1, X_2, \ldots, X_n) = \prod_{i=1}^{n} X_i. \tag{22}$$

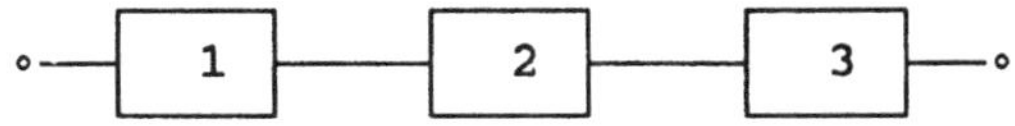

Fig. 4.2.1. Diagram of a series reliability structure

2. *Parallel structure*

The structure of a system <ℂ,φ> is parallel if good performance of at least one component causes the a performance state of a whole system, i.e. the system is failed if and only if all its components are failed (see fig. 4.2.2.):

$$\varphi(\underline{X}) = \max(X_1, X_2, \ldots, X_n) = 1 - \prod_{i=1}^{n} (1 - X_i). \tag{23}$$

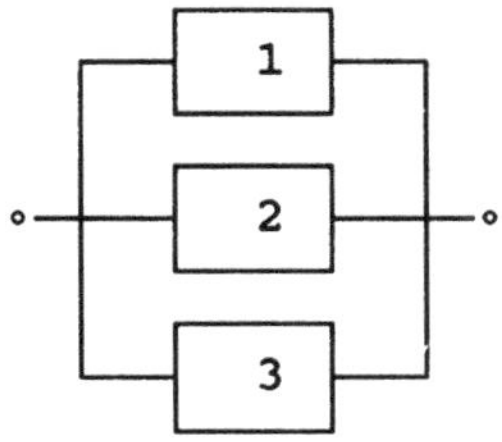

Fig. 4.2.2. Diagram of parallel reliability structure

3. *Structure of the* k-*out*-*of*-n *type*

A system <ℂ,φ> has the k-out-of-n structure (k=1,2,...,n) if it is in the good performance state when at least any k components are in the good performance states:

$$\varphi(\underline{X}) = \begin{cases} 1, & \text{if } \sum_{i=1}^{n} X_i \geq k, \\ 0, & \text{otherwise.} \end{cases} \tag{24}$$

Notice that reliability structure of the system depends on:

a) a functional and constructional structure of the system, i.e. a way of functional and/or construction combination of components;

b) a task that should be performed.

To explain these two kinds of dependencies consider a system of lighting a street. Suppose that street-lamps are uniformly distributed along both sides of the street. Notice that a method of current supply has an effect on reliability structure of such a system. For example, if the left and right sides of the street are supplied independently (in parallel) then reliability structure of this system is in such a case is different than in case of supplying the whole system from one source. On the other hand, reliability structure strongly depends on a task to be performed by the system. Consider two different tasks: first, assume that the system fails to perform if any of street-lamp fails and, secondly, suppose that the lighting system fails if any two neighbouring street-lamps fails. It is easy to verify that reliability structure of the system differs in these two cases.

Reliability structure of complex systems can be analyzed using many methods. Such an analysis is necessary while designing the systems, planning inspections and preventive replacements of system components, fixing spare part to assure uninterrupted functioning of the system etc. On the basis of correctly performed analysis of reliability structure one can improve reliability of the system.

4.3. Fault Tree Analysis

A very useful and popular method for analyzing complex system reliability is so called Fault Tree Analysis (**FTA**). The fault tree itself is a graphical representation of the Boolean failure logic associated with the development of a particular system failure (the **TOP** event) to basic failures (primary events). For example, the TOP event could be the failure of a nuclear power plant guidance control system during its operation with the primary events being the failures of the individual guidance control system components. FTA can be a valuable design tool. It can identify potential accidents in a system design and can help eliminate costly design changes and retrofits. FTA can also be a diagnostic tool. One can predict with it the most likely causes of system failures in the case of system breakdown.

The fault trees are a special cases of decision trees and contain logical gates, (for example, AND, OR, "k-out-of-n", NOR, NAND, NOT) and symbols of top end primary events. The goal of fault tree construction is to model the system conditions that can result in the undesired event. Before construction of the fault tree, the analyst must acquire a thorough understanding of the system. In fact, a system description should be a part of the analysis documentation. The analyst must carefully define the undesired event under consideration, called the "top event".

5. STATISTICAL METHODS IN LIFE TESTING

This paragraph deals mainly with *statistical methods* for solving problems associated with reliability. There are two excellent books: Mann, Schafer and Singpurwalla (1974) and Nelson (1982) that fully describe fundamentals: *data, models, inference* methods and *interpretation* of results. In this paper only examples of some basic ideas of statistical inference in reliability analysis are presented.

The data to which statistical methods are applied in order to estimate reliability measures, are obtained from laboratory or field *life tests*. Life tests are typically applied for prototypes (during the development phase) and manufactured items (during the production and field operation phase). Tested items are subject to stresses and environmental conditions that typify or actually are the intended operating conditions.

The reliability evaluated during design and development phase of item life is usually related to its inherent reliability. This reliability may differ from actually observed reliability because of different operating or/and environmental conditions, different maintenance activity, etc. These influating factors are generally difficult to simulate in a laboratory and to describe using strict, deterministic model. This is the reason why the significant differences are often observed between results of laboratory and field reliability tests. Because of this uncertainty about the mentioned factors the field test is the only way to observe the actual reliability measure but, in practice, there is also a big amount of uncertainty connected with these field conditions. Consequently, random times to failure (for non-repairable items) or successive times between failures (for repairable items) are burdened by this uncertainty.

The next source of uncertainty is a typical for life testing "disorder" in observed data i.e. failure times. The main cause of it is a contradiction between the available test time (usually short) and actual times to failure of tested items (usually long); in most cases it is impossible to collect all failure times (complete data) and censored (incomplete) data are usually registered.

The usual approach to an experimental assessment of item reliability consists in construction of appropriate probabilistic models of an experiment and in applying the statistical inference based on these models. A distribution function such as the Weibull distribution (for non-repairable items) or the Power Law Nonhomogeneous Poisson Process (for repairable items) are good examples of models verified in practice. We will describe these two models more detailly. The point estimation procedures will be shortly described as examples of statistical inference based on the mentioned models.

During life tests the sample of n items from a population of interest is usually observed under the defined stresses. If the life test is terminated at a specified time t_o before all items have failed, such test is called *Type I censoring*. *Type II censoring* occurs when the life test is terminated at the time of a particular r-th failure, $r \le n$. There are essential differences between statistical models in both cases. In Type I censoring the number of failures and all the failure times are random variables, whereas in Type II censoring the number of failures is fixed, and the only random variables are the failure times.

5.1. The Weibull life-time distribution

Recently the Weibull distribution has emerged as the most popular parametric family of failure time distributions. This distribution can be derived either as the asymptotic distribution of the smallest order statistics from a specified probability distribution function or from the failure rate concept.

Let us consider the second approach and use the power function as a very flexible model of failure rate:

$$\lambda(t)=\frac{\beta}{\alpha}\left(\frac{t}{\alpha}\right)^{\beta-1},\alpha,\beta,t>0 \tag{25}$$

It follows from (11) that the two-parameter Weibull distribution function can be written as

$$F(t)=1-\exp\left[-\left(\frac{t}{\alpha}\right)^{\beta}\right] \tag{26}$$

where α is called the *scale* parameter and β is called the shape parameter. The failure rate is decreasing if $\beta<1$, increasing if $\beta>1$ and independent of time for $\beta=1$ (in this case the Weibull distribution reduces to the exponential distribution).

5.2. Estimation

There are several methods of obtaining good estimates of the unknown parameters, α and β. The methods include the iterative solution of the maximum likelihood equations, several types of linear estimation and moment estimators, that has been shown to be less efficient. In the following, maximum likelihood method will be presented.

Suppose that r failure times from a sample of size n were observed. The likelihood function can be expressed for Weibull distribution as

$$L = \frac{n!}{(n-r)!}\prod_{i=1}^{r}\left\{\frac{\beta}{\alpha}\left(\frac{t_i}{\alpha}\right)^{\beta-1}\exp\left[-\left(\frac{t_i}{\alpha}\right)^{\beta}\right]\right\}\left[1-F(t_s)\right]^{n-r} \tag{27}$$

For a specified test termination time $t_s=t_o$ and for the observed time of the r-th failure, at which time testing is terminated, $t_s=t_{(r)}$. The most suitable form of equations, from which the maximum likelihood estimates are obtained is

$$\left[\frac{\sum_{i=1}^{r} t_i^{\beta}\ln t_i+(n-r)t_s^{\beta}\ln t_s}{\sum_{i=1}^{r} t_i^{\beta}+(n-r)t_s^{\beta}}-\frac{1}{\beta}\right]-\frac{1}{r}\sum_{i=1}^{r}\ln t_i = 0$$

$$\frac{1}{r}\left[\sum_{i=1}^{r} t_i^{\beta}+(n-r)t_s^{\beta}\right]-\alpha^{\beta} = 0 \tag{28}$$

The equations (28) are solved iteratively to obtain estimates $\hat{\alpha}$ and $\hat{\beta}$ which make possible the assessment of all reliability characteristics of non-repairable items. For instance, failure rate $\hat{\lambda}(t)$ can be expressed as

$$\hat{\lambda}(t)=\frac{\hat{\beta}}{\hat{\alpha}}\left(\frac{t}{\hat{\alpha}}\right)^{\hat{\beta}-1} \tag{29}$$

The maximum likelihood method has the advantage that it can be applied to life test models in which there are all types of censoring, including the progressive one, in which a portion of survivors (non-failed items) is withdrawn from life test several times during the test.

5.3. Power Law Nonhomogeneous Poisson Process

Systems are mostly *not replaced but repaired* when they fail. This makes a very important difference in statistical modeling of a system life (Ascher and Feingold, 1984; Craw, 1974; Craw, 1990). The usual non-repairable reliability models, such as Weibull one, are not appropriate for repairable systems. Times between failures of such systems are not equally distributed and are dependent in most cases. Consequently, the item failure sequences are modeled by a *stochastic process* with a time-dependent intensity.

A very flexible and practical model for analyzing the failure intensity $\omega(t)$ of repairable systems is a Power Law Nonhomogeneous Poisson Process (PL NHPP) with intensity function

$$\omega(t) = \lambda \beta t^{\beta-1}, \quad t>0, \quad \alpha,\beta>0 \tag{30}$$

where λ,β are model parameters and t is the time from the beginning of functioning of the system.

When system failure intensity does not depend on its age we have a special case of NHPP - Homogeneous Poisson Process (HPP) - equivalent to the widely used Poisson distribution and exponential times between failures.

The parameter β is responsible for a system behaviour:

- for β=1, the failure intensity $\omega(t)$ is time-constant and equal λ, so we have the HPP with stochastically equal (exponential) times between failures;

- for $\beta<1$, $\omega(t)$ is strictly decreasing and the intervals between successive failures are stochastically increasing
- for $\beta>1$, $\omega(t)$ is strictly increasing and the intervals between successive failures are stochastically decreasing.

Because of the mathematical form of (30), which is the same as the failure rate for a Weibull distribution of failure times for non-repairable items some authors call incorrectly (30) the Weibull process.

The PL NHPP has the mean value function

$$E[N(t)]=\lambda t^{\beta} \tag{31}$$

which is the expected number of failures in (0,t), t>0.

The probability that the system fails n times in (0,T) is given by Poisson expression:

$$Pr[N(T)]=n] = \frac{(\lambda T^{\beta})\exp(-\lambda T^{\beta})}{n!} \tag{32}$$

In this paper we will assume that each system under observation is described by the same PL NHPP and is observed in the time interval (0,T). N(T) denotes random number of failures during this time and $0<X_1<X_2<\ldots<X_{N(T)}$ are random successive times of these failures.

In order to estimate a failure intensity of the system, failure data from the k systems operating under the same conditions are collected. In practice there is an uncertainty connected not only with operating or/and environmental conditions, but also with the data themselves, which - for repairable items- are usually more "disorderly". For many reasons an observation may not start at time 0, which is the time of the first operation of the system, but at some later time $S>0$ and it isn't known how many failures occurred before S. For other reasons an observation may not end at time T which is a failure time, but some time after this failure. Four types of censoring are usually distinguished:

A. When the time S is the system failure time we call such observation (data) *failure left censored.*

B. When observation starts at time S which is not a failure time, we call such observation (data) *time left censored.*

C. When observation ends at time T which is a failure time, we call such observation (data) *failure right censored.*

D. When observation ends at time T which is not a failure time, we call such observation (data) *time right censored.*

Suppose now that the k-th system, $k=1,\ldots,K$, is observed continuously from time S_k to time T_k. Let N_k be the number of failures in the period $[S_k, T_k]$ and X_{ik} be the age of this system at the i-th failure occurrence, $i=1,\ldots,N_k$.

In general, the maximum-likelihood estimates (MLE) of the parameters λ and β of the model (30) are (Craw, 1974):

$$\hat{\lambda} = \frac{\sum_{k=1}^{K} N_k}{\sum_{k=1}^{K} (T_k^{\hat{\beta}} - S_k^{\hat{\beta}})} \tag{33a}$$

$$\hat{\beta} = \frac{\sum_{k=1}^{K} N_k}{\hat{\lambda} \sum_{k=1}^{K} (T_k^{\hat{\beta}} \ln T_k - S_k^{\hat{\beta}} \ln S_k) - \sum_{k=1}^{K} \sum_{i=1}^{N_k} \ln X_{ik}} \tag{33b}$$

where $0 \times \ln(0)$ is defined to be 0. The above equations must be solved iteratively.

Once we have estimated λ and β, the MLE of the failure intensity ω is given by:

$$\hat{\omega}(t) = \hat{\lambda}\, \hat{\beta}\, t^{\hat{\beta}-1}, \quad t>0 \tag{34}$$

In some special cases the solution to the equations (33) can be given. For example, when $S_k=0$ and all data are right censored at the same time $T_k= T$ $(k=1,\ldots,K)$ the equations (33) are solved explicitly for $\hat{\lambda}$ and $\hat{\beta}$:

$$\hat{\lambda} = \frac{\sum_{k=1}^{K} N_k}{K T^{\hat{\beta}}} \qquad \hat{\beta} = \frac{\sum_{k=1}^{K} N_k}{\sum_{k=1}^{K} \sum_{i=1}^{N_k} \ln\left(\frac{T}{X_{ik}}\right)} \tag{35}$$

5.4. Classical and Bayesian approach to interval estimation

In Bayes methods one assume that parameters of distribution of interest are random variables Θ. The advantage of Bayes methods is providing a tool to combine previous data, expressed as the prior distribution, with observed data, expressed in density, to obtain better estimates of parameters by using the posterior density.

Let us first consider the classical approach to interval estimation. Suppose that a random sample $(T_1, T_2, ..., T_n)$ of size n is available from a single-parameter probability distribution $f(t,\theta)$. Suppose, further, that the data are adequately described by a statistic $S = g(T_1, T_2, ..., T_n)$, and the probability distribution of S, $f_S(s;\theta)$, depends on θ. This probability distribution is called the sampling distribution of S. The classical approach to constructing the interval is to use $f(s,\theta)$ to determine two numbers s_* and s^* , depending on θ, such that

$$P\ [s_*(\theta) \leq S \leq s^*(\theta)] = 1-\gamma \tag{36}$$

Because we desire a probability statement on the parameter θ and the functions $s_*$and s^* are functions of θ, these functions can be inverted and solved for θ to obtain two new functions, say $h_*(S)$ and $h^*(S)$, such that

$$[h_*(S) \leq \theta \leq h^*(S)] = 1-\gamma \tag{37}$$

In Bayes approach the parameter Θ is considered to be a random variable and assumed to have a prior probability density $g(\theta)$. Applying the Bayes theorem, the posterior density

$$g(\theta|s) = \frac{f(s|\theta)g(\theta)}{f(s)} \tag{38}$$

is well defined at all points where $f(s) \neq 0$. Then, when s has been observed, a probability statement is easily obtained by solving the two equations

$$\int_{-\infty}^{\theta_*} g(\theta|s)d\theta = \frac{\gamma}{2} \qquad \int_{\theta^*}^{\infty} g(\theta|s)d\theta = \frac{\gamma}{2} \tag{39}$$

for θ_* and θ^*, so that

$$P(\theta_* \le \Theta \le \theta^*) = 1-\gamma. \tag{40}$$

The interval (θ_*, θ^*) is a Bayes interval estimate. Equation (40) is a valid probability statement and can be combined with other statements by use of the rules of probability.

REFERENCES

Apostolakis G., and Chu T.L. (1980) The unavailability of system under periodic test and maintenance, Nuclear Technology 50, 5-15.

Ascher H. (1968) Evaluation of repairable system reliability using the "bad-as-old" concept, IEEE Transactions on Reliability R-17, 103-110.

Ascher H., and Feingold H. (1984) Repairable Systems Reliability, Marcel Dekker, New York.

Barlow R.E., and Proschan F. (1965) Mathematical Theory of Reliability, John Wiley & Sons, New York.

Beichelt F., and Franken P. (1983) Zuverlässigkeit und Instandhaltung, VEB Verlag Technik, Berlin.

Cox D.R. (1962) Renewal Theory, Methuen, London.

Craw L.H. (1974) Reliability Analysis for Complex, Repairable Systems, in F.Proschan and R.J.Serfling (Eds.):Reliability and Biometry, SIAM, Philadelphia, pp.379-410,

Craw L.H. (1990) Evaluating the Reliability of Repairable Systems, in: Proceedings Annual Reliability and Maintainability Symposium.

Foglio P.A., and Garriba S. (1980) Reliability analysis of a repairable single unit under general age-dependence, in G. Apostolakis, S. Garriba and S. Volta (Eds.): Synthesis and Analysis Methods for Safety and Reliability Studies, Plenum Press, New York, pp. 251-268.

Garriba S., Reina G., and Volta G. (1976) Availability of repairable units when failure and restoration rates age in real time, IEEE Transactions on Reliability R-25, 88-93.

Henley E.J., and Kumamoto H. (1981) Reliability Engineering and Risk Assessment, Prentice Hall, Englewood Clifs.

Mann N.R., Schafer R.E., and Singpurwalla N.D. (1974) Methods for Statistical Analysis of Reliability and Life Data, John Wiley & Sons, New York.

Nelson W. (1982) Applied Life Data Analysis, John Wiley & Sons, New York.

BRIEF INTRODUCTION TO FUZZY SETS

Mario Fedrizzi* and Janusz Kacprzyk**

* Department of Management and Computer Sciences
University of Trento
Via Inama 5, 38100 Trento, Italy
E-mail: fedrizzi@cs.unitn.it

** Systems Research Institute, Polish Academy of Sciences
ul. Newelska 6, 01-447 Warsaw, Poland
E-mail: kacprzyk@ibspan.waw.pl

Abstract. We present a brief introduction to fuzzy sets theory for the interested readers who has not yet been exposed to this area. Particular emphasis is on basic elements and definitions, and to those which are relevant for the topics covered by this volume. First, the concept of a fuzzy set is presented as a formal means for dealing with the imprecision of meaning, caused mainly by the use of a natural language; such an imprecision is meant to be related to a gradual membership of elements. This is followed by main operations on fuzzy sets, the concepts of cardinality, a fuzzy relation, etc. Then, the concept of a fuzzy number is sketched. Basic operations on fuzzy numbers are dealt with with emphasis on the practically most relevant case of triangular and/or trapezoid fuzzy numbers.

Keywords: fuzzy set, fuzzy relation, fuzzy number.

1. INTRODUCTION

The concept of a *set*, and set theory, are powerful tools in mathematics. Unfortunately, a *sine qua non* condition underlying set theory, i.e. that an element can either belong to a set or not, is often not applicable in real life where many vague terms as "large profit", "high pressure", "moderate temperature", "reliable tools", "safe conditions", etc. are extensively used. Unfortunately, such imprecise descriptions cannot be adequately handled by conventional mathematical tools.

If we wish to maintain the very meaning of imprecise (vague) terms, a crisp differentiation between elements (e.g., pressure values) that are either high or not high may be artificial, and some values may be perceived high to some extent, not fully high and not fully not high.

An attempt to develop a formal apparatus to involve a partial membership in a set was undertaken in the mid-1960's by Zadeh (1965). He introduced the concept of a *fuzzy set* as a collection of objects which might "belong" to it to a degree, from 1 for full belongingness to 0 for full nonbelongingness, through all intermediate values. This was done by employing the concept of a *membership function* assigning to each element of a *universe of discourse* a number from the unit interval to indicate the *intensity* (*grade*) *of belongingness*. The concept of a membership function was evidently an extension of that of a *characteristic function* of a conventional set assigning to the universe of discourse either 0 (nonbelongingness) or 1 (belongingness). Then, basic properties and operations on fuzzy sets were defined by Zadeh (and later by his numerous followers) being essentially extensions (in the above spirit) of their conventional counterparts.

Since its inception, fuzzy sets theory has experienced an unprecedented growth of interest in virtually all fields of science and technology.

The purpose of this paper is to provide an interested reader, who has not yet been exposed to fuzzy sets theory, with a a brief survey of fuzzy sets; the readability and simplicity of exposition will be maintained at the expense of technicalities. We hope that even a novice reader will be able to follow at least most of fuzzy-sets-related arguments in the articles included in this volume. The reader interested more deeply in a particular topic will be referred to an appropriate literature. In principle, we will rather give book references that may be easier to find and provide a more complete coverage, citing only some more important technical articles.

2. FUNDAMENTALS OF FUZZY SETS THEORY

Suppose that $X = \{x\}$ is a *universe of discourse*, i.e. the set of all possible (feasible, relevant,...) elements to be considered with respect to a fuzzy (vague) concept (property). Then a *fuzzy subset* (or a *fuzzy set*, for short) A in X is defined as a set of ordered pairs $\{(x, \mu_A(x))\}$, where $x \in X$ and $\mu_A : X \longrightarrow [0,1]$ is the *membership function* of A; $\mu_A(x) \in [0,1]$ is the *grade of membership* of x in A, from 0 for full nonbelongingness to 1 for full belongingness, through all intermediate values. In some contexts it may be expedient to view the grade of memberships of a particular element as its degree of compatibility with the (vague) concept represented by the fuzzy set. Notice that the degrees of membership are clearly subjective.

Many authors denote $\mu_A(x)$ by $A(x)$. Moreover, a fuzzy set is often equated with its membership function so that both A and $\mu_A(x)$ are often used interchangeably.

Notice that if $[0,1]$ is replaced by $\{0,1\}$, this definition coincides with

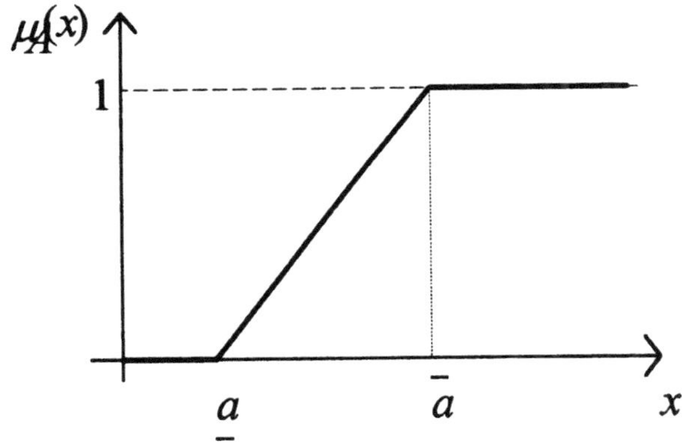

Figure 1: Membership function of a fuzzy set.

the characteristic function based description of an ordinary (nonfuzzy) set. Moreover, the original Zadeh's unit interval is chosen for simplicity, and a similar role may be played by an ordered set, e.g., a lattice (cf. Goguen's concept of an L-fuzzy set that is covered in virtually all books on fuzzy sets, e.g., in Dubois and Prade, 1980).

It is convenient to denote a fuzzy set defined in a finite universe of discourse, say A in $X = \{x_1, \ldots, x_n\}$ as $A = \mu_A(x_1)/x_1 + \cdots + \mu_A(x_n)/x_n$, where "$\mu_A(x_i)/x_i$" (called a singleton) is a pair "grade of membership–element" and "+" is meant in the set-theoretic sense. For instance, if $X = \{1, 2, \ldots, 10\}$, then a fuzzy set "large number" may be given as $A =$ "large number" $= 0.2/6 + 0.5/7 + 0.8/8 + 1/9 + 1/10$ to be meant as: 9 and 10 are surely (to degree 1) "large numbers", 0.8 is a "large number" to degree 0.8, etc. and $1, 2, \ldots, 5$ are surely not "large numbers". Notice that the above degrees of membership are subjective (a "large number" is a subjective concept!) and context-dependent, and – by convention – the singletons with $\mu_A(.) = 0$ are omitted.

In practice it is usually convenient to use a piecewise linear representation of the membership function of a fuzzy set as shown in Fig. 1 since only two values, $\underline{a}$ and $\bar{a}$, are needed.

2.1. Basic properties of fuzzy sets

A fuzzy set A in X is *empty*, $A = \emptyset$, if and only if $\mu_A(x) = 0, \forall x \in X$.

Two fuzzy sets A, B in X are *equal*, $A = B$, iff $\mu_A(x) = \mu_B(x), \forall x \in X$.

A fuzzy set A in X is *included* in (is a *subset* of) a fuzzy set B in X, $A \subseteq B$, if and only if $\mu_A(x) \leq \mu_B(x), \forall x \in X$.

Example 1. Suppose $X = \{1,2,3\}$ and $A = 0.3/1 + 0.5/2 + 1/3$ and $B = 0.4/1 + 0.6/2 + 0.8/3$; then $B \subseteq A$.

An important concept is the *cardinality* of a fuzzy set. If $X = \{x_1, \ldots, x_n\}$, and $A = \mu_A(x_1)/x_1 + \cdots + \mu_A(x_n)/x_n$, then the (nonfuzzy) cardinality of A, the so-called ΣCount (*sigma-count*), is defined as

$$\text{card } A = | A | = \Sigma\text{Count}(A) = \textstyle\sum_{i=1}^{n} \mu_A(x_i) \qquad (1)$$

Example 2. If $X = \{1,2,3,4\}$ and $A = 0.1/1 + 0.4/2 + 0.7/3 + 1/4$, then $\text{card}\, A = 2.2$.

Some other concept of cardinality, mainly defined as a fuzzy number, are also used.

2.2. Basic operations on fuzzy sets

The basic operations are here naturally the complement, intersection and union, as in the conventional set theory.

The *complement* of a fuzzy set A in X, $\neg A$, is defined as

$$\mu_{\neg A}(x) = 1 - \mu_A(x), \forall x \in X \qquad (2)$$

and it corresponds to the negation "not".

The *intersection* of two fuzzy sets A, B in X, $A \cap B$, is defined as

$$\mu_{A \cap B}(x) = \mu_A(x) \wedge \mu_B(x), \forall x \in X \qquad (3)$$

where "$\wedge$" is the minimum, and it corresponds to the connective "and".

The *union* of two fuzzy sets, A, B in X, $A + B$, is defined as

$$\mu_{A \cup B}(x) = \mu_A(x) \vee \mu_B(x), \forall x \in X \qquad (4)$$

where "$\vee$" is the maximum, and it corresponds to the connective "or".

Example 3. If $X = \{1, 2, \ldots, 10\}$, A = "small number"$= 1/1 + 1/2 + 0.8/3 + 0.5/4 + 0.3/5 + 0.1/6$, B = "large number" $= 0.1/5 + 0.2/6 + 0.5/7 + 0.8/8 + 1/9 + 1/10$, then

$$\neg A = \text{"not small number"} = \\ = 0.2/3 + 0.5/4 + 0.7/5 + 0.9/6 + 1/7 + 1/8 + 1/9 + 1/10$$

$$A \cap B = \text{"small number" } \textit{and} \text{ "large number"} = 0.1/5 + 0.1/6$$

$$A \cup B = \text{"small number" } \textit{or} \text{ "large number"} = \\ = 1/1 + 1/2 + 0.8/3 + 0.5/4 + 0.3/5 + 0.2/6 + \\ + 0.5/7 + 0.8/8 + 1/9 + 1/10$$

The above definitions are classic, and have been commonly employed though they are evidently by no means the only ones. For instance, the use of a t-norm for the intersection and an s-norm for the union has often been advocated. They are defined as follows:

- a t-norm is defined as $t : [0,1] \times [0,1] \rightarrow [0,1]$ such that:
  1. $a \, t \, 1 = a$
  2. $a \, t \, b = b \, t \, a$
  3. $a \, t \, b \geq c \, t \, d$ if $a \geq c, b \geq d$
  4. $a \, t \, b \, t \, c = a \, t \, (b \, t \, c) = (a \, t \, b) \, t \, c$

  Some more relevant examples of t-norms are:

  $$\begin{array}{ll} a \wedge b = \min(a,b) - & \text{this is the most popular } t\text{-norm} \\ a \cdot b & \\ 1 - [1 \wedge ((1-a)^p + (1-b)^p)^{1/p}], p \geq 1 & \end{array}$$

- an s-norm (t-conorm) is defined as $s : [0,1] \times [0,1] \rightarrow [0,1]$ such that:
  1. $a \, s \, 0 = a$
  2. as the properties 2–4 for a t-norm

  Some examples of more popular s-norms are:

  $$\begin{array}{ll} a \vee b = \max(a,b) - & \text{this is the most popular } s\text{-norm} \\ a + b - a \cdot b & \\ 1 \wedge (a^p + b^p)^{1/p}, p \geq 1 & \end{array}$$

For other operations on fuzzy sets, see, e.g., Dubois and Prade (1980), Kacprzyk (1983), Zimmermann (1987, 1989, 1991).

2.3. Further properties and related concepts

An *α-cut* (*α-level set*) of a fuzzy set A in X is defined as the ordinary set $A_\alpha \subseteq X$ such that

$$A_\alpha = \{x \in X : \mu_A(x) \geq \alpha\}, \forall \alpha \in [0,1] \tag{5}$$

Example 4. If $A = 1/1 + 0.8/2 + 0.5/3 + 0.1/4$, then $A_{0.1} = \{1,2,3,4\}$, $A_{0.5} = \{1,2,3\}$, $A_{0.8} = \{1,2\}$ and $A_1 = \{1\}$.

The concept of an α-cut of a fuzzy set is crucial for the so-called *decomposition theorem* which states that any fuzzy set A in X may be represented as some (equivalent) operations on conventional sets (subsets of X).

Of fundamental importance is here the so-called *extension principle* which gives a formal apparatus to carry over operations (e.g., arithmetic or algebraic) from sets to fuzzy sets. Namely, if $f : X \to Y$ is a function (operation) and A is a fuzzy set in X, then A induces via f a fuzzy set B in Y given by

$$\mu_B(y) = \begin{cases} \sup_{y=f(x)} \mu_A(x) & \text{if } f^{-1}(x) = \emptyset \\ 0 & \text{otherwise} \end{cases} \tag{6}$$

Example 5. Let $X = \{1, 2, 3, 4\}$, $Y = \{1, 2, \ldots, 6\}$, and $y = x + 2$. If now $A = 0.1/1 + 0.2/2 + 0.7/3 + 1/4$, then $B = 0.1/3 + 0.2/4 + 0.7/5 + 1/6$.

2.4. Fuzzy relations

Fuzzy relations – exemplified by "much larger than", "more or less equal", etc. – are clearly omnipresent in human discourse. Formally, if $X = \{x\}$ and $Y = \{y\}$ are two universes of discourse, then a *fuzzy relation* R is defined as a fuzzy set in the Cartesian product $X \times Y$, characterized by its membership function $\mu_R : X \times Y \to [0, 1]$; $\mu_R(x, y) \in [0, 1]$ reflects the *strength of relation* between $x \in X$ and $y \in Y$.

Example 6. Suppose that $X = \{\text{horse, donkey}\}$ and $Y = \{\text{mule, cow}\}$. The fuzzy relation "similar" may then be defined as

$$\begin{aligned} R = \text{"similar"} \quad &= \quad 0.8/(\text{horse, mule}) + 0.4/(\text{horse, cow}) + \\ &\quad +0.9/(\text{donkey, mule}) + 0.5/(\text{donkey, cow}) \end{aligned}$$

to be read that, e.g., a horse and a mule are similar to degree 0.8, a horse and a cow to degree 0.4, etc.

Notice that for finite, small enough X and Y, a fuzzy relation may be evidently shown in the matrix form.

A crucial concept related to fuzzy relations is their *composition*. If we have two fuzzy relations R in $X \times Y$ and S in $Y \times Z$, then their (*max-min*) *composition* is a fuzzy relation $R \circ S$ in $X \times Z$ defined by

$$\mu_{R \circ S}(x, z) = \sup_{y \in Y} [\mu_R(x, y) \wedge \mu_S(y, z)] \tag{7}$$

Fuzzy relations play a crucial role in virtually all applications, notably in decision making and control, and as good sources of information in this respect we can recommend, say, the articles in Kacprzyk and Fedrizzi (1990).

3. FUZZY NUMBERS

The extension principle defined by (6) is a very powerful tool for extending nonfuzzy relationships to their fuzzy counterparts. It can also be used, e.g., to devise *fuzzy arithmetic*.

A *fuzzy number* is defined (by many, not all authors) as a fuzzy set in the real line, A in R, which is normal (i.e. $\sup_{r \in R} \mu_A(r) = 1$) and bounded convex (i.e. whose all α-cuts are convex and bounded). A fuzzy number may be exemplified by "about five", "a little more than 7", "more or less between 5 and 8", etc.

Notice that function f in (6) may be, say, the sum, product, difference and quotient, and we can extend via (6) the four main arithmetic operations: addition, multiplication, subtraction and division to fuzzy sets, hence obtaining fuzzy arithmetic.

Namely, for the basic four operations we obtain:

- *addition*

$$\mu_{A+B}(z) = \max_{z=x+y} [\mu_A(x) \wedge \mu_B(y)], \forall x, y, z \in R \tag{8}$$

- *subtraction*

$$\mu_{A-B}(z) = \max_{z=x-y} [\mu_A(x) \wedge \mu_B(y)], \forall x, y, z \in R \tag{9}$$

- *multiplication*

$$\mu_{A \cdot B}(z) = \max_{z=x \cdot y} [\mu_A(x) \wedge \mu_B(y)], \forall x, y, z \in R \tag{10}$$

- *division*

$$\mu_{A:B}(z) = \max_{z=x:y, y \neq 0} [\mu_A(x) \wedge \mu_B(y)], \forall x, y, z \in R \tag{11}$$

Unfortunately, the use of the extension principle to define the arithmetic operations on fuzzy numbers is in general numerically inefficient, hence it is usually assumed that a fuzzy number is given in the so-called $L - R$ *representation* whose essence is that the membership function of a fuzzy number is as

$$\mu_A(x) = \begin{cases} L\left(\frac{m-x}{\alpha}\right) & \alpha > 0, \forall x \leq m \\ R\left(\frac{x-m}{\beta}\right) & \beta > 0, \forall x \geq m \end{cases} \tag{12}$$

where function L is such that: (1) $L(-x) = L(x)$, (2) $L(0) = 1$, and (3) L is increasing in $[0, +\infty]$, and similarly function R.

In (12) m is the *mean value* of the fuzzy number A, α is its *left spread*, and β is its *right spread*; notice that when $\alpha, \beta = 0$, then the fuzzy number A boils down to a real number m.

A fuzzy number A can now be written as $A = (m_A, \alpha_A, \beta_A)$, and the arithmetic operations may be defined in terms of the m's, α's and β's. For instance, in the case of *addition*:

$$\begin{aligned} A + B &= (m_A, \alpha_A, \beta_A) + (m_B, \alpha_B, \beta_B) \\ &= (m_A + m_B, \alpha_A + \alpha_B, \beta_A + \beta_B) \end{aligned} \tag{13}$$

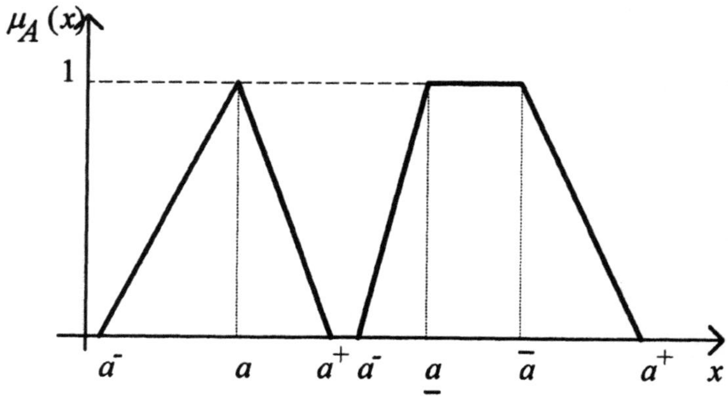

Figure 2: Membership functions of a triangular and trapezoid fuzzy numbers.

and similarly for the other arithmetic operations - see, e.g., Dubois and Prade (1980) or Kaufmann and Gupta (1985) for details.

In practice, however, the $L-R$ representation is further simplified in that the functions L and R are assumed to be linear which leads to *triangular fuzzy numbers* exemplified by the one shown in Fig. 2, and whose membership function is generally given by

$$\mu_A(x) = \begin{cases} (x-a^+)/(a-a^-) & \text{for } a^- \leq x \leq a \\ (a^+-x)/(a^+-a) & \text{for } a < x \leq a_+ \end{cases} \tag{14}$$

Notice that a triangular fuzzy number may be adequate for the formalization of such terms as, say, *around 5* or *much more than 10* (in this case, evidently, a^+ must be a very large number). For the representation of such fuzzy numbers as, e.g., *more or less between 5 and 7* the *trapezoid fuzzy numbers* may be used which are exemplified in Fig. 2 and whose membership function is generally written as

$$\mu_A(x) = \begin{cases} (x-a^+)/(a-a^-) & \text{for } a^- \leq x \leq \underline{a} \\ 1 & \text{for } \underline{a} < x \leq \overline{a} \\ (a^+-x)/(a^+-\overline{a}) & \text{for } \overline{a} < x \leq a^+ \end{cases} \tag{15}$$

The most important issue related to fuzzy numbers is their comparison (ordering) which is a prerequisite for their use in modeling and optimization. In nontrivial cases as, e.g., when the fuzzy numbers are triangular and one is entirely (in the sense of its support) to the left or right of another, this is not a trivial issue. We will not discuss it here in detail due to lack of space, and will refer the reader to the literature (e.g., Kaufmann and Gupta, 1985). Some of these comparison techniques will be used in the papers later on.

4. CONCLUDING REMARKS

The purpose of this short review was to introduce the interested readers to the basic concepts of fuzzy sets theory and fuzzy arithmetic. For more details on the topics covered, the reader is referred to the literature whose list contains more fundamental entries only. Recent publications, whose number is growing, can be found in the publications programs of all major publishers.

BIBLIOGRAPHY

J.C. Bezdek (1980) Pattern Recognition with Fuzzy Objective Function Algorithms, Plenum Press, New York.

D. Dubois and H. Prade (1980) Fuzzy Sets and Systems, Academic Press, New York.

J. Kacprzyk (1983) Multistage Decision-Making under Fuzziness, Verlag TÜV Rheinland, Cologne.

J. Kacprzyk and S.A. Orlovski, Eds. (1987) Optimization Models Using Fuzzy Sets and Possibility Theory, Reidel, Dordrecht.

J. Kacprzyk and M. Fedrizzi, Eds. (1988) Combining Fuzzy Imprecision with Probabilistic Uncertainty in Decision Making, Springer-Verlag, Heidelberg.

J. Kacprzyk and M. Fedrizzi, Eds. (1990) Multiperson Decision Making Models using Fuzzy Sets and Possibility Theory, Kluwer, Dordrecht.

J. Kacprzyk and M. Fedrizzi, Eds. (1992) Fuzzy Regression Analysis. Omnitech Press, Warsaw and Physica-Verlag, Heidelberg.

A. Kandel (1982) Fuzzy Techniques in Pattern Recognition, Wiley, New York.

A. Kandel (1986) Fuzzy Mathematical Techniques with Applications, Addison-Wesley, Reading.

A. Kaufmann and M.M Gupta (1985) Introduction to Fuzzy Arithmetic, Van Nostrand Reinhold, New York.

G.J. Klir and T.A. Folger (1988) Fuzzy Sets, Uncertainty and Information, Prentice-Hall, Englewood Cliffs.

L.A. Zadeh (1965) Fuzzy sets, Information and Control 8, 338–353.

L.A. Zadeh and J. Kacprzyk, Eds. (1992) Fuzzy Logic for the Management of Uncertainty, Wiley, New York.

H.-J. Zimmermann (1985) Fuzzy Sets Theory and its Applications, Kluwer-Nijhoff, Dordrecht.

H.-J. Zimmermann (1987) Fuzzy Sets, Decision - Making, and Expert Systems, Kluwer, Dordrecht.

2

RELIABILITY AND SAFETY ANALYSES UNDER FUZZINESS: GENERAL ISSUES

SYSTEM RELIABILITY FROM THE VIEWPOINT OF EVALUATION AND FUZZY SETS THEORY APPROACH

Takehisa ONISAWA

Institute of Engineering Mechanics
University of Tsukuba
1-1-1, Tennodai, Tsukuba 305, Japan

Abstract – Fuzziness and subjectivity are inherent in a system reliability analysis. This paper mentions the fuzziness and the subjectivity in estimation of a failure probability and an error probability, in modification of the failure probability and the error probability taking the influence of environmental conditions into consideration, in estimation of the dependence level, in evaluation of system reliability and safety. This paper also mentions one of fuzzy sets theory approaches in the system reliability analysis to deal with the fuzziness and the subjectivity. In this approach subjective unreliability measure represented by a fuzzy set is used, and informations on reliability, e.g., reliability estimate, dependence level estimate, are represented by natural language.

Keywords – system reliability, evaluation, subjectivity, fuzzy sets theory, natural language expressions.

1. INTRODUCTION

A main tool of a system reliability analysis is the probability theory and there have been many its applications, e.g., U.S. Atomic Energy Commission (1974) presented WASH-1400 which has been famous as risk assessment of nuclear power plants, and Swain (1963) presented THERP(Technique for Human Error Rate Prediction) which has been famous as a model of human reliability analysis. The probabilistic method was introduced in order to assess system reliability rationally, i.e., objectively. The probability of a system accident obtained by the probabilistic analysis can be compared with the probability of a rare event. That is, risk in a system can be compared with the one in other system quantitatively by the use of the accident probability.

Experts' experienced judgement plays an important role in the system reliability evaluation process even if the probabilistic method is employed. From the viewpoint of the fuzzy sets theory fuzziness and subjectivity are found even in the probabilistic method of the system reliability analysis.

First of all this paper mentions the fuzziness and the subjectivity in the system reliability analysis from the viewpoint of evaluation. This paper also mentions a method to deal with the fuzziness and the subjectivity. The method employs a fuzzy set and can incorporate the subjectivity and the fuzziness into the analysis.

2. SUBJECTIVITY AND FUZZINESS IN SYSTEM RELIABILITY ANALYSIS

Let us consider a system reliability analysis based on system component reliability and human reliability.

2.1 Estimation of Failure Probability and Error Probability

A failure probability is one of equipment measures and an error probability is one of human reliability measures. The failure probability of a system component and the error probability of a human operator in a task are interpreted as evaluation results of system component reliability and human reliability in the task. A database on these probabilities should be established in order to analyze system reliability quantitatively. In practice, however, the database is poor. So the estimation of these probabilities is often dependent on experts' experienced engineering judgement rather than the database. Their judgement is closely related to their subjectivity and fuzziness. The subjectivity and the fuzziness mean that experts cannot estimate these probabilities precisely, exactly and objectively. Reliability measures expressed by numerical values, e.g., the failure probability, the error probability, seem to be objective and exact. As a matter of fact, however, the fuzziness and the subjectivity are latent behind these values.

The estimation of the failure and the error probabilities turns out to be dependent on experts' subjective judgement. Their experienced judgement holds a central position in system reliability assessment as Svenson (1989) mentions.

2.2 Modification of Failure Probability and Error Probability

Hardware reliability and human reliability are dependent on environmental conditions. The failure probability and the error probability should not be a physical constant. They should be modified depending on the environmental conditions. It is hoped that a complete database, which means the database with consideration of conditions in which system components are used or human operators perform tasks, is established immediately. However it is difficult and impossible to establish it because there exist some problems. For example, system components are used in various conditions and it is difficult and impossible to consider all conditions. And the condition in which they are used is not always the same as the condition which is considered at the establishment of the database. It becomes necessary to evaluate the condition in order to modify the failure probability and the error probability.

Usually the modification is performed by multiplying the probabilities by numerical factors. The selection of numerical factors is often dependent on experienced engineering judgement of experts who are familiar with the analyzed system since the database on numerical factors for various conditions cannot be established and since the selection of the numericl factors needs the evaluation of environmental conditions. Their engineering judgement also plays an important role in the evaluation of the environmental conditions and the selection of the numerical factors. The failure probability and the error probability, which are estimated by taking environmental conditions into consideration, are not objective any longer.

It is found that subjectivity and fuzziness are latent behind the numerical values to modify the failure probability and the error probability. The modification of the probabilities is dependent on evaluation results of the condition in which equipment is operated, or in which a human operator performs a task.

2.3 Estimation of Dependence Level

If the dependence exists among subsystems which construct a redundant system, the system does not work redundantly any longer. For example, if failures of subsystems shown in **Fig.1** are dependent on each other, that is, if the failure of the subsystem A has an influence on the failure of the subsystem B, this system does not work as a parallel redundant system any longer.

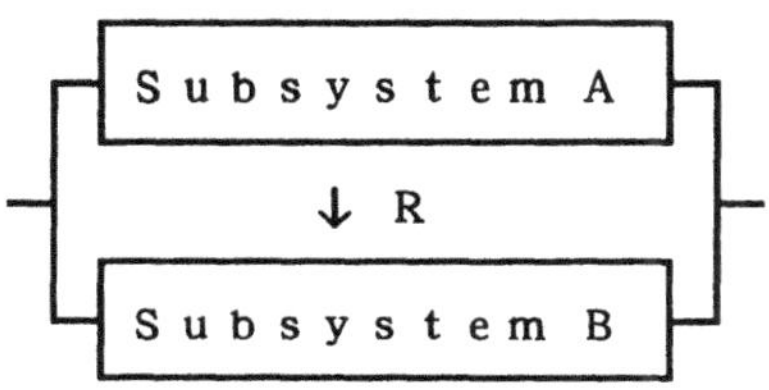

Fig. 1 Dependence

In the probabilistic method a conditional probability is usually used to deal with the dependence. In practice, however, a general database on the conditional probability cannot be established because the data(even if exist) are dependent on the specific system and cannot be used for every system. The information about the dependence is usually obtained by experts who are familiar with an analyzed system. Even if the numerical information about the dependence, e.g., the conditional probability, is obtained, subjectivity and fuzziness are latent behind the numerical information for the same reason in the estimation of the failure probability and the error probability. The dependence level in the analyzed system is estimated by the analysis of the system functional structure. The analysis is highly dependent on experienced engineering judgement of experts who are familiar with the analyzed system. It is said that the construction of a fault tree is dependent on analysts' subjectivity because the fault tree construction is deeply related to the analysis of the system functional structure. Experts' engineering and experienced judgement also holds an important position in the analysis of the system functional structure including the estimation of the dependence level.

2.4 Evaluation of System Safety

Let us consider a parallel system and a series system shown in **Fig. 2**. And the accident probability of each subsystem is assumed to be 10^{-3}. The accident probabilities of the parallel system and the series system are estimated to be 10^{-6} and 2×10^{-3}, respectively.

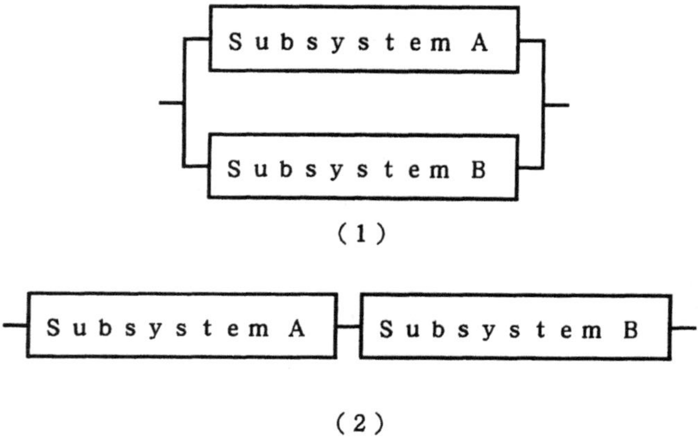

Fig. 2 Parallel System and Series System

In the probabilistic method if accident probabilities of system components are given, the accident probability of a whole system is calculated as shown in the above example. The system accident probability calculated in this way is not objective any longer because the subjectivity and the fuzziness are latent behind the accident probabilities of subsystems or system components. And even if the same accident probability is obtained by the analysis, system reliability and safety is not always evaluated uniquely because subjectivity and fuzziness are inherent in the evaluation. For example, if the accident probability of an analyzed system is estimated to be 10^{-5}[1/year], how do we evaluate safety of the system ? The accident probability, e.g., 10^{-5}[1/year], is usually very very small from the viewpoint of our daily life. The accident probability is usually compared with the probability of a rare event by experts based on their experienced engineering judgement. The evaluation may be dependent on the analyzed system, e.g., whether the system is a nuclear power plant or a chemical plant. The evaluation may be also dependent on a person who evaluates system safety, e.g., he is a person concerned in the system or neighbor of this system. Of course the evaluation may be also dependent on the extent of an accident damage.

What does the very very small probability of the system accident mean ? Does it imply that this accident hardly ever happens ? The small probability of an event does not always mean the low possibility of the event as Dubois and Prade (1980) points out. This is also found by facts that accidents of many high technical developed systems have happened.

Subjectivity and fuzziness are inherent in the evaluation of system reliability and safety. Not only the probabilistic approach but also the possibilistic one are necessary in the evaluation of system safety.

3. FUZZY SETS THEORY APPROACH IN SYSTEM RELIABILITY ANALYSIS

3.1 Natural Language Expressions

As mentioned in the previous section, system reliability is analyzed based on evaluations of system component reliability, human reliability and system functional relation. And in the last stage of the system reliability analysis system reliability and safety is evaluated. The evaluation results cannot be sometimes expressed by numerical values. Even if they are represented by numerical values somehow or other, the subjectivity and the fuzziness, which are latent behind the numerical values, are not communicated into other people.

In these situations natural language expressions are considered since the expressions include subjectivity and fuzziness. It is easier for experts to express system reliability information with linguistic terms than with numerical values when enough data are not obtained. For example, when experts have not complete data about system component reliability, its reliability estimate by numerical vlues such as 10^{-3} [1/hr] is unreasonable and impossible. But they evaluate it easily and reasonably by natural language expressions based on their engineering and experienced judgement such as *the component has standard reliability*, or *the failure probability of the component is about* 10^{-3} *[1/hr]*.

Natural language expressions are appropriate for reliability estimate as Schmucker (1984) points out. As mentioned before, hardware reliability and human reliability are affected by many conditions. The influence of environmental conditions on hardware reliability and human reliability cannot be expressed easily by the use of numerical values. But it can be expressed easily with linguistic terms as presented by Onisawa (1991). The natural language expressions of reliability estimate mean that evaluation results of hardware reliability and human reliability are expressed with natural language taking the influence of the environmental conditions into consideration. And they also mean that results of comparison of hardware or human reliability with experts' standard reliability are expressed with natural language. Of course the expressions include subjectivity.

Experts can express the information about the dependence more easily with linguistic terms than with numerical values. As a matter of fact in human reliability model Swain and Guttmann (1983) use linguistic terms of dependence as shown in **Table 1**.

Natural language expressions are also appropriate for the evaluation of system reliability and safety. As a matter of fact, IAEA and HEA (1992) consider the International Nuclear Event Scale(INES) User's Manual,

Table 1 Conditional Probability of Human Error

Dependence Level	Conditional Probability(Pr)
Complete Dependence	1.0
High Dependence	$(1 + Pr(B))/2$
Moderate Dependence	$(1 + 6 \times Pr(B))/7$
Low Dependence	$(1 + 19 \times Pr(B))/20$
Zero Dependence	$Pr(B)$

$Pr(B)$ is an error probability in a task B assuming that a task A and the task B are indpendent of each other, where a success in the task A or in the task B is assumed to lead to a success in a whole task. The error probability in the whole task is estimated by $Pr(A)\times Pr$, where $Pr(A)$ is an error probability in the task A.

which represents event rating form about nuclear power plant accident by linguistic terms such as level 0 - *below scale*, level 1 - *anomaly*, level 2 - *incident*, level 3 - *serious incident*, level 4 - *accident without significant off-site risk*, level 5 - *accident with off-site risk*, level 6 - *serious accident*, level 7 - *major accident*. It is possible to communicate system reliability and safety to other people by linguistic terms.

A fuzzy set expresses the meanings of the natural language expressions well since their meanings are fuzzy in themselves. The subjectivity and the fuzziness can be communicated to other people well by a fuzzy set. This is one of advantages of the use of the fuzzy set in a system reliability analysis. There have been many approaches of fuzzy sets theory in the area of the system reliability and safety engineering. One of them is a fuzzy probability approach which Noma et al. (1981) and Tanaka et al. (1983) proposed and Singer (1990), Misra et al.(1990) and Kenarngui (1991) employed. In this approach natural language expressions such as *the failure probability of the component is about* 10^{-3} are available. The meaning of *about* 10^{-3} is expressed by a fuzzy set defined on a probability space.

This paper mentions another approach which uses natural language expressions different from the expressions with the fuzzy probability. In the present fuzzy sets theory approach hardware reliability and human reliability are assumed to be expressed by natural language in the form of reliability estimate and its fuzziness. **Table 2** shows natural language expressions of reliability estimate and **Table 3** shows linguistic terms of its fuzziness. Evaluation of the dependence level is also assumed to be expressed by natural language in the form of dependence level estimate and its fuzziness. **Table 4** shows linguistic terms of dependence level estimate, which are referred to Table 1. **Table 5** shows linguistic terms of its fuzziness. Table 5 is the same as Table 3 for simplicity.

Table 2 Natural Language Expressions of Reliability Estimate And Corresponding Parameter x_0

Class	Expressions of Reliability Estimate	Parameter x_0 (Representative Value)
	(System, subsystem, instrument or human operator has)	
1	no reliability	–
2	very low reliability	0.9 – 1.0(0.95)
3	low reliability	0.7 – 0.9(0.8)
4	rather low reliability	0.55 – 0.7(0.625)
5	standard reliability	0.45 – 0.55(0.5)
6	rather high reliability	0.3 – 0.45(0.375)
7	high reliability	0.2 – 0.3(0.25)
8	quite high reliability	0.1 – 0.2(0.15)
9	extremely high reliability	0.05 – 0.1(0.075)
	(Accident, failure or human error is)	
10	next to impossible	0.0 – 0.05(0.025)
11	impossible	–

Table 3 Natural Language Expressions of Fuzziness of Reliability Estimate and Corresponding Parameter m

Class	Expressions of Fuzziness	Parameter m
1	Low Fuzziness	2.0
2	Medium Fuzziness	2.5
3	Rather High Fuzziness	3.0
4	High Fuzziness	3.5

Table 4 Natural Language Expressions of Dependence Level Estimate

Level	Expressions of Dependence Level Estimate
1	Complete Dependence
2	High Dependence
3	Moderate Dependence
4	Low Dependence
5	Zero Dependence

3.2 Translation from Natural Language Information to Quantified Fuzzy Set

The meanings of natural language expressions are represented well by

Table 5 Natural Language Expressions of Fuzziness of Dependence Level Estimate and Corresponding Parameter mr

Class	Expressions of Fuzzinss	Parameter mr
1	Low Fuzziness	2.0
2	Medium Fuzziness	2.5
3	Rather High Fuzziness	3.0
4	High Fuzziness	3.5

a fuzzy set. Every model of human reliability does not employ linguistic terms, for example, Swain and Guttmann (1983) use them as shown in Table 1. Unfortunately, however, the meanings of linguistic terms are represented by numerical factors. They are not expressed well by numerical factors.

In the present approach the translation of linguistic terms of reliability estimation is performed by

$$F(x) = \frac{1}{1 + 20 \times |x - x_0|^m} \tag{1}$$

where x_0 and m are parameters, and $0 \leq x, x_0 \leq 1$. The parameter x_0 gives the maximal grade of $F(x)$ and the parameter m is related to fuzziness. Onisawa (1990) calls a fuzzy set with the membership function (1) as a *failure possibility* in the fuzzy sets theory approach of the system reliability analysis. **Fig. 3** shows an example of the failure possibility. The abscissa in Fig. 3 means subjective unreliability. It is interpreted that the smaller the value of the abscissa is, the lower the subjective unreliability is. The middle point 0.5 means subjective standard unreliability. The ordinate means the grade of membership.

The membership function (1) is selected from the following reasons. Eq. (1) can express that the possibility of subjective unreliability 1 does not necessarily equal to zero even if the value of the parameter x_0 is very small, and that the possibility of subjective unreliability 0 does not necessarily equal to zero even if the value of the parameter x_0 is large.

The parameters x_0 and m correspond to natural language expressions of reliability estimate and its fuzziness, respectively, as shown in Tables 2 and 3. Especially the failure possibility in the class 1 is defined by

$$F(x) = \begin{cases} 1, & x = 1, \\ 0, & x \neq 1, \end{cases} \tag{2}$$

and that in the class 11 is defined by

$$F(x) = \begin{cases} 0, & x \neq 0, \\ 1, & x = 0. \end{cases} \tag{3}$$

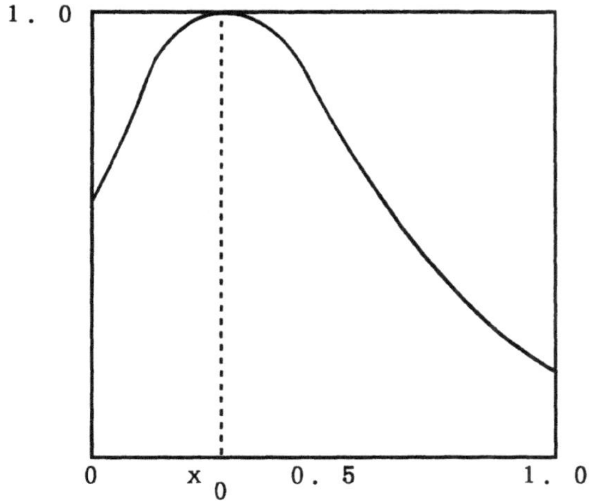

Fig. 3 Failure Possibility

Natural language expressions in the classes 1 and 11 do not take the fuzziness into consideration.

Translation of linguistic terms of dependence information is performed by

$$R(r) = \frac{1}{1 + 20 \times |r - r_0|^{mr}} \tag{4}$$

where r_0 and mr are parameters, and $0 \leq r, r_0 \leq 1$. The parameter mr corresponds to linguistic terms of fuzziness of dependence level estimate as shown in Table 5. The correspondence of the parameter r_0 is mentioned later.

3.3 Comments on Failure Possibility

The failure possibility can be interpreted as the subjective measure of hardware unreliability and human unreliability. When the database on the failure probability and the error probability is not established, reliability estimation is expressed based on the experts' engineering and experienced judgement by natural language better than by numerical values. The natural language expressions are interpreted as the results of the comparison of hardware reliability or human reliability with experts' standard reliability. The natural language expressions are also interpreted as their reliability evaluation results taking the influence of many environmental conditions on reliability into consideration. Hardware unreliability and human unreliability can be treated in the same way by the use of the failure possibility because the meanings of natural language expressions of hardware unreliability and human unreliability are represented by the failure possibility. Then the problem about the hardware reliability measure and the human reliability measure, which Adams (1982) points out, is solved by the use of the failure possibility.

The membership function (1) is defined subjectively. The mappings between linguistic terms and numerical parameters as provided by Tables 2 and 3 are determined subjectively. However behind Eq. (1) there are some meanings as mentioned before. And in the determination of the mappings there are some points to notice. The mappings should be consistent in the following sense. The part of the abscissa of the failure possibility such as $x > 0.5$ means high unreliability, i.e., low reliability. So natural language expressions such as ***high reliability*** should not be mapped into the failure possibility of which parameter x_0 has the value larger than 0.5 such as 0.6. In this sense Table 2 is legitimate. And there are no problems of arbitrariness and subjectivity of the equations and the tables in the failure possibility approach because the equations and the tables are used throughout the system reliability anaysis in order to keep consistency. Of course system reliability should be discussed based on results obtained by experts. The subjectivity and the fuzziness inherent in the analysis are represented openly by the failure possibility. So the system reliability including subjectivity can be discussed openly by the failure possibility approach.

3.4 AND Operation and OR Operation

This approach applies a fault tree analysis. Basic operations are an AND opeation and an OR operation of failure possibilities. The parameterized operations (5) and (6) are employed as the AND operation and the OR operation, respectively.

$$H(x,\ y) = \frac{1}{1+[((1-x)/x)^{1/nH}+((1-y)/y)^{1/nH}]^{nH}}, \tag{5}$$

where $0<x,\ y\leq 1$, $H(x,\ 0)=H(0,\ y)=0$, and nH is a non-negative parameter.

$$G(x,\ y) = \frac{[(x/(1-x))^{nG}+(y/(1-y))^{nG}]^{1/nG}}{1+[(x/(1-x))^{nG}+(y/(1-y))^{nG}]^{1/nG}}, \tag{6}$$

where $0\leq x,\ y<1$, $G(x,\ 1)=G(1,\ y)=1$, and nG is a non-negative parameter. The operations H and G are so called Dombi t-norm and Dombi t-conorm, respectively. They have some properties as Mizumoto (1989) mentions. When nH is infinitive, the operation H becomes a drastic product. When nH is zero, the operation H becomes a min operation. On the other hand when nG is infinitive, the operation G becomes a max operation. When nG is zero, the operation G becomes a drastic sum. From the viewpoint of the reliability analysis, the operations H and G can cover the range of reliability estimate from the most optimistic estimate(when nH and nG are infinitive) to the most pessimistic one(when nH and nG are zero) as Onisawa (1993) points out. The parameterized operations H and G can reflect experts' subjectivity towards an analyzed system. The determination of nH and nG is mentioned later.

3.5 Dependence

Reliability of a parallel system with the dependence as shown in Fig. 1 is analyzed by the following model proposed by Onisawa (1990). Let F_A and F_B be failure possibilities of subsystems A and B, respectively. And let R be a fuzzy set with the membership function (4) representing the dependence level. In this model the following two cases are considered.

(i)The failure of the subsystem A has an influence on that of the subsystem B.

In this case the failure possibility $F_B{}'$ of the whole system is estimated by F_A AND R, i.e.,

$$F_B{}' = H(F_A, R), \tag{7}$$

where the extension principle is employed in this and the following operations.

(ii)The failure of the subsystem A does not have any influence on that of the subsystem B.

As far as the dependence level is not complete, the failure of the subsystem A does not always have any influence on that of the subsystem B. The portion of the failure possibility, which has no influence on the failure of the subsystem B, is estimated by

$$F_A = G(F_A{}', F_B{}') \tag{8}$$

where $F_A{}'$ corresponds to this portion. Eq. (8) implies that the failure of the subsystem A has an influence on that of the subsystem B or not. The failure possibility of the whole system in the second case is estimated by

$$F' = H(F_A{}', F_B). \tag{9}$$

Eq. (9) implies that the failure of the whole system occurs when the failure of the subsystem A and the failure of the subsystem B occur without influence.

The failure possibility of the whole system considering the above two cases is estimated by

$$F = G(F_B{}', F'). \tag{10}$$

Eq. (10) implies that the failure of the whole system occurs when the failure of the subsystem A and the failure of the subsystem B occur with influence or when they occur without influence.

3.6 Natural Language Expressions of Analysis Results

In this approach the analysis result is also expressed by natural language since this approach attaches great importance to natural language expressions. Let F_R be the failure possibility obtained by the analysis and F_S be the failure possibility with the membership function (1). Let α -cuts of F_R and F_S be $(F_R)_\alpha = (x_{1R}(\alpha), x_{2R}(\alpha))$ and $(F_S)_\alpha = (x_{1S}(\alpha), x_{2S}(\alpha))$, respectively. The distance between F_R and F_S is defined by

$$d = \int_0^1 \sqrt{(x_{1R}(\alpha)-x_{1S}(\alpha))^2+(x_{2R}(\alpha)-x_{2S}(\alpha))^2}\,d\alpha \tag{11}$$

The parameters $x_0 (0 \leq x_0 \leq 1)$ and $m(2.0, 2.5, 3.0, 3.5)$ of F_S are selected so as to minimize the distance d and the analysis result is expressed by natural language referring to Tables 2 and 3.

4. DETERMINATION OF PARAMETERS nH, nG and r_0

4.1 Parameters nH and nG

Let us consider the parallel system and the series system as shown in Fig. 2, where the subsystems are independent of each other. The parameters nH and nG are determined depending on the experts' reliability evaluation of the parallel system and that of the series system, respectively, assuming that reliability estimate of each subsystem and its fuzziness are *standard* and *medium*. For example, when an expert evaluates reliability of the parallel system and that of the series system as *high*($x_0 = 0.25$) and *rather low*($x_0 = 0.625$), respectively, the evaluations lead his determination of the parameters nH and nG to 2.05 and 1.0, respectively.

4.2 Parameter r_0

Let reliability estimates of the subsystems A and B shown in Fig. 1 and their fuzziness be *standard* and *medium*, respectively. Let x_{0i} be the parameter x_0 of the failure possibility F_S of a system with dependence level i (i=1:complete dependence, i=2:high dependence, i=3:moderate dependence, i=4:low dependence, i=5:zero dependence). And let r_{0i} be the parameter r_0 of the fuzzy set R representing the dependence level i. When the dependence level is *complete*(i=1), the system has *standard reliability*($x_{01} = 0.5$). On the other hand when the dependence level is *zero*(i=5), reliability of the system is estimated by the operation H with the selected parameter nH. The parameters r_{0i}(i=2, 3, 4) are determined so as to satisfy

$$(x_{0i} - x_{05}) : (x_{01} - x_{0i}) = \begin{cases} 3:1 & (i=2) \\ 1:1 & (i=3) \\ 1:3 & (i=4) \end{cases} \tag{12}$$

For example, let us assume that the parameters nH and nG are determined as 2.05 and 1.0, respectively. In this case $x_{01} = 0.5$ and $x_{05} = 0.25$. The parameters r_{0i}(i=2,3,4) are determined such as

$$\begin{cases} r_{02} = 0.8 & (x_{02} = 0.44) \\ r_{03} = 0.6 & (x_{03} = 0.38) \\ r_{04} = 0.32 & (x_{04} = 0.31) \end{cases}$$

The parameter r_0 is dependent on the parameters nH and nG.

5. EXAMPLE

It is said that a variety of fault trees are obtained for the same analyzed system depending on analysts' subjectvity. In this example it is assumed that the first analyst and the second analyst obtain the fault trees shown in **Fig. 4(1)** and **Fig. 4(2)**, respectively. From the figures it is found that the first analyst considers that this system has no dependence between subsystems and the second one does that this system has dependences between them.

Table 6 Parameters nH and nG Determined by the First and the Second Analysts and the Second Analyst's Dependence Level Estimate and Its Fuzziness

	nH	nG	(Dependence Level Estimate, Fuzziness)
First Analyst	2.05	1.0	–
Second Analyst	2.5	2.5	(Moderate, Medium)

Table 6 shows the parameters nH and nG determined by the two analysts. Assuming that reliability estimate of each subsystem shown in Fig. 2 and its fuzziness are *standard reliability* and *medium*, respectively, the second analyst evaluates reliability of the parallel system and that of the series system as *high reliability*($x_0 = 0.2$) and *standard reliability*($x_0 = 0.55$). On the other hand the first analyst evaluates them as *high reliability*($x_0 = 0.25$) and *rather low reliability*($x_0 = 0.625$).

From Fig. 4 and Table 6 it is found that the first analyst evaluates the system functional relation more optimistically than the second one, but on the other hand the second analyst evaluates reliability of the system more

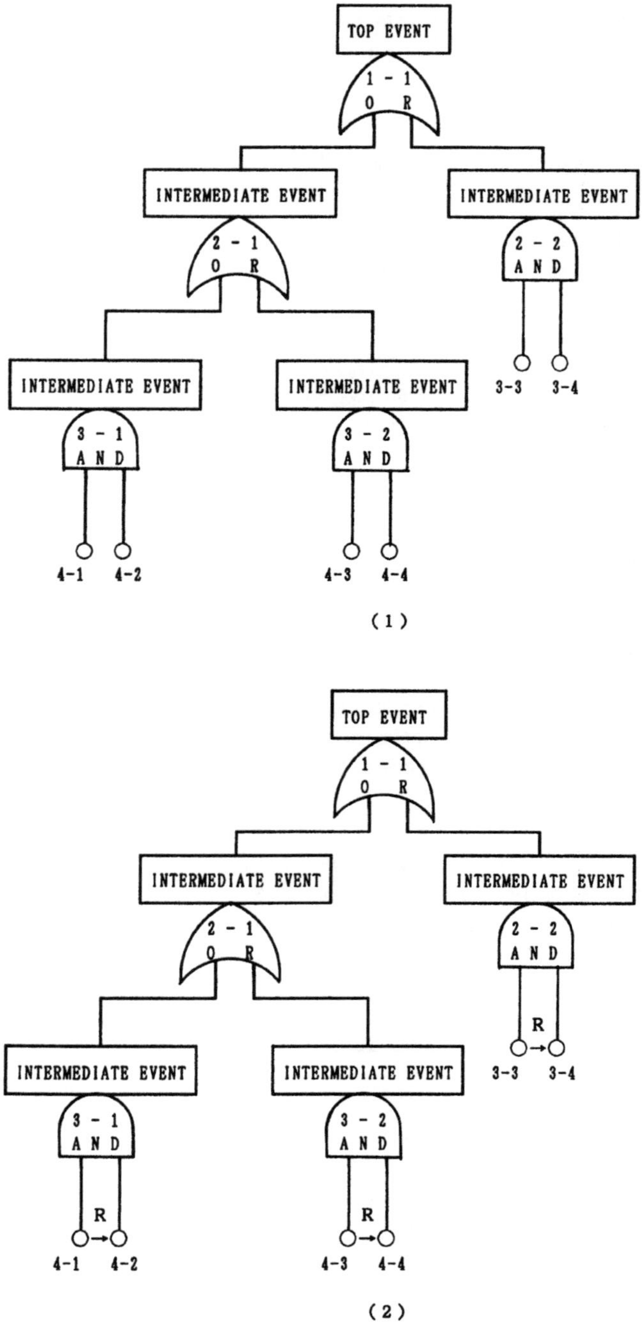

Fig. 4 Analyzed Fault Trees

optimistically than the first analyst.

Let us assume that reliability estimate of each basic event of the fault trees in Fig. 4 and its fuzziness are ***standard reliability*** and ***medium fuzziness***, respectively.

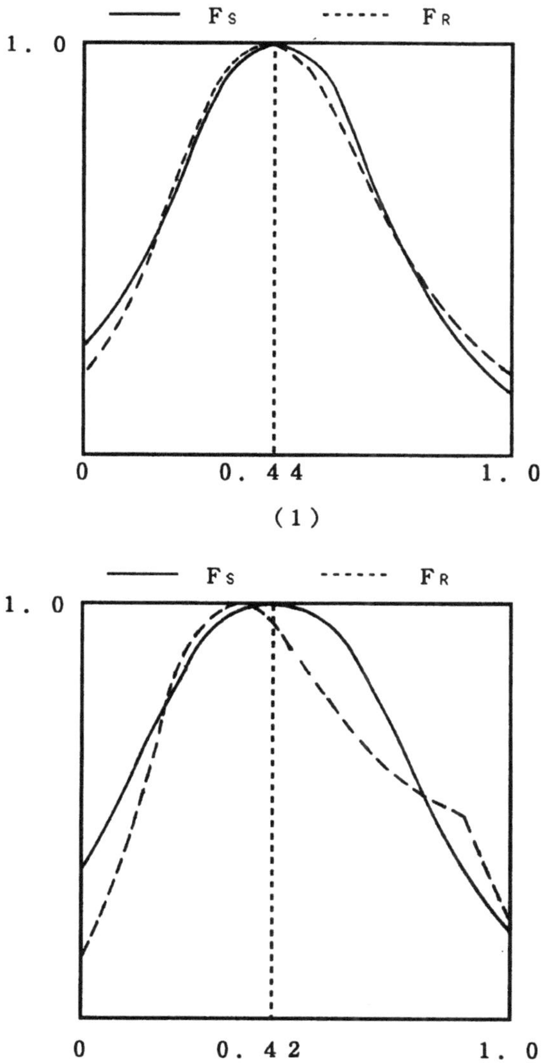

Fig. 5 Results of Analysis

Fig. 5(1) shows failure possibilities obtained by the first analyst and **Fig. 5(2)** shows failure possibilities obtained by the second one. The first analyst evaluates reliability of this system as *rather high* and its fuzziness as *medium*. The second analyst estimates its reliability as *rather high* and its fuzziness as *rather high*.

Subjectivity and fuzziness are inherent in the system reliability analysis. The failure possibility approach can incorporate them into the analysis. Results shown in Fig. 5 are obtained by the use of two analysts' evaluation results of reliability of basic events and the system functional relation, i.e., their subjectivity. In this example both analysts obtain the same result as *this system has rather high reliability*. But there is a possibility that different results are obtained depending on analysts' subjectivity. It is necessary to discuss system reliability based on the obtained results. In the failure possibility approach the obtained results can be examined openly because the evaluation mechanism in this approach is communicated to other people. For example, what if other values of the parameters nH and nG are determined ?, what if reliability estimate of the basic event is not *standard* but *rather high* ?, or what if another estimate of the dependence level is obtained ?, etc.

6. CONCLUSIONS

This paper mentions the fuzziness and the subjectivity inherent in the system reliability analysis from the evaluation point of view. The fuzziness and the subjectivity are found in the estimation of the failure and the error probabilities, in the modification of the failure and the error probabilities with consideration of the influence of the environmental conditions, in the estimation of the dependenec level, in the evaluation of system reliability and safety.

This paper also mentions a fuzzy sets theory approach in the system reliability analysis, i.e., the failure possibility approach, in order to deal with the fuzziness and the subjectivity. The failure possibility is a subjective measure of hardware and human unreliability. This approach can incorporate the subjectivity in the evaluation of system component reliability, human reliability and system reliability, in the evaluation of the dependence level, and in the operations of failure possibilities into the analysis. The subjectivity in the operations of failure possibilities implies the analyst's subjectivity towards the analyzed system, that is, whether its reliability is evaluated optimistically or not. And this approach attaches great importance to the application of natural language expressions in the system reliability analysis because natural language expressions are more appropriate for the expressions of reliability information than numerical expressions when enough data are not obtained.

A main tool of the system reliability analysis is a probabilistic method. There have been many applications of the probabilistic method in this area. The probabilistic method is employed in order to assess system reliability rationally because subjectivity is inherent in the system reliability assessment. So the introduction of subjectivity into the system reliability analysis may be against the reason why the probabilistic method has been employed in this area. But even in the probabilistic method the subjectivity is inherent and the subjectivity is not represented openly by numerical values. On the other hand in the failure possibility approach the subjectivity is represented openly by a fuzzy set. Then experts can discuss openly system reliability based on the fuzzy sets obtained by this approach.

Only one methodology cannot deal with many problems in reliability assessment of a large and complex system. It is hoped that not only the probabilistic approach but also the fuzzy sets theory approach will be applied to the area of system reliability and safety engineering.

REFERENCES

U.S.Atomic Energy Commission (1974), *Reactor Safety Study:An Assessment of Accident Risks in U.S. Commercial Nuclear Power Plants, WASH-1400.*

Swain,A.D. (1963), *A Method for Performing a Human Factors Reliability Analysis*, Report SCR-685, Sandia Corporation.

Svenson, O. (1989), *On Expert Judgement in Safety Analyses in the Process Industries*, Reliability Engineering and System Safety, Vol.25, pp.219-256.

Dubois, D., and Prade, H. (1980), *Fuzzy Sets and Systems; Theory and Applications*, Academic Press, New York.

Schmucker, K.J. (1984), *Fuzzy Sets, Natural Language Computations, and Risk Analysis*, Computer Science Press, Rockville, MD.

Onisawa, T. (1991), *Fuzzy Reliability Assessment Considering the Influence of Many Factors on Reliability*, Int. J. of Approximate Reasoning, Vol.5, pp.265-280.

Swain, A.D., and Guttmann, H.E. (1983), *Handbook of Human Reliability with Emphasis on Nuclear Power Plants Applications, NUREG/CR-1278.*

IAEA and HEA(OECD) (1992), *INES:The International Nuclear Event Scale User's Manual*, Revised and Extended Edition.

Noma, K., Tanaka, H., and Asai, K. (1981), *On Fault Tree Analysis with Fuzzy Probability*, Japanese Journal of Ergonomics, Vol.17, pp.291-297.(in Japanese)

Tanaka, H., Fan, L.T., Lai, F.S., and Toguchi, K. (1983), *Fault Tree Analysis by Fuzzy Probability*, IEEE Trans. on Reliability, Vol.R-32, pp.453-457.

Singer, D. (1990), *A Fuzzy Set Approach to Fault and Reliability Analysis*, Fuzzy Sets and Systems, Vol.34, pp.145-155.

Misra, K.B. and Weber, G.G. (1990), *Use of Fuzzy Set Theory for Level-1 Studies in Probabilistic Risk Assessment*, Fuzzy Sets and Systems, Vol. 37, pp.139-160.

Kenarngui, R. (1991), *Event-tree Analysis by Fuzzy Probability*, IEEE Trans. on Reliability, Vol.R-40, pp.120-124.

Onisawa, T. (1990), *An Application of Fuzzy Concepts to Modelling of Reliability Analysis*, Fuzzy Sets and Systems, Vol.37, pp.269-286.

Adams, J.A. (1982), *Issues on Human Reliability*, Human Factors, Vol.34, pp.1-10.

Mizumoto, M. (1989), *Pictorial Representations of Fuzzy Connectives, Part I:Cases of t-norms, t-conorms, and Averaging Operators*, Fuzzy Sets and Systems, Vol.31, pp.217-242.

Onisawa, T. (1993), *A Model of Subjective Reliability Analysis*, Proc. of Second IEEE International Conference on Fuzzy Systems, San Francisco, pp.756-761.

ISSUES IN POSSIBILISTIC RELIABILITY THEORY

Bart CAPPELLE and Etienne E. KERRE

Applied Mathematics and Computer Science

University of Ghent

Krijgslaan 281-S9

B-9000 Gent, Belgium

Abstract : Recently, alternative models for the classical reliability theory based upon fuzzy set theory and related theories, are proposed to solve the problems of the classical approach. Our model is based upon a combination of multistate structure functions and possibility theory. In this paper, we present a congruence relation on the complete lattice of structure functions, that links several concepts of the theory of multistate structure functions and the possibilistic reliability theory. Some important possibilistic reliability bounds are established and, finally, we apply the theoretic concepts to some concrete examples.

Keywords : Possibilistic reliability functions, multistate structure functions, Dirac measures, possibilistic reliability bounds.

1. INTRODUCTION

Classically, reliability theory has been based upon binary structure functions and probability theory. A binary structure function represents the deterministic relation between the component states and the system state, while probability theory is applied to develop the notion reliability of both components and systems.

Some obvious problems arise while applying this theory. A binary structure function allows only two states: a perfect functioning or a complete failure. Since many real systems are more complex by nature, intermediate states must be allowed. This is the topic of multistate structure functions which is, in a way, closely related to fuzzy set theory, since many real life problems simply cannot be represented by a dichotomous model.

By allowing intermediate states, we must extend the classical notion of reliability based upon the probability of failure or functioning of a component or system. Recent research, however, showed that probability theory is not the only possible way of representing imprecision and uncertainty. Possibility theory and fuzzy set theory, e.g., provide useful alternatives to the probabilistic approach of reliability.

First, we discuss some useful concepts of the theory of multistate structure functions on complete lattices. In the third and fourth section, we introduce the possibilistic reliability functions and discuss a special case, which links several concepts of the theory of multistate structure functions and possibilistic reliability functions. Afterwards, we give some examples to exemplify the usefulness of the introduced concepts.

2. MULTISTATE STRUCTURE FUNCTIONS

In this section we introduce some basic notions and concepts, that we apply furtheron. The binary structure functions are too restrictive to model many real life situations, since the concepts of failure or functioning are not always well-defined or since a binary approach is too restrictive. Hence, in the multistate approach we allow intermediate states. We assume that a system has a finite number of components n and that the state space for both the components and the system is an arbitrary complete lattice (Birkhoff, 1967). In Montero, Cappelle and Kerre (1993), the need for complete lattices in the theory of structure functions is explained, while in Cappelle (1991a) some fundamental results about the lattice theoretic approach are proven. First, we introduce the notion structure function, which is the deterministic relationship between the component states and the system state.

Definition 1 (Cappelle, 1991a) *Let* $(L_i, \leq)$, $1 \leq i \leq n$, *and* $(L, \leq)$ *be* $n+1$ *complete lattices. An* $L_1 \times \cdots \times L_n - L$*-mapping* ϕ, *satisfying*

(i) $\phi(0, \ldots, 0) = 0$ *and* $\phi(1, \ldots, 1) = 1$,

(ii) ϕ *is isotone, that is,*

$$(\forall(\mathbf{x}, \mathbf{y}) \in (L_1 \times \cdots \times L_n)^2)(\mathbf{x} \leq \mathbf{y} \Rightarrow \phi(\mathbf{x}) \leq \phi(\mathbf{y})),$$

is a structure function from $(L_1 \times \cdots \times L_n, \leq)$ *to* $(L, \leq)$.

We silently assumed that $L_1 \times \cdots \times L_n$ has been provided with its product ordering. For a justification of both conditions we refer to Montero, Cappelle and Kerre (1993). These conditions transform the following intuitive ideas into a mathematical form:

(i) whenever a component assumes a better state, the system must not deteriorate,

(ii) if every component assumes its optimal (worst) state, the system assumes its optimal (worst) state.

As shown in Montero, Tejada and Yáñez (1988) and Cappelle (1991a), definition 1 covers most multistate models introduced by, e.g., El-Neweihi, Proschan and Sethuraman (1978) and Baxter (1984). The set of the structure functions from $(L_1 \times \cdots \times L_n, \leq)$ to $(L, \leq)$ is denoted by $\mathcal{M}(L_1 \times \cdots \times L_n, L)$. On the set $\mathcal{M}(L_1 \times \cdots \times L_n, L)$ we define the pointwise order relation $\preceq$, that is, for any two structure functions ϕ and ψ from $(L_1 \times \cdots \times L_n, \leq)$ to $(L, \leq)$

$$\phi \preceq \psi \Leftrightarrow (\forall \mathbf{x} \in L_1 \times \cdots \times L_n)(\phi(\mathbf{x}) \leq \psi(\mathbf{x})).$$

Theorem 1 (Cappelle, 1991a) $(\mathcal{M}(L_1 \times \cdots \times L_n, L), \preceq)$ *is a complete lattice.*

We will not dwell upon the properties of the complete lattice of structure functions. For an exhaustive overview, we refer to Cappelle (1991a). However, we will discuss one additional important property. First, we introduce the notion congruence relation on a lattice.

Definition 2 (Grätzer, 1978) *Let* $(L, \leq)$ *be a lattice and* θ *a binary relation on* L; θ *is a congruence relation if and only if*

(i) θ *is an equivalence relation on* L,

(ii) *for any elements* x_1, x_2, y_1 *and* y_2 *of* L

$$x_1 \in [y_1]_\theta \text{ and } x_2 \in [y_2]_\theta \Rightarrow x_1 \wedge x_2 \in [y_1 \wedge y_2]_\theta \text{ and } x_1 \vee x_2 \in [y_1 \vee y_2]_\theta.$$

In definition 2, $\wedge$ ($\vee$) is the infimum (supremum) operator on the lattice $(L, \leq)$ and $[x]_\theta$ is the equivalence class of θ, generated or represented by x. The infimum (supremum) operator on the set of structure functions is denoted by $\frown$ ($\smile$). For two arbitrary structure functions ϕ_1 and ϕ_2,

$$\phi_1 \frown \phi_2 : L_1 \times \cdots \times L_n \to L : \mathbf{x} \mapsto \phi_1(\mathbf{x}) \wedge \phi_2(\mathbf{x}). \tag{1}$$

The operator $\smile$ is defined analogously. Now, we introduce the relation Θ_A on the set of structure functions.

Theorem 2 *Let* A *be a subset of* $L_1 \times \cdots \times L_n$ *and consider the binary relation* Θ_A *on* $\mathcal{M}(L_1 \times \cdots \times L_n, L)$, *satisfying*

$$(\forall(\phi_1, \phi_2) \in \mathcal{M}(L_1 \times \cdots \times L_n, L)^2)(\phi_1 \Theta_A \phi_2 \Leftrightarrow (\forall \mathbf{x} \in A)(\phi_1(\mathbf{x}) = \phi_2(\mathbf{x}))), \tag{2}$$

then Θ_A *is a congruence relation.*

Proof. Let A be a subset of $L_1 \times \cdots \times L_n$. Since the equality relation on L is an equivalence relation, Θ_A is an equivalence relation as well. We prove the second condition of definition 2. Let ϕ_1, ϕ_2, ψ_1 and ψ_2 be four structure functions from $(L_1 \times \cdots \times L_n, \leq)$ to $(L, \leq)$ and assume that

$$\phi_1 \Theta_A \phi_2 \text{ and } \psi_1 \Theta_A \psi_2. \tag{3}$$

Taking into account (2), we must prove that

$$(\phi_1 \frown \psi_1) \Theta_A (\phi_2 \frown \psi_2)$$

or equivalently

$$(\forall \mathbf{x} \in A)((\phi_1 \frown \psi_1)(\mathbf{x}) = (\phi_2 \frown \psi_2)(\mathbf{x})). \tag{4}$$

Formula (4) is deduced immediately from (2) and assumption (3). In an analogous way $(\phi_1 \smile \psi_1)\Theta_A(\phi_2 \smile \psi_2)$ follows from assumption (3). Hence, Θ_A is a congruence relation on the complete lattice of structure functions. □

For the sake of simplicity, in the sequel we denote $\Theta_{\{\mathbf{x}\}}$ by $\Theta_{\mathbf{x}}$, for any element $\mathbf{x}$ of $L_1 \times \cdots \times L_n$.

Corollary 1 *Let ϕ be a structure function from $(L_1 \times \cdots \times L_n, \leq)$ to $(L, \leq)$.*

(i) $[\phi]_{\Theta_\emptyset} = \mathcal{M}(L_1 \times \cdots \times L_n, L)$,

(ii) $[\phi]_{\Theta_\mathbf{0}} = [\phi]_{\Theta_\mathbf{1}} = [\phi]_{\Theta_{\{\mathbf{0},\mathbf{1}\}}} = \mathcal{M}(L_1 \times \cdots \times L_n, L)$,

(iii) $[\phi]_{\Theta_{L_1 \times \cdots \times L_n}} = \{\phi\}$.

Proof. The statements easily follow from the definition of the congruence relation Θ_A, A an arbitrary subset of $L_1 \times \cdots \times L_n$, and the boundary conditions of structure functions. □

Corollary 2 *Let A be a subset of $L_1 \times \cdots \times L_n$, ϕ and ψ two arbitrary structure functions from $(L_1 \times \cdots \times L_n, \leq)$ to $(L, \leq)$, then*

$$\psi \in [\phi]_{\Theta_A} \Leftrightarrow (\forall \mathbf{x} \in A)(\psi(\mathbf{x}) = \phi(\mathbf{x})).$$

Proof. Immediate from (2) and the definition of equivalence class of an equivalence relation. □

Theorem 3 *Let A be a subset of $L_1 \times \cdots \times L_n$ and ϕ a structure function from $(L_1 \times \cdots \times L_n, \leq)$ to $(L, \leq)$, then $([\phi]_{\Theta_A}, \preceq)$ is a complete sublattice of $(\mathcal{M}(L_1 \times \cdots \times L_n, L), \preceq)$.*

Proof. Let $(\phi_i \mid i \in I)$ be a non-empty family in $[\phi]_{\Theta_A}$. Since

$$(\forall \mathbf{x} \in A)(\inf_{i \in I} \phi_i(\mathbf{x}) = \phi(\mathbf{x}) = \sup_{i \in I} \phi_i(\mathbf{x})),$$

both $\inf_{i \in I} \phi_i$ and $\sup_{i \in I} \phi_i$ belong to $[\phi]_{\Theta_A}$. Hence, $([\phi]_{\Theta_A}, \preceq)$ is a complete sublattice of $(\mathcal{M}(L_1 \times \cdots \times L_n, L), \preceq)$. □

One easily proves that $[\phi]_{\Theta_A}$ is a closed interval, denoted by $[l(A, \phi), u(A, \phi)]$. The structure functions $l(A, \phi)$ and $u(A, \phi)$ can explicitly be determined by applying the concepts of ideal and dual ideal structure functions. For more details we refer to Cappelle (1993a). Before explaining the importance of the relation Θ_A to practical reliability engineering, we prove two important properties.

Property 1 *Let A and B be two subsets of $L_1 \times \cdots \times L_n$ and ϕ a structure function from $(L_1 \times \cdots \times L_n, \leq)$ to $(L, \leq)$, then*

$$[\phi]_{\Theta_{A \cup B}} = [\phi]_{\Theta_A} \cap [\phi]_{\Theta_B}. \tag{5}$$

Proof. Let ψ be an arbitrary element of $[\phi]_{\Theta_{A \cup B}}$, then

$$(\forall \mathbf{x} \in A \cup B)(\psi(\mathbf{x}) = \phi(\mathbf{x})).$$

Hence,

$$[\phi]_{\Theta_{A\cup B}} \subseteq [\phi]_{\Theta_A} \text{ and } [\phi]_{\Theta_{A\cup B}} \subseteq [\phi]_{\Theta_B},$$

which leads to

$$[\phi]_{\Theta_{A\cup B}} \subseteq [\phi]_{\Theta_A} \cap [\phi]_{\Theta_B}.$$

Now, we prove the reverse inclusion. Let ψ be an element of $[\phi]_{\Theta_A} \cap [\phi]_{\Theta_B}$, then

$$(\forall \mathbf{x} \in A)(\psi(\mathbf{x}) = \phi(\mathbf{x})) \text{ and } (\forall \mathbf{x} \in B)(\psi(\mathbf{x}) = \phi(\mathbf{x})).$$

From the latter formula, we deduce that (5) holds. □

Property 2 *Let A and B be two subsets of $L_1 \times \cdots \times L_n$ and ϕ a structure function from $(L_1 \times \cdots \times L_n, \leq)$ to $(L, \leq)$, then*

$$A \subseteq B \Rightarrow [\phi]_{\Theta_B} \subseteq [\phi]_{\Theta_A}. \tag{6}$$

Proof. Since for any two subsets A and B of $L_1 \times \cdots \times L_n$,

$$A \subseteq B \Leftrightarrow A \cup B = B,$$

(6) follows from property 1. □

Obviously, the reverse of implication (6) does not necessarily hold. We explain the importance of this highly theoretic concept to practical reliability engineering.

In a concrete problem, first of all the physical system must be modeled by a structure function. In practice, we often must approximate the structure function from a set of *observations*. An observation is an ordered couple $(\mathbf{x}, y)$, $\mathbf{x}$ an element of $L_1 \times \cdots \times L_n$ and y an element of L. Thus, an observation contains some basic information of the structure function ϕ that models the physical system. Hence, the meaning of an observation $(\mathbf{x}, y)$ is that $\phi(\mathbf{x}) = y$. In reality, it is almost impossible to determine or know all observations. Therefore, only for a limited amount of component configurations, that is, elements of $L_1 \times \cdots \times L_n$, we know the corresponding system state. From theorem 3, we can deduce the existence of a smallest and greatest structure function, that satisfy the conditions imposed by the set of observations. All suitable structure functions for the system are situated between this lower and upper bound. Property 2 is even more important. It states that adding more information, that is, enlarge the set of observations, decreases the number of structure functions that satisfy the given information. Since $[\phi]_{\Theta_{L_1 \times \cdots \times L_n}} = \{\phi\}$, this procedure of reducing the set of possible structure functions, satisfying the conditions imposed by the observations, converges to the unique structure function that models the system properly. The structure functions $l(A, \phi)$ and $u(A, \phi)$ will lead to some useful reliability bounds.

3. THE POSSIBILISTIC RELIABILITY FUNCTIONS

In classical reliability, probability theory is considered as the unifying model to represent uncertainty. This assumption is not surprising, since classical reliability theory was developed at the beginning of the thirties and mainly after the second world war as an application of probability theory and quality control. Later on, reliability theory became a new, mainly a probabilistic, field of interest. Since probabilistic models proved successful, one may wonder if there is really a need in reliability theory for alternative uncertainty models, in general, and possibilistic uncertainty models, in particular.

Some points of interest must be mentioned here in order to formulate a proper answer to the problem mentioned above. On the first hand, the reliability of a system deals with the confidence we have that the system will function, or at least will function at a minimal performance. Therefore, a reliable system is a system we can trust. The confidence that the system will function properly at a certain level is classically defined in a probabilistic way, and leads to the well-known definition that the reliability of a system is the probability that the system functions during a certain time period. As mentioned above, reliability theory was developed systematically from the early fifties. At that time, non-probabilistic uncertainty models were not available or at least not very popular. Nowadays, we have plenty of alternatives for probability theory.

On the other hand, in the early sixties some important deficiencies of the probabilistic approach became apparant. NASA, e.g., developed alternative models to analyse the reliability aspects of the Saturnus V missile, since a classical approach failed (Cohen, 1984). There were some reasons why a probabailistic approach was not successful. There was, e.g., an accumulation of errors due to the lack of sufficient statistical information about the failure aspects of the components, hence, there was an overestimation of the probability of failure. A qualitative approach was more appropriate. Since the introduction of fuzzy sets and possibility theory, new tools became available to model uncertainty. They are more qualitative by nature and can therefore be applied to situations where a quantitative approach is very unlikely or even impossible.

Recently, several models, based upon fuzzy set theory were proposed to solve the problems mentioned above. Amongst others, we mention the introduction of fuzzy probabilities by Tanaka, Fan, Lai and Toguchi (1983), the fuzzification of the classical reliability function by Singer (1990), the combination of fuzzy states and fuzzy probabilities by Cai and Wen (1991) and the introduction of human error possibilities by Onisawa (1988,1990). In this paper, however, we develop some basic properties of the possibilistic counterpart, introduced by Cappelle and Kerre (1991), of the classical reliability function. In Cappelle (1991a) this concept was further explored. In both cases, however, a product possibility measure was considered, hence, the components were assumed to be possibilisticly independent. In many cases, this assumption is a serious oversimplification of reality since the failure of one component may affect the possibility of failure of another component.

For any mapping ϕ from a universe X to a complete lattice $(L, \leq)$, that is, an L-fuzzy set, we denote by $S(\alpha, \phi)$ and $S^d(\alpha, \phi)$ the sets

$$S(\alpha, \phi) = \{x \mid x \in X \text{ and } \alpha \leq \phi(x)\}$$

and

$$S^d(\alpha, \phi) = \{x \mid x \in X \text{ and } \phi(x) \leq \alpha\}.$$

The set $S(\alpha, \phi)$ is called the α-level set of ϕ, while the set $S^d(\alpha, \phi)$ is called the dual α-level set of ϕ. A related set is $N(\alpha, \phi)$,

$$N(\alpha, \phi) = \{x \mid x \in X \text{ and } \phi(x) = \alpha\}.$$

Obviously,

$$(\forall \alpha \in L)(N(\alpha, \phi) = S(\alpha, \phi) \cap S^d(\alpha, \phi)). \tag{7}$$

In the sequel, by Π we denote a possibility measure on $L_1 \times \cdots \times L_n$, which is not necessarily a product measure. Hence, we allow possibilisticly dependent components.

Definition 3 (Cappelle, 1991a; Cappelle and Kerre, 1993) *Let ϕ be a structure function from $(L_1 \times \cdots \times L_n, \leq)$ to $(L, \leq)$. The possibilistic reliability function $\rho(\Pi, \phi)$ of ϕ is the $L - [0,1]$ mapping*

$$\rho(\Pi, \phi) : L \to [0,1] : \alpha \mapsto \Pi(S(\alpha, \phi))$$

and the dual possibilistic reliability function $\rho^d(\Pi, \phi)$ of ϕ is the $L - [0,1]$ mapping

$$\rho^d(\Pi, \phi) : L \to [0,1] : \alpha \mapsto \Pi(S^d(\alpha, \phi)).$$

Hence, for any α of L, the system represented by ϕ performs at least at level α with possibility $\rho(\Pi,\phi)(\alpha)$, while it performs at most at level α with possibility $\rho^d(\Pi,\phi)(\alpha)$. The possibilistic reliability function deals with the functioning aspects of a system, while the dual possibilistic reliability function deals with the failure aspects of a system. Amongst the important properties of these functions, we mention that $\rho(\Pi,\phi)$ is a decreasing mapping and that $\rho^d(\Pi,\phi)$ is an increasing mapping, i.e.,

$$(\forall(\alpha,\beta)\in L^2)(\alpha\leq\beta\Rightarrow\rho(\Pi,\phi)(\alpha)\geq\rho(\Pi,\phi)(\beta)$$
$$\text{and } \rho^d(\Pi,\phi)(\alpha)\leq\rho^d(\Pi,\phi)(\beta)).$$

They satisfy the boundary conditions

$$\rho(\Pi,\phi)(0)=1=\rho^d(\Pi,\phi)(1).$$

Property 3 *Let ϕ be a structure function from $(L_1\times\cdots\times L_n,\leq)$ to $(L,\leq)$, then*

$$(\forall\alpha\in L)(\Pi(N(\alpha,\phi))\leq\min(\rho(\Pi,\phi)(\alpha),\rho^d(\Pi,\phi)(\alpha))). \tag{8}$$

Proof. Taking into account (7), we easily obtain that for any α of L

$$\Pi(N(\alpha,\phi))\leq\rho(\Pi,\phi)(\alpha) \text{ and } \Pi(N(\alpha,\phi))\leq\rho^d(\Pi,\phi)(\alpha).$$

Formula (8) easily follows. □

By calculating the minimum of both possibilistic reliability functions, one obtains information about the possibility that the system assumes the state α.

Theorem 4 *Let ϕ be a structure function from $(L_1\times\cdots\times L_n,\leq)$ to $(L,\leq)$, then*

$$\rho(\Pi,\phi):L\mapsto[0,1]:\alpha\mapsto\sup_{\mathbf{x}\in S(\alpha,\phi)}\Pi([\mathbf{x},\mathbf{1}])$$

and

$$\rho^d(\Pi,\phi) : L \mapsto [0,1] : \alpha \mapsto \sup_{\mathbf{x}\in S^d(\alpha,\phi)} \Pi([\mathbf{0},\mathbf{x}]).$$

We do not prove this important result, since it is a straightforward extension of theorem 1 in Cappelle and Kerre (1993), proven for the so-called n-ary structure functions, that is, structure functions from $([0,1]^n,\leq)$ to $([0,1],\leq)$.

Theorem 5 *Let A be a subset of $L_1 \times \cdots \times L_n$, then*

$$\rho(\Pi, l(A,\phi)) \preceq \rho(\Pi,\phi) \preceq \rho(\Pi, u(A,\phi)) \tag{9}$$

and

$$\rho^d(\Pi, u(A,\phi)) \preceq \rho^d(\Pi,\phi) \preceq \rho^d(\Pi, l(A,\phi)). \tag{10}$$

Proof. Since $l(A,\phi) \preceq \phi \preceq u(A,\phi)$ implies that for any α of L

$$S(\alpha, l(A,\phi)) \subseteq S(\alpha,\phi) \subseteq S(\alpha, u(A,\phi))$$

and

$$S^d(\alpha, u(A,\phi)) \subseteq S^d(\alpha,\phi) \subseteq S^d(\alpha, l(A,\phi)),$$

(9) and (10) immediately follow from the isotonicity of a possibility measure. □

Theorem 5 provides important and useful reliability bounds for the possibilistic reliability function, that we will explore furtheron in the examples.

4. DIRAC MEASURES

To gain insight into the possibilistic reliability functions, we consider a special case, the Dirac measures. These measures were introduced by De Cooman et al. (1992).

Definition 4 *Let X be a universe and x_o an element of X. The mapping $D(x_o, \cdot)$ from $\mathcal{P}(X)$ to $[0,1]$,*

$$D(x_o, \cdot) : \mathcal{P}(X) \to [0,1] : A \mapsto \begin{cases} 1 & ; \quad x_o \in A \\ 0 & ; \quad x_o \in \mathrm{co}A \end{cases}$$

is a Dirac measure in the element x_o.

One easily verifies that Dirac measures are both possibility and necessity measures. This fact greatly affects the properties of these measures. For more details we refer to De Cooman et al. (1992).

Lemma 1 *Let ϕ be a structure function from $(L_1 \times \cdots \times L_n, \leq)$ to $(L, \leq)$ and $\mathbf{x}_o$ an arbitrary element of $L_1 \times \cdots \times L_n$, then*

$$\rho(D(\mathbf{x}_o, \cdot), \phi) : L \to [0,1] : \alpha \mapsto \begin{cases} 1 & ; \quad \alpha \leq \phi(\mathbf{x}_o) \\ 0 & ; \quad \textit{elsewhere} \end{cases} \tag{11}$$

and

$$\rho^d(D(\mathbf{x}_o, \cdot), \phi) : L \to [0,1] : \alpha \mapsto \begin{cases} 1 & ; \quad \phi(\mathbf{x}_o) \leq \alpha \\ 0 & ; \quad \textit{elsewhere} \end{cases} \tag{12}$$

Proof. Taking into account the analogy, we only prove (11). From the definition of the reliability function and a Dirac measure we deduce that

$$(\forall \alpha \in L)(\rho(D(\mathbf{x}_o, \cdot), \phi)(\alpha) = 1 \Leftrightarrow \mathbf{x}_o \in S(\alpha, \phi))$$

and

$$(\forall \alpha \in L)(\rho(D(\mathbf{x}_o, \cdot), \phi)(\alpha) = 0 \Leftrightarrow \mathbf{x}_o \in \mathrm{co}S(\alpha, \phi)).$$

Since,

$$(\forall \alpha \in L)(\mathbf{x}_o \in S(\alpha, \phi) \Leftrightarrow \alpha \leq \phi(\mathbf{x}_o) \text{ and } \mathbf{x}_o \in S^d(\alpha, \phi) \Leftrightarrow \phi(\mathbf{x}_o) \leq \alpha),$$

we immediately obtain (11), which completes the proof. □

To establish the relationship among the possibilistic reliability functions, the congruence relations of the previous section and Dirac measures, we introduce the relations (ρ, Π) and (ρ^d, Π), based upon the possibilistic reliability functions.

Definition 5 *Let ϕ and ψ be two structure functions from $(L_1 \times \cdots \times L_n, \leq)$ to $(L, \leq)$, then*

$$\phi(\rho, \Pi)\psi \Leftrightarrow \rho(\Pi, \phi) = \rho(\Pi, \psi) \text{ and } \phi(\rho^d, \Pi)\psi \Leftrightarrow \rho^d(\Pi, \phi) = \rho^d(\Pi, \psi).$$

Theorem 6 *The relations (ρ, Π) and (ρ^d, Π) are equivalence relations.*

Proof. Obvious, taking into account definition 5. □

Two structure functions are (ρ, Π)-equivalent, (ρ^d, Π)-equivalent respectively, if their possibilistic reliability functions, their dual possibilistic reliability functions respectively, equal. The equivalence classes, associated to these equivalence relations, generated by a structure function ϕ, are denoted by $[\phi]_{(\rho,\Pi)}$ and $[\phi]_{(\rho^d,\Pi)}$. When two structure functions are both (ρ, Π)- and (ρ^d, Π)-equivalent, the systems behave similar w.r.t. their reliability aspects. Obviously, the equivalence relations are highly dependent upon the possibility distribution Π. Now, we link the notions (ρ, Π)- and (ρ^d, Π)-equivalence, Dirac measures and the relation Θ_A. To this effect we prove theorem 7 and two corollaries. They reflect the remarkable relationship among these notions.

Theorem 7 *Let ϕ be a structure function from $(L_1 \times \cdots \times L_n, \leq)$ to $(L, \leq)$ and $\mathbf{x}_o$ an arbitrary element of $L_1 \times \cdots \times L_n$, then*

$$\begin{aligned} [\phi]_{(\rho,D(\mathbf{x}_o,\cdot))} &= \{\psi \mid \psi \in \mathcal{M}(L_1 \times \cdots \times L_n, L) \text{ and } \psi(\mathbf{x}_o) = \phi(\mathbf{x}_o)\} \\ &= [\phi]_{(\rho^d,D(\mathbf{x}_o,\cdot))}. \end{aligned} \tag{13}$$

Proof. From lemma 1, we deduce that

$$\rho(D(\mathbf{x}_o,\cdot),\phi) : L \to [0,1] : \alpha \mapsto \begin{cases} 1 & ; \quad \alpha \leq \phi(\mathbf{x}_o) \\ 0 & ; \quad \text{elsewhere} \end{cases}$$

and

$$\rho(D(\mathbf{x}_o,\cdot),\psi) : L \to [0,1] : \alpha \mapsto \begin{cases} 1 & ; \quad \alpha \leq \psi(\mathbf{x}_o) \\ 0 & ; \quad \text{elsewhere} \end{cases}$$

Hence, both reliability functions equal if and only if $\phi(\mathbf{x}_o) = \psi(\mathbf{x}_o)$. Analogously, we prove that the dual reliability functions of ϕ and ψ equal if and only if $\phi(\mathbf{x}_o) = \psi(\mathbf{x}_o)$. Hence, (13) holds. □

Corollary 3 *Let ϕ be an arbitrary structure function from $(L_1 \times \cdots \times L_n, \leq)$ to $(L, \leq)$.*

$$[\phi]_{(\rho,D(\mathbf{0},\cdot))} = [\phi]_{(\rho,D(\mathbf{1},\cdot))} = \mathcal{M}(L_1 \times \cdots \times L_n, L)$$
$$= [\phi]_{(\rho^d,D(\mathbf{0},\cdot))} = [\phi]_{(\rho^d,D(\mathbf{1},\cdot))}$$

Proof. Immediate from theorem 7 and the boundary conditions for structure functions. □

Corollary 4 *Let ϕ be a structure function from $(L_1 \times \cdots \times L_n, \leq)$ naar $(L, \leq)$ and $\mathbf{x}_o$ an element of $L_1 \times \cdots \times L_n$, then*

$$[\phi]_{(\rho,D(\mathbf{x}_o,\cdot))} = [\phi]_{\Theta_{\mathbf{x}_o}} = [\phi]_{(\rho^d,D(\mathbf{x}_o,\cdot))}.$$

Proof. Immediate from theorem 7. □

Dirac measures are very crisp. Indeed, when the uncertainty on $L_1 \times \cdots \times L_n$ is represented by the Dirac measure $D(\mathbf{x}_o,\cdot)$, we know for sure that component i assumes the state $\mathrm{pr}_i(\mathbf{x}_o)$, $1 \leq i \leq n$, that is, the ith projection of $\mathbf{x}_o$. In reality, the most optimal situation appears when all components assume their optimal state. Hence, the most ideal possibility measure is $D(\mathbf{1},\cdot)$. Taking into account corollary 3 and theorem 7 all systems behave similar, which is not very surprising, considering the boundary conditions of the structure functions. From lemma 1, we deduce that $\rho(D(\mathbf{1},\cdot),\phi)$ is the constant mapping on 1 and $\rho^d(D(\mathbf{1},\cdot),\phi)$ equals $\chi_{\{1\}}$, the characteristic mapping of the singleton $\{1\}$. In an analogous way $D(\mathbf{0},\cdot)$ is the representation of a complete failure, and hence, is the worst possible situation. There is no hope that a system will show some functioning at any level; $\rho(D(\mathbf{0},\cdot),\phi)$ is the characteristic mapping $\chi_{\{0\}}$ and $\rho^d(D(\mathbf{0},\cdot),\phi)$ is the constant mapping on 1. In general we conclude that when the possibilistic reliability functions are more similar to those derived from $D(\mathbf{0},\cdot)$, the system tends to fail. Analogously, when the reliability functions are similar to those stemming from $D(\mathbf{1},\cdot)$, we can conclude that the system tends to function perfectly. In the next section some examples will clarify the different concepts.

5. SOME CONCRETE EXAMPLES

In this section, we give some possibilities of how to apply the different concepts, introduced in the previous sections. Therefore, consider the structure function ϕ from $([0,1]^2,\leq)$ to $([0,1],\leq)$,

$$\phi : [0,1]^2 \rightarrow [0,1] : (x_1,x_2) \mapsto \frac{x_1+x_2}{2},$$

and the possibility distribution π

$$\pi : [0,1]^2 \rightarrow [0,1] : (x_1,x_2) \mapsto 1- \mid x_1 - \frac{1}{2} \mid - \mid x_2 - \frac{1}{2} \mid . \qquad (14)$$

In Cappelle and Kerre (1993), we proved that

$$\rho(\Pi,\phi) : [0,1] \rightarrow [0,1] : \alpha \mapsto \min(1, 2-2\alpha)$$

and

$$\rho^d(\Pi,\phi) : [0,1] \rightarrow [0,1] : \alpha \mapsto \min(1, 2\alpha).$$

First, assume that, besides the boundary conditions, we know the value of ϕ in only two points, e.g.,

$$\phi(\frac{1}{4},\frac{3}{4}) = \frac{1}{2} \text{ and } \phi(\frac{1}{3},\frac{2}{3}) = \frac{1}{2}.$$

For the sake of simplicity, we will denote the set $\{(0,0),(\frac{1}{4},\frac{3}{4}),(\frac{1}{3},\frac{2}{3}),(1,1)\}$ by A. Hence, the set $[\phi]_{\Theta_A}$ is the closed interval $[l(A,\phi),u(A,\phi)]$. Taking into account the isotonicity of ϕ and after some calculations, we obtain explicitly

$$l(A,\phi):[0,1]^2 \to [0,1]:(x_1,x_2) \mapsto \begin{cases} 1 & ; \quad x_1 = x_2 = 1 \\ \frac{1}{2} & ; \quad (x_1,x_2) \in [\frac{1}{4},1] \times [\frac{3}{4},1] \cup \\ & \qquad [\frac{1}{3},1] \times [\frac{2}{3},1] \setminus \{(1,1)\} \\ 0 & ; \quad \text{elsewhere} \end{cases}$$

and

$$u(A,\phi):[0,1]^2 \to [0,1]:(x_1,x_2) \mapsto \begin{cases} 0 & ; \quad x_1 = x_2 = 0 \\ \frac{1}{2} & ; \quad (x_1,x_2) \in [0,\frac{1}{4}] \times [0,\frac{3}{4}] \cup \\ & \qquad [0,\frac{1}{3}] \times [0,\frac{2}{3}] \setminus \{(0,0)\} \\ 1 & ; \quad \text{elsewhere} \end{cases}$$

One easily verifies that

$$l(A,\phi) \preceq \phi \preceq u(A,\phi).$$

From theorem 5, we calculate the boundaries for the possibilistic reliability functions. We find that

$$\rho(\Pi,l(A,\phi)):[0,1] \to [0,1]:\alpha \mapsto \begin{cases} 1 & ; \quad \alpha = 0 \\ \frac{5}{6} & ; \quad 0 < \alpha \leq \frac{1}{2} \\ 0 & ; \quad \frac{1}{2} < \alpha \leq 1 \end{cases}$$

and $\rho(\Pi,u(A,\phi))$ is the constant mapping on 1, while the dual possibilistic reliability functions are

$$\rho^d(\Pi,u(A,\phi)):[0,1] \to [0,1]:\alpha \mapsto \begin{cases} 0 & ; \quad 0 \leq \alpha < \frac{1}{2} \\ \frac{5}{6} & ; \quad \frac{1}{2} \leq \alpha < 1 \\ 1 & ; \quad \alpha = 1 \end{cases}$$

and $\rho^d(\Pi, l(A, \phi))$ is the constant mapping on 1. Although very few information is known about the structure function ϕ, we may conclude that the system probably assumes a value of about $\frac{1}{2}$. Therefore, we calculated the minimum of the approximations $\rho(\Pi, u(A, \phi))(\alpha)$ and $\rho(\Pi, l(A, \phi))(\alpha)$, which is a lower bound for $\Pi(N(\alpha, \phi))$, $\alpha \in [0, 1]$. Obviously, adding more information generates more accurate possibilistic reliability bounds.

Finally, we consider the structure function ϕ from $([0,1]^2, \leq)$ to $(\{0,1\}^2, \leq)$,

$$\phi : [0,1]^2 \to \{0,1\}^2 : (x_1, x_2) \mapsto \begin{cases} (1,1) & ; \quad \frac{1}{4} \leq x_1 \leq 1 \text{ and } \frac{1}{4} \leq x_2 \leq 1 \\ (1,0) & ; \quad 0 \leq x_1 < \frac{1}{4} \text{ and } \frac{1}{4} \leq x_2 \leq 1 \\ (0,1) & ; \quad \frac{1}{4} \leq x_1 \leq 1 \text{ and } 0 \leq x_2 < \frac{1}{4} \\ (0,0) & ; \quad \text{elsewhere} \end{cases}$$

ϕ models, e.g., a system of two components subject to two kinds of failure. On $[0,1]^2$ we consider once more the possibility distribution π defined in (14). Hence, the possibilistic reliability functions are mappings from $\{0,1\}^2$ to $[0,1]$; $\rho(\Pi, \phi)$ is the constant mapping on 1 and

$$\rho^d(\Pi, \phi) : \{0,1\}^2 \to [0,1] : \begin{cases} (0,0) & \mapsto & \frac{1}{2} \\ (0,1) & \mapsto & \frac{3}{4} \\ (1,0) & \mapsto & \frac{3}{4} \\ (1,1) & \mapsto & 1 \end{cases}$$

Taking into account the remarks of section 3 and 4, we may conclude that the system probably assumes its optimal state, that is, $(1,1)$, or, less probably, a degraded state but not the total failure, that is, $(0,1)$ or $(1,0)$, since the reliability functions are more similar to $\rho(D(\mathbf{1}, \cdot), \phi)$ and $\rho^d(D(\mathbf{1}, \cdot), \phi)$ than to $\rho(D(\mathbf{0}, \cdot), \phi)$ and $\rho^d(D(\mathbf{0}, \cdot), \phi)$.

6. CONCLUSION

The congruence relation, defined on the complete lattice of structure functions, provides the reliability engineer, with some powerful tools to model a physical system and to obtain some important possibilistic reliability bounds from a set of observations. Adding more observations, diminishes the number of appropriate structure functions, satisfying all conditions imposed by the set of observations and improves the possibilistic reliability bounds. From the approximation of a structure function, some interesting possibilistic reliability bounds are derived. By Dirac measures, a measure that is both a necessity and a possibility measure, we were able to link several concepts of multistate structure functions and the theory of possibilistic reliability functions. In a few examples, we have shown how these highly theoretic concepts can be applied to many concrete problems.

Acknowledgement: The authors whish to thank the National Fund of Scientific Research (Belgium) for the support of this research. Bart Cappelle is currently a research assistant of the NFSR.

REFERENCES

L.A. Baxter (1984), *Continuum Structures I*, J. of Appl. Prob. **21**, 802–815.

G. Birkhoff (1967) "Lattice Theory," AMS Colloquium Publication Volume 24, Providence, Rhode Island.

Cai Kai-Yuan and Wen Chuan-Yuan (1990), *Street-lighting Lamp Replacement: a Fuzzy Viewpoint*, Fuzzy Sets and Systems **37**, 161–172.

B. Cappelle, E.E. Kerre and F.R. Vanmassenhove (1991) *A Possibilistic Uncertainty Model Applied to the Barlow-Wu Class of Multistate Structures* in "Proceedings of IFSA '91 - Daychair mathematics", Brussels, July 7-10, 1991 p25-28.

B. Cappelle (1991a) *Multistate Structure Functions and Possibility Theory: an Alternative Approach to Reliability* in E.E. Kerre, Ed.:Introduction to the Basic Principles of Fuzzy Set Theory and Some of its Applications, Communication and Cognition Gent, 1991, 252–293.

B. Cappelle and E.E. Kerre (1993) *On a Possibilistic Approach to Reliability Theory* in B.M. Ayyub, Ed.: Proceedings of the Second International Symposium on Uncertainty Modelling and Analysis (ISUMA 93), April25–28 1993, Maryland M.D.; IEEE Computer Society Press Los Alamitos, California, 415–418.

B. Cappelle (1993a) *Benaderingen van een structuurfunctie met behulp van ideaal- en duale ideaalstructuurfuncties,* Internal Report (in Dutch), Gent, 1993, 16 pages.

H. Cohen (1984), *Space Reliability Technology: A Historical Perspective,* IEEE Transactions on Reliability **33**, 36–40.

G. De Cooman, E.E. Kerre and F.R. Vanmassenhove (1992), *Possibility Theory: An Integral Theoretic Approach,* Fuzzy Sets and Systems **46**, 287–300.

E. El–Neweihi, F. Proschan and J. Sethuraman (1978), *Multistate Coherent Systems,* J. of Appl. Prob. **15**, 675–688.

G. Grätzer (1978) "General Lattice Theory," Birkhäuser Verlag, Basel.

J. Montero, J. Tejada and J. Yáñez (1988), *General Structure Functions,* Proceedings Workshop on Knowledge-Based Systems and Models of Logical Reasoning, Dec 26–31 1988, Cairo (Egypt).

J. Montero, B. Cappelle and E.E. Kerre (1993), *The Usefulness of Complete Lattices in Reliability Theory,* In this volume.

T. Onisawa (1988), *An approach to Human Reliability in Man-Machine Systems using Error Possibility,* Fuzzy Sets and Systems **27**, 87–103.

T. Onisawa (1990), *An Application of Fuzzy Concepts to modelling of Reliability Analysis,* Fuzzy Sets and Systems **37**, 268–286.

D. Singer (1990), *A Fuzzy Set Approach to Fault Tree and Reliability Analysis,* Fuzzy Sets and Systems **34**, 145–155.

H. Tanaka, L.T. Fan, F.S. Lai, K. Toguchi (1983), *Fault Tree Analysis by Fuzzy Probability,* IEEE Transactions on Reliability **32**, 453–457.

COHERENT SYSTEMS IN PROFUST RELIABILITY THEORY

CAI Kai-Yuan *
WEN Chuan-Yuan
ZHANG Ming-Lian

Department of Automatic Control
Beijing University of Aeronautics and Astronautics
Beijing 100083, China

ABSTRACT Profust reliability theory is based on the probability assumption and the fuzzy-state assumption. In an attempt to provide a uniform foundation for profust reliability theory, in this paper we introduce the concept of coherent systems and distinguish two classes of systems: closely coherent systems and loosely coherent systems. To formulate the relationships between system reliabilities and component reliabilities, we present a number of general results, results concerning convexity, and results concerning unimodality. It is argued that for a coherent system, component reliability improvements don't certainly enhance the system reliability.

Keywords Profust reliability, Coherent system, Convexity, Unimodality

1 INTRODUCTION

It's well-known that conventional (i.e., probist) reliability theory is based on the probability assumption and the binary-state assumption. Although these two assumptions sound reasonable for a large variety of systems, it has been strongly argued that they are unacceptable in extensive cases [1-11] and thus various kinds of fuzzy reliability theories can be developeed, including profust reliability theory, posbist reliability theory, and posfust reliability theory Among them, profust reliability theory is based on the following two assumptions:

A1. *Fuzzy-State assumption*: The meaning of system failure is defined in a fuzzy way and thus at any time the system is in fuzzy failure state or in fuzzy success state to some extent.

A2. *Probability assumption:* The system failure behavior can be fully characterized in the context of probability measures.

It's generally thought of that the concept of coherent system provides a uniform foundation for probist reliability theory. Similarly, in order to provide a foundation for profust reliability theory, in this paper we introduce the concept of coherent systems in the context of profust reliability theory. However we must note profust reliability behavior of coherent systems may exhibit in extremely varying ways, and thus we should distinguish closely coherent systems and loosely coherent systems. This will be explained in Section 2. For coherent systems, the central

---

* Present Address: **Centre for Software Reliability, City University, Northampton Square, London EC1V 0HB, U.K.**

problem is to establish relationships among system reliabilities and component reliabilitis. For this purpose, Section 3 gives general results for coherent systems. Section 4 presents results concerning convexity of lifetime distributions. Section 5 is devoted to developing results concerning unimodality of lifetime distributions. Finally, concluding remarks are contained in Section 6.

2 COHERENT SYSTEMS

Let $U=\{u\}$ be the state universe of discourse, discrete or continuous, where u represents a particular non-fuzzy system state and U comprises all non-fuzzy states of the system. On this non-fuzzy universe we define a fuzzy system failure state

$$F=\{u,\mu_F(u)\} \tag{2.1}$$

and a fuzzy system success state

$$S=\{u,\mu_S(u)\} \tag{2.2}$$

where μ_F and μ_S are failure and sucess membership functions, respectively. In general, we take the postulation that

$$\mu_S(u)=1-\mu_F(u).........\forall u\in U \tag{2.3}$$

Let

$$T_{SF}=\{m_{xy},\mu_{T_{SF}}(m_{xy})\} \tag{2.4}$$

where m_{xy} represents the transition from state x to state y. Then T_{SF} is interpreted as the transition from S to F, and is just a fuzzy event.

Profust system reliability is defined as

$$R(t)=P\{\,T_{SF} \text{ doesn't occur during the time interval } (0,t)\} \tag{2.5}$$

Then it can be shown that [8]

$$R(t)=1-E\mu_F=E.\mu_S \tag{2.6}$$

or

$$R(t)=1-F(t)=\overline{F}(t) \tag{2.7}$$

where

$$F(t)=E.\mu_F \tag{2.8}$$

where E is the expectation operator in snese of Lebesegue-Stieljes integral.

Suppose a system is of order n with failure membership function μ_{F_i} and success membership function μ_{S_i} for the ith component. We assume μ_F is uniquely determined by $\mu_{F_1},\ldots,\mu_{F_n}$ and μ_S is uniquely determined by $\mu_{S_1},\ldots,\mu_{S_n}$. We denote

$$\mu_F=\Phi_F(\mu_{F_1},\ldots,\mu_{F_n}) \tag{2.9}$$

$$\mu_S=\Phi_S(\mu_{S_1},\ldots,\mu_{S_n}) \tag{2.10}$$

DEFINITION 2.1 $\Phi_F\,/\,\Phi_S$ is referred to as failure/success membership structure function.

DEFINITION 2.2 μ_{F_i} / μ_{S_i} is irrelevant to Φ_F / Φ_S if Φ_F / Φ_S is constant in μ_{F_i} / μ_{S_i}. Otherwise μ_{F_i} / μ_{S_i} is relevant to Φ_F / Φ_S.

DEFINITION 2.3 Φ_F is coherent if

(1). Φ_F is increasing in each argument μ_{F_i}.

(2). Each μ_{F_i} is relevant to Φ_F.

(3). $\Phi_F(0,\ldots,0)=0, \Phi_F(0,\ldots,0)=1$.

DEFINITION 2.4 Φ_S is coherent if

(1). Φ_S is increasing in each argument μ_{S_i}.

(2). Each μ_{S_i} is relevant to Φ_S.

(3). $\Phi_S(0,\ldots,0)=0, \Phi_S(0,\ldots,0)=1$

DEFINITION 2.5 A system, denoted by $\Phi=(\Phi_F,\Phi_S)$, is coherent iff Φ_F and Φ_S are each coherent.

We see that the system reliability is $E\mu_S = E\Phi_S(\mu_{S_1},\ldots,\mu_{S_n})$ and the component reliabilities are $E\mu_{S_1},\ldots,E\mu_{S_n}$. Then it is worthy to wonder what the relationships are between the system reliability and the component reliabilities, and whether, given the component reliabilities, the system reliability is uniquely determined.

EXAMPLE 2.1 In conventional (i.e., probist) reliability theory[12], the reliability of a coherent system is uniquely determined by its component reliabilities if all the components are independent, and an opposite situation arises if dependence underllies among the components.

EXAMPLE 2.2 Consider a mixture model (system) defined as follows [9]

$$\mu_F = \sum_{i=1}^{n} c_i \mu_{F_i}$$

$$\mu_S = 1 - \sum_{i=1}^{n} c_i \left(1-\mu_{S_i}\right)$$

where $c_i \geq 0$ for each $i=1,\ldots,n$ and $\sum_{i=1}^{n} c_i = 1$. Apparantly the system reliability is uniquely determined by component reliabilities.

EXAMPLE 2.3 Consider a system with two independent components and the following relationships

$$\mu_F = \mu_{F_1}^2 \mu_{F_2}$$

$$\mu_S = 1-\left(1-\mu_{S_1}\right)^2\left(1-\mu_{S_2}\right)$$

Obviously, $E\mu_{F_1}, E\mu_{F_2}$ cannot uniquely determine $E\mu_F$, and $E\mu_{S_1}, E\mu_{S_2}$ cannot uniquely determine $E\mu_S$ either.

From the above examples we see that whether the reliability of a coherent system is uniquely determined by its component reliabilities is heavily involved in the function forms of Φ_F and Φ_S, the distributions of $\mu_{F_1},\ldots,\mu_{F_n}$ and $\mu_{S_1},\ldots,\mu_{S_1}$, and the dependence of components. So it is valuable to distinguish two classes of coherent systems: closely coherent systems and loosely coherent systems.

DEFINITION 2.6 A coherent system $\Phi=(\Phi_F,\Phi_S)$ of order n is said to be close if $E\Phi_F$ can be uniquely determined by $E\mu_{F_1},\ldots,E\mu_{F_n}$, and $E\Phi_S$ can be uniquely determined by $E\mu_{S_1},\ldots,E\mu_{S_n}$.

DEFINITION 2.7 A coherent system $\Phi=(\Phi_F,\Phi_S)$ is loose if it is not close.

3 GENERAL RESULTS

In the sequel, from time to time, we use a uniform form $g(X_1,\ldots,X_n)$ to represent $\Phi_F(\mu_{F_1},\ldots,\mu_{F_n})$ and $\Phi_S(\mu_{S_1},\ldots,\mu_{S_n})$. We only give general results concerning $g(X_1,\ldots,X_n)$. Results on $\Phi_F(\mu_{F_1},\ldots,\mu_{F_n})$ and $\Phi_S(\mu_{S_1},\ldots,\mu_{S_n})$ can be easily obtained by taking substitutions of $\Phi_F(\mu_{F_1},\ldots,\mu_{F_n})$ and $\Phi_S(\mu_{S_1},\ldots,\mu_{S_n})$ for $g(X_1,\ldots,X_n)$, respectively. Thus in the sequel we don't present results on $\Phi_F(\mu_{F_1},\ldots,\mu_{F_n})$ and $\Phi_S(\mu_{S_1},\ldots,\mu_{S_n})$ seperately. Further, $\vec{X}=(X_1,\ldots,X_n)$, and $\vec{X}^{(1)}\le\vec{X}^{(2)}$ means $X_i^{(1)}\le X_i^{(2)}$ for each $i=1,\ldots,n.g(X_1,\ldots,X_n)$ may be signified $g(\vec{X})$.

THEOREM 3.1 Let $g(\vec{X})$ be a *Borel*-measurable function and nondecreasing in each argument. Let $\vec{X}^{(1)}\le\vec{X}^{(2)}$. Then $Eg(\vec{X}^{(1)})\le Eg(\vec{X}^{(2)})$.

PROOF: Obvious.

This theorem implies that if the component performance is improved in a deterministic way, the system reliability is certainly improved.

EXAMPLE 3.1 Consider a system with n components, where failure and success membership functions of the ith component are, respectively, determined by

$$\mu_{F_i}=\mu_{F_i}^{(1)}\mu_{F_i}^{(2)}$$

and

$$\mu_{S_i}=1-\left(1-\mu_{S_i}^{(1)}\right)\left(1-\mu_{S_i}^{(2)}\right)$$

That is, each component has two subcomponents, whose failure and success memberhsip functions are indexed with upperscripts.

Then we have

$$E\mu_F\le E\mu_F^{(j)}\ldots\ldots\ldots j=1,2$$

$$E\mu_S\ge E\mu_S^{(j)}\ldots\ldots\ldots j=1,2$$

Nevertheless, if the component performance is improved is a probabilistic way, that is, the component reliabilities are improved, it is not sufficient to assert that the system reliability is improved accordingly. This observation differs from the corresponding one in probist reliability theory, which indicates that component reliability improvements certainly enhance reliability of a coherent system.

EXAMPLE 3.2 Consider a system with two independent components and

$$\mu_F = 1-(1-\mu_{F_1})^2(1-\mu_{F_2})^2$$
$$\mu_S = \mu_{S_1}^2\mu_{S_2}^2$$

Let $\mathrm{var}(X)$ denote the variance of random variable X. Then

$$E\mu_S = E\mu_{S_1}^2 . E\mu_{S_2}^2$$
$$=\left[(\mathrm{var}(\mu_{S_1}))^2+(E\mu_{S_1})^2\right]\left[(\mathrm{var}(\mu_{S_2}))^2+(E\mu_{S_2})^2\right]$$

In this way we can denote

$$E\mu_S^{(1)}=\left[(\mathrm{var}(\mu_{S_1}^{(1)}))^2+(E\mu_{S_1}^{(1)})^2\right]\left[(\mathrm{var}(\mu_{S_2}^{(1)}))^2+(E\mu_{S_2}^{(1)})^2\right]$$
$$E\mu_S^{(1)}=\left[(\mathrm{var}(\mu_{S_1}^{(2)}))^2+(E\mu_{S_1}^{(2)})^2\right]\left[(\mathrm{var}(\mu_{S_2}^{(2)}))^2+(E\mu_{S_2}^{(2)})^2\right]$$

From the above equalities we see that $E\mu_{S_1}^{(1)} \le E\mu_{S_1}^{(2)}$ and $E\mu_{S_2}^{(1)} \le E\mu_{S_2}^{(2)}$ don't certainly facilate $E\mu_S^{(1)} \le E\mu_S^{(2)}$.

However we should note that under some adequate conditions, the component reliability improvements may enhance the system reliability. The related implications may be revealed, more or less, from the following lemma [13].

LEMMA 3.1 If $X=(X_n,\Im_n)$ is a submartingale and $g(X)$ is a convex downward and nondecreasing with $E|g(X_n)| \prec \infty$ for all $n \succ 0$. Then $(g(X_n),\Im_n)$ is also a submartigale.

At this point we should point out that what the necessary and sufficient conditions are to ensure the system reliability to be increasing in the component reliabilities is just an open problem. Here we take no assertions about the conditions for system reliability improvement under the improvements of component reliabilities.

LEMMA 3.2 [14] Let X be a random variable and let g be a nonnegative *Borel*-measurable function such that $Eg(X) \prec \infty$. Suppose that g is even and nondecreasing on $[0,\infty)$. Then, for every $\varepsilon \succ 0$

$$P\{|X| \ge \varepsilon\} \le \frac{Eg(X)}{g(\varepsilon)} \tag{3.1}$$

Further, if $a.s.\sup g(X) \prec \infty$, (Note: $a.s.\sup|X| = \inf\{c: 0 \le c \le \infty, P\{|X| \succ c\}=0\}$), then

$$P\{|X| \ge \varepsilon\} \ge \frac{Eg(X)-g(\varepsilon)}{a.s.\sup g(X)} \tag{3.2}$$

holds for every $\varepsilon \succ 0$.

LEMMA 3.3 [13] (Fubini's Theorem) Let $\xi=\xi(\omega_1,\omega_2)$ be an $\Im_1 \otimes \Im_2$-measurable function, integrable with respect to the measure $\mu_1 \times \mu_2$,

$$\int_{\Omega_1\times\Omega_2} |\xi(\omega_1,\omega_2)| d(\mu_1 \times \mu_2) \prec \infty$$

Then the integrals $\int_{\Omega_1} \xi(\xi_1,\xi_2)\mu_1(d\omega_1)$ and $\int_{\Omega_2} \xi(\omega_1,\omega_2)\mu_2(d\omega_2)$

(1). are defined for all ω_1 and ω_2.

(2). are respectively $\mathfrak{I}_2$- and $\mathfrak{I}_1$-measurable function with

$$\mu_2\left\{\omega_2: \int_{\Omega_1} |\xi(\omega_1,\omega_2)|\mu_1(d\omega_1)=\infty\right\}=0$$

$$\mu_1\left\{\omega_1: \int_{\Omega_2} |\xi(\omega_1,\omega_2)|\mu_2(d\omega_2)=\infty\right\}=0$$

and (3).

$$\int_{\Omega_1\times\Omega_2} \xi(\omega_1,\omega_2)d(\mu_1\times\mu_2)$$

$$=\int_{\Omega_1}\left[\int_{\Omega_2} \xi(\omega_1,\omega_2)\mu_2(d\omega_2)\right]\mu_1(d\omega_1) \tag{3.3}$$

$$=\int_{\Omega_2}\left[\int_{\Omega_1} \xi(\omega_1,\omega_2)\mu_1(d\omega_1)\right]\mu_2(d\omega_2)$$

THEOREM 3.2 Let $X_1,\ldots,X_n$ be independent and $g(X_1,\ldots,X_2)$ be a nonnegative *Borel* function such that $Eg(X_1,\ldots,X_n)\prec\infty$. Suppose g is even and nondecreasing in each argument on $[0,\infty)$. Then

$$Eg(X_1,\ldots,X_2)\geq g(\varepsilon_1,\ldots,\varepsilon_2).\prod_{i=1}^{n}P\{|X_i|\geq\varepsilon_i\} \tag{3.4}$$

holds for every n-tuple $(\varepsilon_1,\ldots,\varepsilon_n)$, where $\varepsilon_i\succ 0$ for each $i=1,\ldots,n$.

PROOF: Since $X_1,\ldots,X_n$ are independent, from Fubini's Theorem, we have

$$Eg(X_1,\ldots,X_n)=E_{X_n}\left[E_{X_{n-1}}\left[\ldots\left[E_{X_1}g(X_1,\ldots,X_n)\right]\ldots\right]\right]$$

However Lemma 3.2 implies that for every $\varepsilon_1\succ 0$,

$$E_{X_1}g(X_1,\ldots,X_n)\geq g(\varepsilon_1,X_2,\ldots,X_n).P\{|X_1|\geq\varepsilon_1\}$$

Thus

$$E_{X_2}\left[E_{X_1}g(X_1,\ldots,X_n)\right]\geq E_{X_2}\left[g(\varepsilon_1,X_2,\ldots,X_n).P\{|X_1|\geq\varepsilon_1\}\right]$$
$$=E_{X_2}g(\varepsilon_1,X_2,\ldots,X_n).P\{|X_1|\geq\varepsilon_1\}$$
$$\geq g(\varepsilon_1,\varepsilon_2,X_3,\ldots,X_n).P\{|X_2|\geq\varepsilon_2\}.P\{|X_1|\geq\varepsilon_1\}$$

In an inductive way, inequality (3.4) can be obtained and the theorem is thus proved.

THEOREM 3.3 Let $X_1,\dots,X_n$ be independent and $g(X_1,\dots,X_n)$ be a nonnegative *Borel* function such that $Eg(X_1,\dots,X_n)\prec\infty$. Suppose g is even and nondecreasing in each argument on $[0,\infty)$. Further $a.s.\sup g(X)\le 1$. Then

$$Eg(X_1,\dots,X_n)\le g(\varepsilon_1,\dots,\varepsilon_n)+\sum_{i=1}^{n}P\{|X_i|\ge\varepsilon_i\} \tag{3.5}$$

holds for every n-tuple $(\varepsilon_1,\dots,\varepsilon_n)$, where $\varepsilon_i\succ 0$ for each $i=1,\dots,n$.

PROOF: Because of the indepedence of $X_1,\dots,X_n$ and Fubini's Theorem, we have

$$Eg(X_1,\dots,X_n)=E_{X_n}\left[E_{X_{n-1}}\left[\dots\left[E_{X_1}g(X_1,\dots,X_n)\right]\dots\right]\right]$$

From Lemma 3.2, we obtain

$$E_{X_1}g(X_1,\dots,X_n)\le g(\varepsilon_1,X_2,\dots,X_n)+P\{|X_1|\ge\varepsilon_1\}$$

Thus

$$\begin{aligned}E_{X_2}\left[E_{X_1}g(X_1,\dots,X_n)\right]&\le E_{X_2}g(\varepsilon_1,X_2,\dots,X_n)+P\{|X_1|\ge\varepsilon_1\}\\ &\le g(\varepsilon_1,\varepsilon_2,X_3,\dots,X_n)+P\{|X_2|\ge\varepsilon_2\}+P\{|X_1|\ge\varepsilon_1\}\end{aligned}$$

In an inductive way inequality (3.5) can be obtained and the theorem is thus proved.

LEMMA 3.4 Suppose $(\Omega,\Im,P)$ is a probability space. Then P can be expressed as a convex mixture $\theta_1P_1+\theta_2P_2+\theta_3P_3$.

PROOF: Suppose A_1,A_2,A_3 construct a decomposition of Ω. Let

$$\theta_i=\int_{A_i}dP(\omega)\dots\dots\dots i=1,2,3$$

Then $\theta_1+\theta_2+\theta_3=1$. Let

$$P_i(B)=\frac{1}{\theta_i}.P(B\cap A_i)\dots\dots\dots\forall B\subset\Im$$

In this way P_i is a probability measure and

$$P=\theta_1P_1+\theta_2P_2+\theta_3P_3$$

The lemma is thus proved.

THEOREM 3.4 Suppose X is a random variable and $g(X)$ is a nonnegative, *Borel*-measurable function, Further $g(X)$ is even and nondecreasing on $[0,\infty)$. Then, for every $\varepsilon\succ 0$, there exists $r^*\succ 1$ such that

$$P\{|X|\ge\varepsilon\}\le\frac{r^*.Eg(X)-g(\varepsilon)}{(r^*-1)g(\varepsilon)} \tag{3.6}$$

PROOF: Let $A_\varepsilon=\{x:|x|\ge\varepsilon\}$. Consider two cases.

Case 1. Suppose $P(A_\varepsilon)=1$. Then

$$Eg(X)=\int_{A_\varepsilon} g(x)dP(x)=\int_{|x|\ge\varepsilon} g(x)dP(x)$$
$$\ge\int_{|x|\ge\varepsilon} g(\varepsilon)dP(x)=g(\varepsilon)$$

So for every $r^* \succ 1$, we obtain

$$\frac{r^* Eg(X)-g(\varepsilon)}{(r^*-1)g(\varepsilon)}\ge 1=P(A_\varepsilon)$$

Case 2. Suppose $P(A_\varepsilon)\prec 1$. Then, from Lemma 3.4, we can express P as a convex mixture $\theta_1 P_1+\theta_2 P_2+\theta_3 P_3$, where P_1 has support $[-\varepsilon,\varepsilon]$, P_2 has support $[\varepsilon,\infty)$, and P_3 has support $(-\infty,-\varepsilon]$. Write $\tau_i=\int g(x)dP_i(x), i=1,2,3$. For $i=1$,

$$\tau_1=\int g(x)dP_1(x)$$
$$\le\int_{-\varepsilon}^{\varepsilon} g(x)dP(x)=g(\varepsilon)$$

Thus we can choose some $r^* \succ 1$ such that

$$r^*\tau_1\ge g(\varepsilon)$$

Hence

$$P_1(A_\varepsilon)=0\le\frac{r^*\tau_1-g(\varepsilon)}{(r^*-1)g(\varepsilon)}$$

Further, if $i=2,3$, then $P_i(A_\varepsilon)=1$ and so *Case 1* above shows that

$$P_i(A_\varepsilon)\le\frac{r^*\tau_i-g(\varepsilon)}{(r^*-1)g(\varepsilon)}\ Eg(X)\prec g(\varepsilon)$$

So inequality (3.6) is obtained and the theorem is thus proved.
REMARK Denote

$$f(r^*)=\frac{r^* Eg(X)-g(\varepsilon)}{(r^*-1)g(\varepsilon)}$$

It is easy to verify that $f(r^*)\ge 1$ and $f(r^*)$ is decreasing in r^* when $Eg(X)\ge g(\varepsilon)$. Further $f(r^*)\le\dfrac{Eg(X)}{g(\varepsilon)}\prec 1$ and $f(r^*)$ is increasing in $r*$ when $Eg(X)\prec g(\varepsilon)$.

THEOREM 3.5 Let $X_1,\ldots,X_n$ be independent and $g(X_1,\ldots,X_n)$ be a nonnegative *Borel* function. Suppose $g(X_1,\ldots,X_n)$ is even and nondecreasing in each argument on $[0,\infty)$. Then, for every n-tuple $(r_1^*,\ldots,r_n^*)$, where $r_i^*\succ 1$ for each $i=1,\ldots,n$, such that

$$Eg(X_1,\ldots,X_n)\ge g(\varepsilon_1,\ldots,\varepsilon_n)\cdot\prod_{i=1}^{n} q_i(\varepsilon_i,r_i^*) \tag{3.7}$$

where

$$q_i(\varepsilon_i, r_i^*) = \frac{(r^* - 1)P\{|X_i| \ge \varepsilon_i\} + 1}{r_i^*} \tag{3.8}$$

PROOF: Since $X_1, \ldots, X_n$ are independent, from Fubini's Theorem, we have

$$Eg(X_1, \ldots, X_n) = E_{X_n}\left[E_{X_{n-1}}\left[\cdots\left[E_{X_1} g(X_1, \ldots, X_n)\right]\cdots\right]\right]$$

From Theorem 3.4, there holds

$$E_{X_1} g(X_1, \ldots, X_n) \ge q_1(\varepsilon_1, r_1^*) . g(\varepsilon_1, X_2, \ldots, X_n)$$

Thus

$$E_{X_2}\left[E_{X_1} g(X_1, \ldots, X_n)\right] \ge q_1(\varepsilon_1, r_1^*) . E_{X_2} g(\varepsilon_1, X_2, \ldots, X_n)$$

$$\ge q_1(\varepsilon_1, r_1^*) . q_2(\varepsilon_2, r_2^*) . E_{X_3} g(\varepsilon_1, \varepsilon_2, X_3, \ldots, X_n)$$

In an inductive way, inequality (3.7) can be obtained and the theorem is thus obtained.

4 RESULTS CONCERNING CONVEXITY

LEMMA 4.1 [15] (Jensen's Inequality) If g is convex and EX finite, then

$$g(EX) \le Eg(X)$$

THEOREM 4.1 Suppose $X_1, \ldots, X_n$ are independent and $EX_1, \ldots, EX_n$ finite. Further, $g(X_1, \ldots, X_n)$ is coordinatewise convex. Then

$$Eg(X_1, \ldots, X_n) \ge g(EX_1, \ldots, EX_n) \tag{4.1}$$

PROOF: Since $X_1, \ldots, X_n$ are independent, from Fubini's Theorem, we arrive at

$$Eg(X_1, \ldots. X_n) = E_{X_n}\left[E_{X_{n-1}}\left[\cdots\left[E_{X_1} g(X_1, \ldots, X_n)\right]\cdots\right]\right]$$

Futher, $g(X_1, \ldots, X_n)$ is coordinatewise convex, from Lemma 4.1, we have

$$E_{X_1} g(X_1, \ldots, X_n) \ge g(EX_1, X_2 \ldots, X_n)$$

Thus

$$E_{X_2}\left[E_{X_1} g(X_1, \ldots, X_n)\right] \ge E_{X_2} g(EX_1, X_2, \ldots, X_n)$$

$$\ge g(EX_1, EX_2, X_3, \ldots, X_n)$$

In an inductive way, inequality (4.1) can be obtained and the theorem is thus proved.

LEMMA 4.2 [16] (Markov inequality) Let $X \ge 0$ be a random variable with positive finite expectation EX. Then, for every $\varepsilon \succ 0$

$$P\{X \ge \varepsilon . EX\} \le 1/\varepsilon$$

THEOREM 4.2 Suppose nonnegative random variables $X_1, \ldots, X_n$ are independent. Suppose $g(X_1, \ldots, X_n)$ is nonnegative and coordinatewise convex. Then

$$\max_{1\le i\le n} P\left\{X_i \ge \frac{EX_i}{g(EX_1,\dots,EX_n)}\right\} \le Eg(X_1,\dots,X_n) \tag{4.2}$$

PROOF: From Markov inequality, we have

$$P\left\{X_i \ge \frac{EX_i}{g(EX_1,\dots,EX_n)}\right\} \le g(EX_1,\dots,EX_n)\dots\dots\dots;i=1,\dots,n$$

Thus

$$\max_{1\le i\le n} P\left\{X_i \ge \frac{EX_i}{g(EX_1,\dots,EX_n)}\right\} \le g(EX_1,\dots,EX_n)$$

Since $X_1,\dots,X_n$ are independent and $g(X_1,\dots,X_n)$ is coordinatewise convex, from Theorem 4.1, there holds

$$g(EX_1,\dots,EX_n) \le Eg(X_1,\dots,X_n)$$

Hence inequality (4.2) holds and the theorem is thus proved.

5 RESULTS CONCERNING UNIMODALITY

DEFINITION 5.1 A real random varaible X or its distribution function F is called unimodal about mode (or vertex) v if F is convex on $(-\infty, v)$ and concave on (v,∞)

THEOREM 5.1 Let random variable X be unimodal about model M and $g(X)$ be a nonnegative and *Borel* -measurable function. Suppose $g(X)$ is nondecreasing on $\Re$ (the real number line) and $g^{\leftarrow}(y)=\max_{y=g(x)}\{x\}$ be convex on $(-\infty,g(M))$ and concave on $(g(M),\infty)$. Further, $g^{\leftarrow}(g(M))=M$. Then $g(X)$ is unimodal about mode $g(M)$.

PROOF: Denote distribution function of X as F_X and distribution function of $g(X)$ as F_g. Then we have

$$F_g(y)=P\{g(X)\le y\}$$
$$=P\{X\le g^{\leftarrow}(y)\}=F_X(g(y))$$

Since X is unimodal about mode M, X is convex on $(-\infty,M)$. Further, it is known that $g^{\leftarrow}(y)$ is convex on $(-\infty,g(M))$ and F_X is a monotone increasing function. Hence, on $(-\infty,g(M))$, given arbitary $0\le\alpha_1=1-\alpha_2\le 1$, we have

$$F_g(\alpha_1 y_1+\alpha_2 y_2)=F_X(g^{\leftarrow}(\alpha_1 y_1+\alpha_2 y_2))$$
$$\le F_X(\alpha_1 g^{\leftarrow}(y_1)+\alpha_2 g^{\leftarrow}(y_2))$$
$$\le \alpha_1 F_X(g^{\leftarrow}(y_1))+\alpha_2 F_X(g^{\leftarrow}(y_2))=\alpha_1 F_g(y_1)+\alpha_2 F_g(y_2)$$

That is to say, $F_g(y)$ is convex on $(-\infty, g(M))$. Similarly, it can be proved $F_g(y)$ is concave on $(g(M), \infty)$. So $g(X)$ is unimodal about mode $g(M)$.

LEMMA 5.1 [17] Let X have a unimodal distribution with mode M. For $a \in \Re$ and $r \succ 0$, let $\tau_r = E(|X-a|^r)$. Then

$$|a-M|^r \le r^* \tau_r$$

where r^* is given by

$$r^* \succ (r+1) \ldots and \ldots r^*(r^*-r-1)^r = r^r$$

THEOREM 5.2 Let X_i be a nonnegative random variable and have a unimodal distribution with mode M_i for each $i=1,\ldots,n$. Let $g(X_1,\ldots,X_n)$ be a nonnegative *Borel* function. Suppose $g(X_1,\ldots,X_n)$ is coordinatewise convex and nondecreasing on $[0,\infty)$. Further, $X_1,\ldots,X_n$ are independent. Then

$$Eg(X_1,\ldots,X_n) \ge g\left(\frac{M_1}{1+\sqrt{2}},\ldots,\frac{M_n}{1+\sqrt{2}}\right) \tag{5.1}$$

PROOF: First, from Lemma 5.1, we have

$$EX_i \ge \frac{M_i}{1+\sqrt{2}} \ldots\ldots\ldots ; i=1,\ldots,n$$

However Theorem 4.1 implies

$$Eg(X_1,\ldots,X_n) \ge g(EX_1,\ldots,EX_n)$$

Since $g(X_1,\ldots,X_n)$ is coordinatewise nondecreasing, inequality (5.1) is thus obtained and the theorem is proved.

LEMMA 5.2 [17] Let X have a distribution which is unimodal about zero. Then, for every $r \succ 0$ and $\varepsilon \succ 0$

$$P\{|X| \ge \varepsilon\} \le \left(\frac{r}{r+1}\right)^r \cdot \frac{E(|X|^r)}{\varepsilon^r}$$

THEOREM 5.3 Suppose nonnegative random variable $X_1,\ldots,X_n$ are independent and all are unimodal about zero. Let $g(X_1,\ldots,X_n)$ be coordinatewise convex and nondecreasing on $[0,\infty)$. Then, for every n-tuple $(\varepsilon_1,\ldots,\varepsilon_n)$, where $\varepsilon_i \succ 0$ for $i=1,\ldots,n$,

$$Eg(X_1,\ldots,X_n) \ge g(\varepsilon_1^*,\ldots,\varepsilon_n^*) \tag{5.2}$$

where

$$\varepsilon_i^* = 2\varepsilon_i P\{X_i \ge \varepsilon_i\}$$

PROOF: First, from Lemma 5.2, we have

$$EX_i \ge 2\varepsilon_i P\{X_i \ge \varepsilon_i\}$$

However Theorem 4.1 implies

$$Eg(X_1,\ldots,X_n) \ge g(EX_1,\ldots,EX_n)$$

Since $g(X_1,\ldots,X_n)$ is coordinatewise nondecreasing, inequality (5.2) is thus obtained and the theorem is proved.

LEMMA 5.3 [17] Suppose X is a unimodal random variable with variance σ^2 and mode M. Then

$$(EX-M)^2 \le 3\sigma^2$$

THEOREM 5.4 Suppose $X_1,\ldots,X_n$ are nonnegative and independent, and X_i is unimodal with mode M_i and variance σ^2 for each $i=1,\ldots,n$. Let $g(X_1,\ldots,X_n)$ be coordinatewise convex and nondecreasing on $[0,\infty)$. Then

$$Eg(X_1,\ldots,X_n) \ge g(\alpha_1,\ldots,\alpha_n) \tag{5.3}$$

where

$$\alpha_i = \max\left(M_i - \sqrt{3}\sigma_i, 0\right)\ldots\ldots\ldots; i=1,\ldots,n$$

PROOF: First, from Lemma 5.3, we have

$$EX_i \ge M_i - \sqrt{3}\sigma_i$$

Since X_i is nonnegative,

$$EX_i \ge \max\left(M_i - \sqrt{3}\sigma_i, 0\right)$$

However Theorem 4.1 implies

$$Eg(X_1,\ldots,X_n) \ge g(EX_1,\ldots,EX_n)$$

Further, $g(X_1,\ldots,X_n)$ is coordinatewise nondecreasing, thus inequality (5.3) is obtained and the theorem is proved.

THEOREM 5.5 Suppose $X_1,\ldots,X_n$ are nonnegative and independent, and X_i is unimodal with mode M_i and variance σ^2 for each $i=1,\ldots,n$. Let $g(X_1,\ldots,X_n)$ be a nonnegative *Borel* function. Let $g(X_1,\ldots,X_n)$ be coordinatewise convex and nondecreasing on $[0,\infty)$. Then

$$Eg(X_1,\ldots,X_n) \ge g(\beta_1,\ldots,\beta_n) \tag{5.4}$$

where

$$\beta_i = \max\left(\frac{M_i}{1+\sqrt{2}}, M_i - \sqrt{3}\sigma_i\right)$$

PROOF: From Lemma 5.1, we have

$$EX_i \ge \frac{M_i}{1+\sqrt{2}}$$

Further, Lemma 5.1 indicates

$$EX_i \ge M_i - \sqrt{3}\sigma_i$$

Thus

$$EX_i \ge \max\left(\frac{M_i}{1+\sqrt{2}}, M_i - \sqrt{3}\sigma_i\right)$$

However Theorem 4.1 implies

$$Eg(X_1,\ldots,X_n) \geq g(EX_1,\ldots,EX_n)$$

Since $g(X_1,\ldots,X_n)$ is coordinatewise nondecreasing, inequality (5.4) is thus obtained and the theorem is proved.

6 CONCLUDING REMARKS

It is well-known that the concept of coherent systems provides a uniform foundation for probist reliability theory. In this paper we have defined the concept of coherent systems for profust reliability theory and distinguish two classes of them: closely coherent systems and loosely coherent systems. For a closely coherent system, the system reliability can be uniquely determined by its component reliabilities. However this is not the case for a loosely coherent system. For coherent systems, the central problem is to formulate the relationships between system reliabilities and component reliabilities. To this end, we have presented a number of general results, results concerning convexity and results concerning unimodality. A surprising assertion is that component reliability improvements don't certainly enhance the system reliability. This is contrary to the corresponding assertion in probist reliability theory. Further we point out that what the necessary and sufficient conditions are for a coherent system to be close or loose is just an open problem.

REFERENCES

1. K.Y.Cai, C.Y.Wen, M.L.Zhang, "A Critical Review on Software Reliability Modeling", *Reliability Engineering and System Safety,* Vol.32, 1991, pp357-371.
2. K.Y.Cai, C.Y.Wen, "Uncertainties of Operating Time in System Reliability", *Acta Aeronautica et Astronautica Sinica,* Vol.11, No.1, 1990, pp98-105.
3. K.Y.Cai. C.Y.Wen, "Streeting-Lighting Lamps Replacement: A Fuzzy Viewpoint", *Fuzzy Sets and Systems,* Vol.37, 1990, pp161-172.
4. K.Y.Cai, C.Y.Wen, M.L.Zhang, "Fuzzy Reliability Modeling of Gracefully Degradable Computing Systems", *Reliability Engineering and System Safety,* Vol.35, 1991, pp141-157.
5. K.Y.Cai, C.Y.Wen, Z.J.Chen, "A Preliminary Analysis of CIMS Reliability with Fuzzy Viewpoint", *Proc. Beijing International Conference on System Simulation and Scientific Computing,* 1989, pp771-775.
6. K.Y.Cai, C.Y.Wen, M.L.Zhang, "Survivability Index for CCNs: A Measure of Fuzzy Reliability", *Reliability Engineering and System Safety,* Vol.33, 1991, pp71-99.
7. K.Y.Cai, C.Y.Wen, M.L.Zhang, "Fuzzy Variables as a Basis for a Theory of Fuzzy Reliability in the Possibility Context", *Fuzzy Sets and Systems,* Vol.42, 1991, pp145-172.
8. K.Y.Cai. C.Y.Wen, M.L.Zhang, "Fuzzy States as a Basis for a Theory of Fuzzy Reliability", *Microelectronics and Reliability,* Vol.33, No.15, 1993, pp2253-2263.
9. K.Y.Cai, C.Y.Wen, M.L.Zhang, "Mixture Models in Profust Reliability Theory", *to appear.*

10. K.Y.Cai, C.Y.Wen, M.L.Zhang, "Posbist Reliability Behavior of Typical Systems with Two Types of Failures", *Fuzzy Sets and Systems*, Vol.43, 1991, pp17-32.
11. K.Y.Cai, C.Y.Wen, M.L.Zhang, "Posbist Reliability Behavior of Fault-Tolerant Systems", *to appear.*
12. R.E.Barlow, F.Proschan, *Statistical Theory of Reliability and Life Testing: Probability Models,* Holt, Rinehart and Winston, Inc., 1975.
13. A.N.Shiryayev, *Probability*, Springer-Verlag, 1984.
14. R.G.Laha, V.K.Rohatgi, *Probability Theory*, John Wiley & Sons, 1979,
15. P.R.Bhat, *Modern Probability Theory*, John Wiley & Sons, 1981.
16. J.Galambos, *Advanced Probability Theory,* Marcel Dekker, Inc., 1988.
17. S.Dharmahikari, K.Joag-dev, *Unimodality, Convexity, and Applications,* Academic Press, 1988.

THE USEFULNESS OF COMPLETE LATTICES IN RELIABILITY THEORY

Javier MONTERO[1], Bart CAPPELLE[2] and Etienne E. KERRE[2]

[1]Department of Statistics and O.R.
Faculty of Mathematics
Universidad Complutense de Madrid
28040 Madrid, Spain

[2]Applied Mathematics
and Computer Science
University of Ghent
Krijgslaan 281-S9
B-9000 Gent, Belgium

Abstract : The main aim of this paper is to show how lattice theory in the very next future will be a useful tool in analysing complex real reliability problems, not properly modelled within classical reliability theory. The introduction of a complete lattice as a state space appears not only of theoretical importance that allows to understand several phenomena with respect to reliability theory better, but as a need claimed from practical engineering. Two important topics are discussed in this general framework: incomparability of component and system states and the duality principle. The strong relationship between the ideas of fuzzy set theory and the ideas that led to the introduction of the theory of multistate structure functions will become clear.

Keywords : Complete lattices, multistate structure functions, non-binary reliability theory.

1. INTRODUCTION

When developing a theory of structure functions defined on arbitrary complete lattices, one often meets some criticism against this choice. Some say that complete lattices provide a too abstract framework and that there are no obvious arguments to introduce them. In this paper, we shall try to show that complete lattices apply to real life situations and that they make possible to study rather complex phenomena with respect to reliability theory and practice. Seminal ideas can be found in Montero et al. (1988) and Cappelle (1991).

Complete lattices have been successfully applied to other fields apart from reliability theory. We can at least trace two important applications in the past. First of all, recent developments in the theory of mathematical morphology show that the general framework of lattice theory is almost indispensable in explaining complex phenomena in an easy way. For an excellent overview and outline of this theory we refer to Ronse (1989). Second, the introduction of L-fuzzy sets (Goguen, 1967) provides a general framework for Zadeh's fuzzy set theory (Kerre, 1991).

In general, complete lattices apply whenever *ordinal* information must be represented. In the theory of mathematical morphology we must manipulate and classify images, while in fuzzy set theory we must check the fulfillment of some property (classical examples are the properties "young" or "beautiful"). Sentences such as "the aircraft is more similar to a Boeing 747 than to a Tupolev 159-M" or "our production quality is higher than the competitor's" make sense in pattern recognition and information representation. Analogous sentences are common among engineering practitioners.

Some claim that complete chains, i.e., totally ordered sets, can be applied in most cases. But very often this assumption is an oversimplification of reality, since we can not deal with incomparable elements. We encounter difficulties that are hard to solve when dealing with complex non-single criterion based problems, where the system state itself is evaluated according to various criteria (Montero et al., 1992).

Hence, when we have a closer look at real life problems, the claim for comparability is not so obvious. Let us suggest some concrete examples:

(i) consider a genealogical tree and the set of John's ancestors; the *natural* order relation on the set of ancestors is $x \leq y$ if and only if y is an ancestor of x, x and y two arbitrary elements of the set of John's ancestors; the order relation on the set of John's ancestors is not linear, i.e., generally for each pair of ancestors neither (x, y) "x is an ancestor of y" nor "y is an ancestor of x" holds;

(ii) suppose we must judge if one weather condition is more dangerous than another weather condition while driving a car; most of us will judge close and thick fog far more dangerous than ordinary rain, but will most of us be able to decide which weather condition is worse: thick fog or icy and snowy weather; some just will say that these weather circumstances are dangerous in a different way and, hence, somehow incomparable;

(iii) consider a particular manufactured product; it can be very difficult to define a linear order when comparing the "quality" of products made by different people, since "quality" refers to many aspects of the manufactured product.

The state space representation in practical reliability has a deep link with preference relations in decision making and we all know that decision making relations are rarely linear. Hence, linearly ordered sets are not the most suitable tool to order objects. The application of partially ordered sets (posets for short) to reliability theory is not surprising from this point of view, since we must be able to sort the states from "bad" to "good" states. Even the applied uncertainty model to study the probability or possibility of failure or degradation, is mainly based upon an order relation. Indeed, the uncertainty model allows to sort the system states from most probable (most possible) to least probable (least possible). Hence, the relationship between reliability theory and lattice theory is not artificial.

A quite different approach tries to introduce some kind of fuzzy uncertainty in the basic binary model, from the existence of non-random (possibilistic) uncertainties attached to real systems and the fuzzy nature of the idea of "performance" and "failure". The term "safety" or "reliability" can then be modelled in a fuzzy framework (see Zimmermann (1983)). In the particular area of reliability, fuzziness can lead to at least two possible basic models:

(i) characterize system reliability behaviour in a possibilistic context,

(ii) assume that the system failure is defined in a fuzzy way.

Following Cai's notation (see Cai, 1991 for a personal report on the research done at the Bejing University of Aeronautics and Astronautics), combination leads to at least three types of fuzzy reliability theories:

(i) PROFUST reliability: a theory based upon PRObability theory and a FUzzy STate assumption,

(ii) POSBIST reliability: a theory based upon POSsibility theory and a BInary STate assumption,

(iii) POSFUST reliability: a theory based upon POSsibility Theory and a FUzzy STate assumption.

An alternative combined approach has been developed by Onisawa (see Onisawa, 1989 for a personal review of his research), by considering simultaneously equipment "failure possibility" and human "error possibility", both derived from the estimation of failure/error probabilities based on a safety criterion.

Our paper deals with multistate structure functions. Hence, we consider the case that each component or system may assume one of many states. Since we assume that a state space can be any complete lattice, one notices the deep link with the L-fuzzy sets of Goguen (1967). Indeed, a structure function can be viewed as a kind of fuzzy set that models the relationship between the fuzzy notions failure, almost failure, almost functioning and perfect functioning between the components and the system. From that point of view, a state of a component is a kind of membership degree in the fuzzy set of "good states." The main aim of this paper is to show how complete lattices provide a general framework to explain complex phenomena in non-classical reliability theory. One will notice the deep link with the main ideas of fuzzy set theory.

2. COMPLETE LATTICES IN RELIABILITY THEORY

When studying classical reliability theory two parts can be distinguished: a general theory of binary structure functions and a probabilistic (time-dependent) uncertainty model. By means of a structure function the deterministic relationship between the states of the components and the system state is modelled, while probability theory provides a general framework to define the notion *reliability* properly: the reliability of an *item* (component or system) is the probability that the item functions properly. According to classical reliability theory, the items must always assume either one of two possible states: perfect functioning or total failure. The system reliability is then calculated from the (basic) component reliabilities.

Since many real life situations are simply not binary in nature, a dichotomous approximation, initially introduced by Birnbaum et al. (1961), sometimes is far too inaccurate. The model has been extended systematically from 1978 on in order to allow intermediate states between perfect functioning and total failure[1]. Among others we mention the finite models of El-Neweihi et al. (1978), Griffith (1980) and Natvig (1982) and the infinite or continuous models introduced by Block and Savits (1984) and Baxter (1984, 1986). Some technical difficulties arise depending on the complexity of the state space, and a variety of solutions to particular problems can be found in the literature. We remind the reader, e.g., of the overwhelming number of notions of the coherence of structure functions, most of them reduce to the well-known coherence property when applied to binary structure functions (Ohi and Nishida (1984) mention at least fourteen different notions).

By assuming that the state space is a complete lattice, we are not just extending the notion of binary structure function to a multistate model, we are also modelling many real life systems where more restrictive approaches do not apply and, moreover, creating a general framework with notions that allow a better comprehension of the basic concepts initially defined for particular systems only.

In the sequel two problems are discussed in order to exemplify our approach: the possible *incomparability of states* and a general *duality principle.* The first problem requires a complete lattice as a state space, while the second one can be properly justified by considering the complete lattice of structure functions.

3. A STATE SPACE FOR COMPONENTS AND SYSTEMS

Let us assume that each component provides a set of *parameters* that can be evaluated. The evaluation of these parameters is called a *state*, that is, a characteristic for the behaviour of the component at a certain time. The performance level of a copy machine, e.g., can be tested by making a copy of a special grey chart; this grey chart copy can be an excellent parameter for the evaluation of the performance of

[1] The ideas and first attempts to formulate multistate structure functions are much older, cf., Premo (1963), and some notes in Gnedenko et al. (1972).

the copy machine. Hence, an *evaluation* is a mapping from the set of parameters into a set L, the set of all the possible states of the item. In order to be able to distinguish between worse and better states, L must be provided with an order relation $\leq$ (Birkhoff, 1967). For any states a and b, $a \leq b$ implies that state b is better than state a or, equivalently, that state a is worse than state b.

The couple $(L, \leq)$ is called a poset, and as pointed out above, in general the order relation is not linear. The state space must be a *complete lattice*, i.e., given any subset in L, at least its greatest lower bound, called infimum, and its smallest upper bound, called supremum, exist. Although an extensive justification of the completeness is far beyond the scope of this paper, this demand allows, e.g., the development of general reliability bounds. A *chain* or totally ordered set is a poset where any two elements are *comparable*, i.e.,

$$(\forall(a, b) \in L^2)(a \leq b \text{ or } b \leq a).$$

When $(\{0, 1\}, \leq)$ is the state space for both the system and its components, where $\leq$ is the well-known order relation, we obtain the classical binary model, introduced by Birnbaum et al. (1961). When $L = \{0, 1, \ldots, M\}$, with $M > 2$, the finite state models of El-Neweihi et al. (1978) appear. When both the state space of components and system equal $([0, 1], \leq)$, Baxter's continuous infinite model is obtained (Baxter, 1984). It must be pointed out that Block and Savits (1984) also considered an infinite continuous model based on the chain $(\mathbb{R}^+, \leq)$. However, this chain is not complete and therefore this model has not been considered here[2]. Obviously under our general approach, the state space for each component is not required to be always the same. Hence, the following definition applies (see Montero et al., 1988, and Cappelle, 1991):

---

[2] From our point of view, a greatest element, called infinity, must be added as the perfect functioning state to transform $(\mathbb{R}^+, \leq)$ into a complete lattice.

Definition 1 *Let* $(L_i, \leq), 0 \leq i \leq n$, *be n arbitrary complete lattices and assume that* $(L_1 \times \cdots \times L_n, \leq)$ *has been provided with its product ordering, i.e., for an arbitrary* $\mathbf{x}$ *and* $\mathbf{y}$ *in* $L_1 \times \cdots \times L_n$

$$\mathbf{x} \leq \mathbf{y} \Leftrightarrow (\forall i \in \{1, \ldots, n\})(x_i \leq y_i).$$

An $L_1 \times \cdots \times L_n - L_o$ *mapping* ϕ *is a structure function if and only if*

(i) ϕ *is isotone, i.e.,*

$$(\forall (\mathbf{x}, \mathbf{y}) \in (L_1 \times \cdots \times L_n)^2)(\mathbf{x} \leq \mathbf{y} \Rightarrow \phi(\mathbf{x}) \leq \phi(\mathbf{y})),$$

(ii) $\phi(0, \ldots, 0) = 0$ *and* $\phi(1, \ldots, 1) = 1$.

Since a system that assumes a worse state when every component assumes a better state is quite unusual or poorly designed (Barlow, 1975), the isotonicity is justified. The boundary condition is also widely accepted and applied. It reflects the fact that whenever each component assumes it worst (best) state, the system must assume its worst (best) state.

When $L_i = \{0, 1\}, 0 \leq i \leq n$, the structure functions are called binary structure functions (Barlow and Proschan, 1975). An El-Neweihi et al. (1978) structure function ϕ appears when $L_i = \{0, 1, ..., M\}, 0 \leq i \leq n$, and ϕ is idempotent, i.e.,

$$(\forall x \in \{0, 1, \ldots, M\})(\phi(x, \ldots, x) = x).$$

4. INCOMPARABLE STATES

In section 3 we have introduced arbitrary complete lattices as the state space for both components and systems. It has been pointed out that not every two states are comparable. For an arbitrary poset $(P, \leq)$, two elements a and b of P are *incomparable* if and only if $\neg(a \leq b)$ and $\neg(b \leq a)$ holds (Birkhoff, 1967), and it is denoted by $a \parallel b$.

As pointed out above, incomparable states are of particular interest in the theory of structure functions. Very often we cannot determine if state a is better or worse than another state b, or when both states are similar. When dealing with systems subject to n different kinds of failure, e.g., where a transition diagram indicates the possible transitions between all the states, the diagram often suggests a poset with $n + 1$ elements. Element 1 represents the perfect functioning state and the elements a_i, $1 \leq i \leq n$, represent the failure of type i (see for example Proctor (1976), Elsayed (1979) and Yamashiro (1980)). Hence, the order relation $\leq$ imposed on the set $\{a_1, \ldots, a_n, 1\}$ is obvious:

$$(\forall i \in \{1, \ldots, n\})(a_i \leq 1) \text{ and } (\forall (i,j) \in \{1, \ldots, n\}^2)(i \neq j \Leftrightarrow a_i \parallel a_j).$$

It is quite useless to consider a failure of type i better or worse than another different failure of type j, since different types of failure do not reflect a degradation in several stages. Adding a smallest element 0, i.e.,

$$(\forall x \in \{0, a_1, \ldots, a_n, 1\})(0 \leq x),$$

transforms the given state space into the complete lattice $(\{0, a_1, \ldots, a_n, 1\}, \leq)$. In fact, the structural model of a component subject to n different types of failure is more complex if only structural information had been taking into account. However, it has been simplified by considering some extra (probabilistic) information: having n binary failures a_i, $1 \leq i \leq n$, a binary space $\{0, 1\}$ is associated to each failure, such that the initial state space for the system is the complete lattice $(\{0,1\},\leq)$, whereas 1 and 0, respectively, are the perfect functioning and the total failure state, respectively. If we additionally know that failure times are continuously distributed and that the system completely fails when the first failure is observed, all states within the lattice $(\{0,1\}^n,\leq)$ except $n+1$ will have zero probability. Therefore, only $n+1$ states can be observed: no failure (1) and the *first* failure of type i, a_i, $1 \leq i \leq n$. Hence, the proposed state space simplifies the model, since many non-observable states are excluded.

Let us consider a motor opened valve, MOV for short. The MOV can fail to open when it is closed or fail to close when it is open. Hence, the MOV is subject to two kinds of failure which can hardly be compared. Practical considerations suggest a complete lattice with five elements as a state space:

(i) state 1: the MOV is in optimal condition,

(ii) state a: the MOV must be maintained but still functions,

(iii) state b: the MOV can be opened but can no longer be closed,

(iv) state c: the MOV can be closed but can no longer be opened,

(v) state 0: a total failure of the MOV is observed.

Just like in the previous example, this state space is a simplified version of a more complex state space, by considering two failure types each having a three element evaluation set: $\{0,\frac{1}{2},1\}$, the intermediate state $\frac{1}{2}$ represents functioning with maintenance. One must notice that in the proposed state space, the associated order relation is not the product ordering, i.e., $(\{0,\frac{1}{2},1\}^2,\leq)$, since the engineer—according to the chosen state space—prefers functioning though any maintenance is needed of both opening and closing mechanisms to complete failure of one of the mechanisms, despite the other is perfectly functioning.

5. A GENERAL DUALITY PRINCIPLE

In the previous sections we have pointed out that the order relation on the set of possible states determines when states are similar, better, worse or incomparable and that the assumption of a linear ordering of the state space is not appropriate for many practical problems. In this section we shall exemplify how our general approach will gain insight into the standard concepts in reliability theory.

A duality principle deals with a *reversed* order on the set of states. The "better" states become the "worse" states, and *vice versa*. In fact, many systems subject to two dual failure types (Barlow and Proschan, 1975) can be represented as the conjunction of two dual systems. Structural duality, however, very often is defined without any special consideration. From our point of view, this particular property of the state space, i.e., the possibility of exchanging the good and the bad states, is a special case of the selfduality of a poset. An arbitrary poset $(P, \leq)$ is selfdual if and only if there exists an order reversing permutation δ on P (Birkhoff, 1967), that is,

$$(\forall (a,b) \in P^2)(a \leq b \Leftrightarrow \delta(b) \leq \delta(a)).$$

Therefore, the following definition is proposed:

Definition 2 (Montero 1988, Cappelle 1991) *Let δ_i be an order reversing permutation of $(L_i, \leq)$, $1 \leq i \leq n$, and δ_o be an order reversing permutation of $(L_o, \leq)$. The dual (w.r.t. $\delta_1, \ldots, \delta_n$ and δ_o) of a structure function ϕ, denoted by $\Delta(\phi)$, is defined as*

$$\Delta(\phi) : L_1 \times \cdots \times L_n \to L_o : \mathbf{x} \mapsto \delta_o(\phi(\delta_1(x_1), \ldots, \delta_n(x_n))).$$

The mapping Δ on the set of structure functions is called a duality.

By $\mathcal{M}(L_1 \times \cdots \times L_n, L)$ we denote the set of all $L_1 \times \cdots \times L_n - L_o$ structure functions, and we provide it with a pointwise order relation $\preceq$, such that for any two arbitrary structure functions ϕ_1 and ϕ_2

$$\phi_1 \preceq \phi_2 \iff (\forall \mathbf{x} \in L_1 \times \cdots \times L_n)(\phi_1(\mathbf{x}) \leq \phi_2(\mathbf{x})).$$

It is easily verified that $(\mathcal{M}(L_1 \times \cdots \times L_n, L), \preceq)$ is a complete lattice (Cappelle 1991) and that Δ is an order reversing permutation of $(\mathcal{M}(L_1 \times \cdots \times L_n, L), \preceq)$.

From this point of view, structural duality is an order reversing permutation of $(\mathcal{M}(L_1 \times \cdots \times L_n, L), \preceq)$. As a special case we find that the classical duality in the binary case and linearly ordered finite multistate models is unique. Indeed, there exists a unique order reversing permutation δ on the chain $(\{0, \ldots, M\}, \leq)$,

$$\delta : \{0, \ldots, M\} \to \{0, \ldots, M\} : x \mapsto M - x.$$

Hence, the duality Δ transforms any structure function ϕ into a structure function $\Delta(\phi)$,

$$\Delta(\phi) : \{0, \ldots, M\}^n \to \{0, \ldots, M\} : \mathbf{x} \mapsto M - \phi(M - x_1, \ldots, M - x_n).$$

This is exactly the duality principle introduced in the past, for the linearly ordered finite state systems.

Problems arise when considering more general state spaces, e.g., when dealing with continuous systems, i.e., the complete chain $([0,1], \leq)$ as the state space for both the system and its components. It is obvious that there is more than one order reversing permutation on $([0,1], \leq)$. Hence, an enormous variety of order reversing permutations on $(\mathcal{M}([0,1]^n, [0,1]), \preceq)$ can be constructed, making Baxter's duality principle,

$$\Delta(\phi) : [0,1]^n \to [0,1] : \mathbf{x} \mapsto 1 - \phi(1 - x_1, \ldots, 1 - x_n),$$

for any $[0,1]^n - [0,1]$ structure function ϕ, not so evident. Still Montero et al. (1988) provide a very important argument to consider Baxter's choice: the order reversing permutation δ,

$$\delta : [0,1] \to [0,1] : x \mapsto 1 - x,$$

is the only order reversing permutation that is measure preserving with respect to the Lebesgue measure[3]. However, when a possibilistic uncertainty model is applied, there is no reason to consider only Baxter's definition (cf. Cappelle, 1991).

6. CONCLUSION

On the one hand, the existence of components and systems with incomparable states requires a lattice-based reliability approach in order to capture all the essential information without artificial oversimplification of the mathematical model.

On the other hand, it has been pointed out that besides a real increase in potential applications, a general approach is useful in order to clarify basic concepts and notions. In particular, it has been shown that the idea of a duality must be related to the structure of $(\mathcal{M}(L_1 \times \cdots \times L_n, L_o), \preceq)$, and although the standard duality principle for the finite chain model is unique, this result does not apply in general. Hence, in the near future, lattice theory must have a deep influence on system reliability, from a theoretical as well as from a practical point of view.

We mention once again the strong link with fuzzy set theory. Throughout the paper we have shown that the intermediate states stem from the fuzziness of the terms 'perfect functioning' and 'total failure'. Moreover the lattice that is the state space of a component or the system may be, e.g., a lattice of fuzzy numbers. Hence,

[3] The result is more general, being the key assumption to the existence of an σ-finite measure associated to the space of states.

our approach allows the possibility of both crisp states and fuzzy states. The fuzzy states, e.g., can be applied to model non-stochastic uncertainty.

Finally, it must be pointed out that in this paper we have focused our attention strictly on the problem of the state space representation. We did not discuss the possibilistic uncertainty models (see, e.g., Cai , Chuan and Zhang (1991), and Cappelle (1991)).

Acknowledgement: The authors whish to thank the National Fund of Scientific Research (Belgium) and DGICYT (national grant number PB91-0389) for the support of this research, and the referees for their helpful comments. Bart Cappelle is currently a research assistant of the NFSR.

REFERENCES

R. E. Barlow and F. Proschan (1975) "Statistical Theory of Reliability and Life Testing," Holt, Rinehart and Winston, New York.

L. A. Baxter (1984), *Continuum Structures I*, J. Appl. Prob. **21**, 802–815.

L. A. Baxter (1986), *Continuum Structures II*, Math. Proc. Camb. Phil. Soc. **99**, 331–338.

G. Birkhoff (1967) "Lattice Theory," AMS Colloquium Publication Volume 24, Providence, Rhode Island.

Z. W. Birnbaum, J. D. Esary and S. C. Saunders (1961), *Multicomponent systems and structures, and their reliability*, Technometrics **3**, 55–77.

H. W. Block and T. H. Savits (1984), *Continuous Multistate Structure Functions*, Operations Research **32**, 703–714.

K. Y. Cai (1991), *Fuzzy Reliability Theories*, Fuzzy Sets and Systems **40**, 510-511.

K. Y. Cai, C. Y. Chuan and M. L. Zhang (1991), *Posbist Reliability Behaviour of Typical Systems with two Types of Failure*, Fuzzy Sets and Systems **43**, 17-32.

B. Cappelle (1991) *Multistate structure functions and Possibility Theory: an alternative approach to reliability* in E.E. Kerre, Ed.: Introduction to the Basic Principles of Fuzzy Set Theory and Some of its Applications, Communication and Cognition, Gent, 1991, 252–293.

E. El–Neweihi, F. Proschan and J. Sethuraman (1978), *Multistate Coherent Systems*, J. Appl. Prob. **15**, 675–688.

E. A. Elsayed and A. Zebib (1979), *A Repairable Multistate Device*, IEEE Trans. Rel. **28**, 81–82.

B. Gnedenko, Y. Beliaev and A. Soloviev (1972) "Méthodes Mathématiques en Théorie de la Fiabilité," Mir, Moscow.

J. A. Goguen (1967), *L-Fuzzy Sets*, J. Math. An. Appl. **18**, 145–174.

W. S. Griffith (1980), *Multistate Reliability Models*, Journal of Applied Probability **17**, 735–744.

E. E. Kerre (1991) *Basic Principles of Fuzzy Set Theory for the Representation and Manipulation of Imprecision and Uncertainty* in E.E. Kerre, Ed.: Introduction to the Basic Principles of Fuzzy Set Theory and Some of its Applications, Communication and Cognition, Gent, 1991,1–158.

J. Montero, J. Tejada and J. Yáñez (1988), *General Structure Functions*, Proceedings Workshop on Knowledge-Based Systems and Models of Logical Reasoning, Dec 26–31 1988 Cairo (Egypt).

J. Montero, J. Tejada and J. Yáñez (1992), *Multivalued Structure Functions*, European Journal of Operational Research (in press).

B. Natvig (1982), *Two suggestions of how to define a multistate coherent system*, Advances in Applied Probability **14**, 434–455.

F. Ohi and T. Nishida (1984), *On Multistate Coherent Systems*, IEEE Trans. Rel. **33**, 284–288.

T. Onisawa (1989), *Fuzzy Set Theory in Reliability Analysis*, Fuzzy Sets and Systems **30**, 361-363.

A. F. Premo (1963), *The use of Boolean Algebra and a Truth Table in the Formulation of a Mathematical Model of Success*, IEEE Trans. Rel. **12**, 45–49.

C. L. Proctor and B. Singh (1976), *A Repairable 3–State Device*, IEEE Trans. Rel. **25**, 210–211.

C. Ronse (1989), *Introduction to the algebraic basis of morphological operations*, 5th International Workshop on Stereology, Stochastic Geometry and Image Analysis, 20 pages.

M. Yamashiro (1980), *A Repairable Multi-State Device with General Repair Time*,

IEEE Trans. Rel. **29**, 276.

H. J. Zimmermann (1983), *Using Fuzzy Sets in Operational Research*, European Journal of Operational Research **13**, 201-216.

3

FAULT TREE ANALYSIS USING FUZZY SETS AND POSSIBILITY THEORY

MULTI STATE FAULT TREE ANALYSIS USING FUZZY PROBABILITY VECTORS AND RESOLUTION IDENTITY

KRISHNA B. MISRA* and K.P.SOMAN **

* Disaster Prevention and Impact Minimization Division,
NEERI, Nehru Marg, Nagpur - 440 020, India

** Reliability Engineering Centre,
Indian Institute of Technology, Kharagpur - 721 302, West Bengal, India

Abstract : In this paper, we propose a method of estimating top event fuzzy probability of a fault tree in case of a system consisting of multistate elements. To the best of our knowledge, no such attempt has ever been made in this direction in the past. Beta fuzzy probability vectors, as proposed by Stein [9], are used to model the joint-possibility distribution of multistate elements. The use of resolution identity keeps the computational requirement at its minimum. However, the estimation procedure is based on Zadeh's extension principle.

Keywords : *Fuzzy probability, resolution identity, multistate elements*

1. INTRODUCTION

Fault tree analysis continues to remain an area of research in the broad discipline of reliability and safety engineering. The primary objective of fault tree construction and evaluation is to determine whether or not the system design offers an acceptable level of safety and reliability, that minimizes the probability of occurrence of top event (an undesired system event) and is within a specified bounds. If the system design is found inadequate, the design is modified by first identifying those critical events that significantly contribute to the top occurrence of event. When all design changes have been incorporated, the fault tree is re-evaluated to determine if the revised design provides an acceptable level of safety.

As discussed in [3, 7 and 8], there are inherent weaknesses in the current probabilistic risk assessment or safety analysis and these mainly stem from the existence of a large amount of uncertainty in the results due to the use of imprecise failure data and secondly due to uncertainty of modelling in case of dependent failures, common cause failures, human operator's failures, aging of plants etc. To mathematically treat some of these uncertainties, Tanaka [1] and Misra [2,3] proposed fuzzy sets' approach. These approaches are based on fault trees with two state elements and have certain weaknesses (elaborated in section 4) and are actually computationally intensive. The computations for arithmetic operations can be considerably simplified if L-R representation of fuzzy numbers due to Dubois and

Prade [4] is used. Mizumoto and Tanaka [5] also derived some algebraic properties of fuzzy numbers.

The fuzzy numbers with continuous membership functions have been considered as extensions of the concept of confidence interval by Kaufmann and Gupta [6]. Instead of considering a confidence interval at a unique level, it is considered more generally at all levels between 0 and 1. It has been also proved that arithmetic operations on fuzzy numbers based on the extension principle can be decomposed into operations on confidence interval at various levels from 0 to 1. Thus the problem of fuzzy field is transformed into a problem in a nonfuzzy field without loss of information. Kaufmann and Gupta [7] provided simplified computations for triangular fuzzy numbers. Recently, Soman and Misra [14] proposed a new method of computing top event's fuzzy probability, which is computationally efficient and is free from the weaknesses of previous methods. In the present paper, we extend this approach to systems consisting of multistate elements. Extended fuzzy numbers, resolution identity, extension principle and fuzzy probability vectors form the basis of the methodology of this paper. Formulas for arithmetic operation on extended fuzzy numbers with continuous membership function have been derived using the operation on bound points of all α-level set intervals instead of using general set operations. Based on this framework, top event's fuzzy probability has been evaluated.

2. EXTENDED FUZZY NUMBER

Let x be a continuous real variable restricted by a possibility distribution function $\mu(x) \varepsilon [0,1]$ such that

(i) $\mu(x)$ is piecewise continuous;
(ii) $\mu(x)$ is a convex fuzzy set; and
(iii) there exists a region for which $\mu(x)$ has a flat top, i.e. corresponds to $\mu(x)=1$.

A fuzzy set which satisfies the above requirements is called an extended fuzzy number. Figure 1 shows graphical representation of such an extended fuzzy number

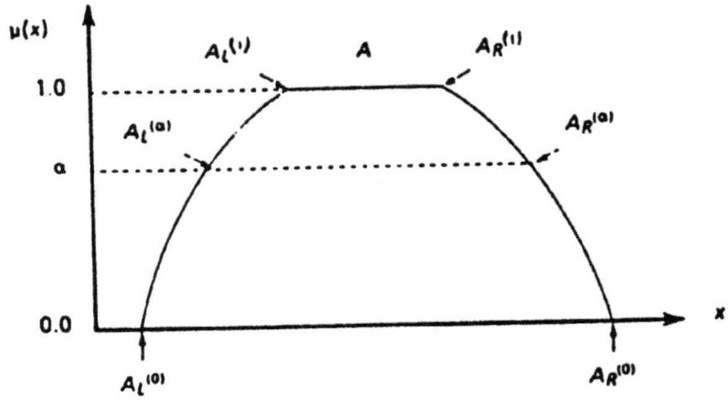

Fig. 1 Bound Points for Alpha Level Set Interval of $\mu_A(x)$

In Kaufmann and Gupta [6], it was shown that the computational effort involved with operations on fuzzy numbers based on the extension principle can be considerably reduced by decomposing the membership functions into α-levels and conducting mathematical operations directly on the confidence intervals. But in general, it cannot be used directly for α - level calculations.

For any fuzzy number **A** (either a general fuzzy number or an extended fuzzy number), which has the membership function $\mu_A(x)$, an interval bounded by two points at each α-level ($0 \leq \alpha \leq 1$) can be obtained by using the α-cut method. The symbols $A_L^{(\alpha)}$ and $A_R^{(\alpha)}$ have been used in this paper to represent the left-end-point and the right-end-point of this interval (as is shown in Fig. 1). When $A_L^{(\alpha)}$ and $A_R^{(\alpha)}$ are considered for all α- levels, due to monotonicity and normality of the membership functions, we can express a general or an extended fuzzy number **A**, using the following form :

$$A \rightarrow [A_L^{(\alpha)}, A_R^{(\alpha)}], \qquad 0 \leq \alpha \leq 1$$

This was defined as a resolution identity by Zadeh [10]. Using the expression for the membership function at the α-level instead of usual $\mu(x)$, the algebraic properties of fuzzy numbers can be described more conveniently.

2.1 Arithmetic Operations on Extended Fuzzy Numbers

The arithmetic operations on two extended fuzzy numbers can be represented more compactly using resolution identity. The properties of the results can be analyzed more conveniently based on interval arithmetic [4,5]. Using the concept of confidence interval, the following arithmetic operations, similar to those defined for interval arithmetic, on two extended fuzzy numbers **A** and **B** can be easily performed.

$$\mathbf{C} = \mathbf{A} + \mathbf{B} \rightarrow [\mathbf{C}_L^{(\alpha)}, \mathbf{C}_R^{(\alpha)}] = [\mathbf{A}_L^{(\alpha)} + \mathbf{B}_L^{(\alpha)}, \mathbf{A}_R^{(\alpha)} + \mathbf{B}_R^{(\alpha)}]$$

$$\mathbf{C} = \mathbf{A} - \mathbf{B} \rightarrow [\mathbf{C}_L^{(\alpha)}, \mathbf{C}_R^{(\alpha)}] = [\mathbf{A}_L^{(\alpha)} - \mathbf{B}_R^{(\alpha)}, \mathbf{A}_R^{(\alpha)} - \mathbf{B}_L^{(\alpha)}]$$

$$\mathbf{C} = \mathbf{A} \,.\, \mathbf{B} \rightarrow [\mathbf{C}_L^{(\alpha)}, \mathbf{C}_R^{(\alpha)}]$$

$$= [\min(\mathbf{A}_L^{(\alpha)} . \mathbf{B}_L^{(\alpha)}, \mathbf{A}_R^{(\alpha)} . \mathbf{B}_L^{(\alpha)}, \mathbf{A}_L^{(\alpha)} . \mathbf{B}_R^{(\alpha)}, \mathbf{A}_R^{(\alpha)} . \mathbf{B}_R^{(\alpha)}),$$
$$\max(\mathbf{A}_L^{(\alpha)} . \mathbf{B}_L^{(\alpha)}, \mathbf{A}_R^{(\alpha)} . \mathbf{B}_L^{(\alpha)}, \mathbf{A}_L^{(\alpha)} . \mathbf{B}_R^{(\alpha)}, \mathbf{A}_R^{(\alpha)} . \mathbf{B}_R^{(\alpha)})]$$

When **A** and **B** are positive,

$$\mathbf{C} = [\mathbf{C}_L^{(\alpha)}, \mathbf{C}_R^{(\alpha)}] = [\mathbf{A}_L^{(\alpha)} . \mathbf{B}_L^{(\alpha)}, \mathbf{A}_R^{(\alpha)} . \mathbf{B}_R^{(\alpha)}]$$

Finally,

$$C = A \div B \rightarrow [C_L^{(\alpha)}, C_R^{(\alpha)}] = [A_L^{(\alpha)}, A_R^{(\alpha)}] \times \left[\frac{1}{\mathbf{B}_R^{(\alpha)}}, \frac{1}{\mathbf{B}_L^{(\alpha)}}\right]$$

$$\text{if } 0 \notin [\mathbf{B}_L^{(\alpha)}, \mathbf{B}_R^{(\alpha)}]$$

In general if $\mathbf{C} = f(\mathbf{A},\mathbf{B})$, then

$$\mathbf{C} \rightarrow [\mathbf{C}_L^{(\alpha)}, \mathbf{C}_R^{(\alpha)}]$$
$$= [\min\{f(\mathbf{A}_L^{(\alpha)}, \mathbf{B}_L^{(\alpha)}), f(\mathbf{A}_L^{(\alpha)}, \mathbf{B}_R^{(\alpha)}), f(\mathbf{A}_R^{(\alpha)}, \mathbf{B}_L^{(\alpha)}), f(\mathbf{A}_L^{(\alpha)}, \mathbf{B}_R^{(\alpha)})\},$$
$$\max\{f(\mathbf{A}_L^{(\alpha)}, \mathbf{B}_L^{(\alpha)}), f(\mathbf{A}_L^{(\alpha)}, \mathbf{B}_R^{(\alpha)}), f(\mathbf{A}_R^{(\alpha)}, \mathbf{B}_L^{(\alpha)}), f(\mathbf{A}_L^{(\alpha)}, \mathbf{B}_R^{(\alpha)})\}] \quad [1]$$

3. FUZZY PROBABILITY

Fuzzy probability, i.e, a fuzzy set defined in probability space, represents a fuzzy number, between zero and one, assigned to the probability of an event. One can choose, depending upon the suitability, a wide variety of membership functions for fuzzy probability. Tanaka [1] used a trapezoidal membership function, which is a special case of an extended fuzzy number. It is represented by a set of four parameters, viz.,
$\mu_{P_i} = T(\alpha_{i1}, \alpha_{i2}, \beta_{i2}, \beta_{i1})$, where T stands for a trapezoidal membership function and is shown in Fig. 2.

The membership function of such a fuzzy number is given by :

$$\mu_{P_i}(p) = \begin{cases} 0 & ; \text{ for } 0 \leq p \leq \alpha_{i1} \\ 1 - \dfrac{\alpha_{i2} - p}{\alpha_{i2} - \alpha_{i1}} & ; \text{ for } \alpha_{i1} \leq p \leq \alpha_{i2} \\ 1 & ; \text{ for } \alpha_{i2} \leq p \leq \beta_{i2} \\ 1 - \dfrac{p - \beta_{i2}}{\beta_{i1} - \beta_{i2}} & ; \text{ for } \beta_{i2} \leq p \leq \beta_{i1} \\ 0 & ; \text{ for } \beta_{i1} \leq p \leq 1 \end{cases}$$

Another membership function that is suitable for representing a fuzzy probability is shown in Fig. 3, and is given by

$$\mu_{P_i}(p) = \begin{cases} \exp(-((p - m_{iL}) / \sigma_{iL})^2) & ; \text{ for } 0 \leq p \leq m_{iL} \\ 1 & ; \text{ for } m_{iL} \leq p \leq m_{iR} \\ \exp(-((p - m_{iR}) / \sigma_{iR})^2) & ; \text{ for } m_{iR} \leq p \leq 1 \end{cases}$$

Symbolically, such a fuzzy probability is written as $G(\mu_{iL}, \sigma_{iL}, \mu_{iR}, \sigma_{iR})$, where G stands for a Gaussian membership function. Note the similarity of the expression with that of the standard Gaussian probability distribution.

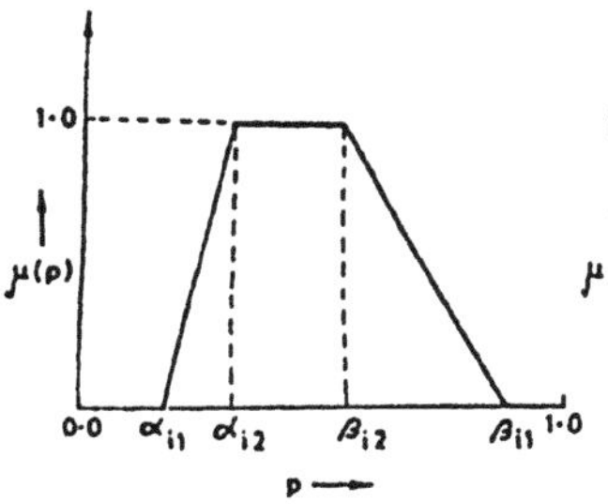

Fig. 2 **A Trapezoidal Fuzzy Number**

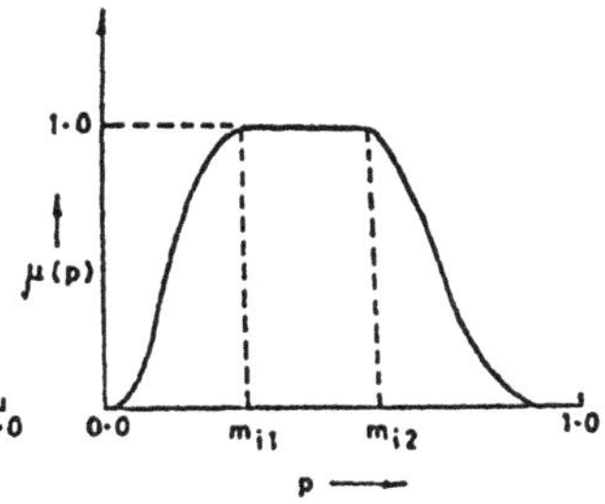

Fig. 3 **A Gaussian Fuzzy Number**

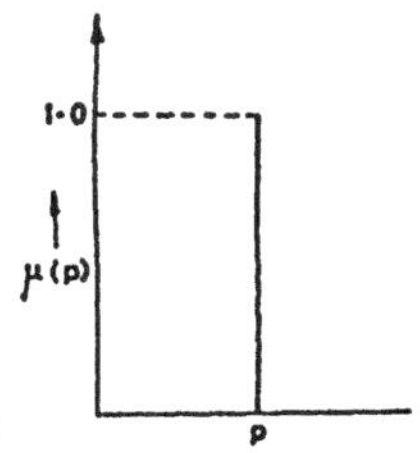

Fig. 4 **A Crisp Fuzzy Number**

Additionally, we may also have crisp probability, which is a special case of fuzzy probability. The membership function is as shown in Fig. 4. Note that α- cut is just a point and is same for all $\alpha \varepsilon$ [0,1]

3.1 Fuzzy Probability Vectors

Fuzzy Variable

Let π be a possibility measure defined on the class of all subsets of a set Γ Nahmias [13] defined a real valued fuzzy variable **X** as a function $\mathbf{X} : \Gamma \rightarrow \mathbf{R}$. Its membership function is computed from

$$\mu_x(x) = \pi(\gamma \varepsilon \Gamma : \mathbf{X}(\gamma) = x), \quad x \ \varepsilon \ \mathbf{R}$$

With this approach μ is not primitive, rather it is computed from π. The membership function is usually thought of as defining a fuzzy set. However, here the fuzzy variable **X** determines a fuzzy set and from that a μ is derived. This situation is analogous to the role of random variables and probability distribution in probability theory.

Non-interactive and Interactive Fuzzy Variables

Two fuzzy variable **X** and **Y** are said to be non-interactive if

$$\mu_{x,y}(x,y) = \mathbf{min}\ [\ \mu_x(x),\ \mu_y(y)\ ] \quad \text{for all } x, y\ \varepsilon\ \mathbf{R}$$

The term $\mu_{x,y}(x,y)$ is called joint possibility distribution ie $\pi(\mathbf{X}=x, \mathbf{Y}=y)$. In multistate fault trees, we have to deal with fuzzy state probabilities $\mathbf{P}_1, \mathbf{P}_2, \mathbf{P}_3, ..., \mathbf{P}_n$ such that $\mathbf{P}_1+\mathbf{P}_2+\mathbf{P}_3+...+\mathbf{P}_n=1$. This is an example of interactive fuzzy numbers, for which the relation given above does not hold good.

In literature, there are two methods to find the joint possibility distribution of interactive fuzzy variables. In the first method [11], it is assumed that if any one of p_i is eliminated, then the remaining become non-interactive on Δ_n, where $\Delta_n = \{(p_1, p_2, p_3, ..., p_n) : p_i \geq 0,\ \Sigma\, p_i = 1\}$. Because of this assumption, their membership functions may be individually specified. Thus we can obtain joint possibility distribution from the marginal possibility distribution of constituent elements as :

$$\mu_p(p_1, p_2, .., p_i, ..., p_n) = \min [\ \mu_1(p_1), .., \mu_i(p_i), .., \mu_{n-1}(p_{n-1})\ ],$$
$$(p_1, ..., p_n) \in \Delta_n$$

In the second method [12], all the $\mathbf{P}_i$ are assumed to be non- interactive on Δ_n, the domain of interest. The membership functions of constituent elements are individually specified and the joint possibility distribution is derived from the relation :

$$\mu_p(p_1, p_2, .., p_i, ..., p_n) = \min [\ \mu_1(p_1), .., \mu_i(p_i), .., \mu_n(p_n)\ ],$$
$$(p_1, ..., p_n) \in \Delta_n$$

These two methods will not always give the same result [9], and this raises a question of which method to use? One condition that any self consistent method must satisfy is that the marginal distribution must be obtainable from joint distribution via supremum [10]. For example, the marginal of $\mathbf{P}_i$ is

$$\mu_i(p_i) = \sup [\ \mu_p(p_1, ..., p_i, ..., p_n)\ ]$$

where the supremum is over all $(p_1,...,p_i,...,p_n)$ with p_i being fixed, such that $(p_1,..., p_i,..., p_n) \in \Delta_n$. Stein [9] has shown that both these methods are not consistent. According to him, the solution lies in directly specifying the joint possibility distribution of $(p_1, p_2,..,p_i,..,p_n)$. A class of joint possibility distribution suitable for interactive fuzzy probabilities is a multivariate fuzzy beta distribution. In this paper, since all the discussion is based on fuzzy beta distribution we give a brief review of it.

Fuzzy Beta Distribution

The fuzzy vector $\underline{\mathbf{P}} = (\mathbf{P}_1, \mathbf{P}_2, .. \mathbf{P}_i, .., \mathbf{P}_n)$ has a fuzzy beta possibility distribution if

$$\mu_p(p_1, ..., p_i, ..., p_n) = C(\beta_1, ..., \beta_i, ..., \beta_n)\, p_1^{\beta 1} .. p_i^{\beta i} .. p_n^{\beta n} \quad (2)$$

for all $(p_1, ..., p_i, ..., p_n) \in \Delta_n$ and $\beta_i \geq 0$, $i = 1, 2, ..., n$. The constant C is chosen such that μ_p is normalized (i.e., the supremum is 1). The value of C which normalizes the fuzzy beta is given by

$$C(\beta_1, .., \beta_i, .., \beta_n) = (\Sigma \beta_i)^{\Sigma \beta_i} / \prod (\beta_i^{\beta_i}) \quad (3)$$

where the sums and products are carried over all i such that $\beta_i > 0$. The derivation of above formula is given by Stein [9]. If all $\beta_i = 0$, then C =1.

The marginal possibility distribution of k-*th* variable of the fuzzy beta distribution given by (2) is again a fuzzy beta distribution (for n =2) with membership function given by

$$\mu_k(p) = C(\beta_k, \beta_r)\, p^{\beta_k}(1-p)^{\beta_r}, \text{ where } \beta_r = \sum_{i \neq k} \beta_i . \quad (4)$$

Example : A component in a system has three mutually exclusive states with approximate probabilities of being in these states as 0.5, 0.4, and 0.1 respectively, then $\beta_1 = 5\beta$, $\beta_2 = 4\beta$, $\beta_3 = \beta$, where β is any positive number. This will give marginals with modes at 0.5, 0.4 and at 0.1. The value of β controls the degree of spread around these values. As β approaches 0, each marginal fuzzy probability approaches a crisp number. The choice of β is purely subjective.

4. FUZZY FAULT TREE ANALYSIS

A fault tree is a logical representation showing combination of events starting with basic events that may lead to the occurrence of top event (undesired event). Tanaka et. al [1] were the first to propose fuzzy sets approach to fault tree analysis with two state elements for handling various uncertainties in data. They used possibilities of failure, viz., a fuzzy set defined in probability space. Using this concept, a degree of uncertainty can be allocated to each value of the probability of failure. Thus different aspects of the uncertainty i.e., probability and possibility are simultaneously treated. The approach is based on extension principle [10]. However it suffers from following disadvantages:

(a) It cannot be applied to fault trees with repeated events. In this method the cutsets are treated as if they are disjoint.

(b) If a fault tree contains an event and its complementary event, the method would provide an erroneous result as it treats possibility distribution of complementary and uncomplementary events as independent.

(c) The formula derived is restricted to trapezoidal shaped membership functions.

(d) If a fault tree contains more than two events, one must resort to approximate method of evaluation.

Later on, in 1989, Misra et. al [2,3] proposed two different methods using fuzzy set concepts but these also suffered from the first two drawbacks mentioned above.

Recently, the present authors [14] proposed a new method based on resolution identity and extension principle for fault trees with two state elements. This method is computationally efficient and overcomes the drawbacks mentioned above and provides fault analysis with multistate elements using fuzzy probability.

4.1 Fault Tree Analysis with Multistate Elements

Unlike fault tree with two state elements, the analysis of fault trees with multistate elements is complicated on account of the following reasons:

i) modeling of components' failure possibilities by appropriate joint possibility distribution. As far as the present authors know, so far only one family of joint possibility distribution is appeared in literature [9]. Therefore modeler is restricted in his choice of distribution.

ii) For a given α-level, ordinary fuzzy probability has a unique infimum and a supremum, which facilitate fast computation of α- cuts of top event's possibility distribution of probability. In case of a joint possibility distribution for multistate elements, we have an infinite number of probability vectors with a given possibility of α. From these infinite number of vectors, we have to choose the one which minimizes (maximizes) the top event probability expression. This selection can be done only by optimizing the top event probability considering all multistate elements' probability vectors together. This certainly increases our computational requirements.

Example: Let us illustrate the methodology by taking a simple example. Consider the example of an electronic filter given in [7] on page 777. The circuit under consideration is shown in Fig. 5. It consists of a diode with three states, a resistor with two states, and a capacitor with three states. The output can be normal or one which is false but not dangerous or false and dangerous (if an a.c appears as an output, already certain components connected to the output of the filter would have been destroyed. Thus three outputs are distinguished which are taken as system states.

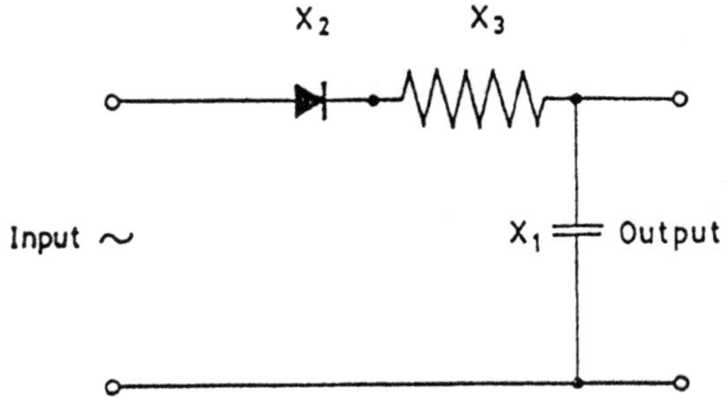

Fig 5 An Electronic Filter Circuit

The components and system states are distinguished as follows :

Component		States	Definition
Capacitor	:	X_{11}	short circuit fault
		X_{12}	open circuit fault
		X_{13}	working normally
Diode	:	X_{21}	short circuit fault
		X_{22}	open circuit fault
		X_{23}	working normally
Resistor	:	X_{31}	open circuit fault
		X_{32}	working normally

Further, the system states are recognized as :

T_1	Normal output
T_2	False but safe output signal
T_3	False and dangerous output signal

The fault trees leading to system states T_2 and T_3 are shown in Fig. 6 and Fig. 7, respectively.

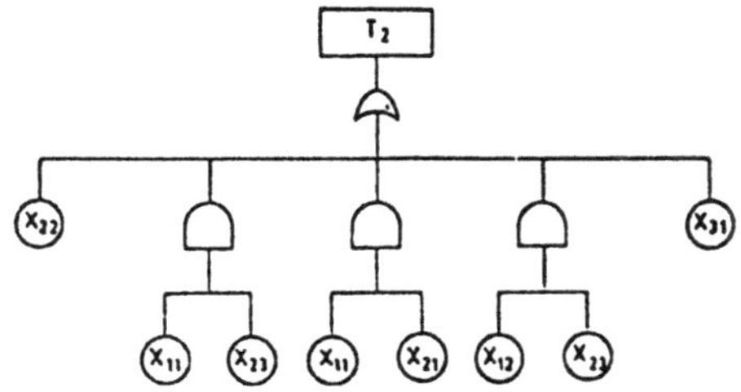

Fig 6 Fault Tree for the Top Event T_2

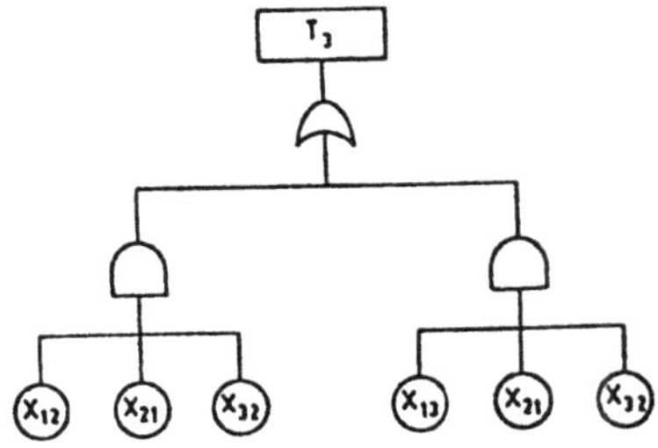

Fig. 7 Fault Tree for the Top Event T_3

We have $T_2 = X_{22} + X_{31} + X_{11}X_{23} + X_{11}X_{21} + X_{12}X_{23}$

On disjointing the terms, we obtain

$$T_2 = X_{22} + X_{31}\overline{X}_{22} + X_{12}X_{23}\overline{X}_{31} + X_{11}X_{21}\overline{X}_{31}$$
$$T_3 = X_{12}X_{21}X_{32} + X_{13}X_{21}X_{32}$$

Let the top event probabilities associated with T_2 and T_3 be represented by Q_2 and Q_3, respectively.

$$Q_2 = p_{22} + p_{31}(1-p_{22}) + (1-p_{31})(p_{12}p_{23} + p_{11}p_{23} + p_{11}p_{21}) \qquad (5)$$

and $\quad Q_3 = p_{12}p_{21}p_{32} + p_{13}p_{21}p_{32}$

Let the joint possibility distribution of the components' state probabilities be given by

$$\mu_{P1}(p_{11},p_{12},p_{13}) = C(0.5,1,3.5)(p_{11})^{0.5}(p_{12})^{1.0}(p_{13})^{3.5}$$

$\mu_{P2}(p_{21},p_{22},p_{23}) = C(1,1,8)(p_{21})^1(p_{22})^1(p_{23})^8$

Resistor is modeled as two state device and let its possibility distribution corresponding to its failure state be a trapezoidal number given by

$\mu_{P3}(p_{31}) = T(.05, .1, .2, .25)$

Now, we can formulate the problem of estimation of α-cuts of top event fuzzy probability through a nonlinear mathematical optimization. Let $Q_{2L}^{(\alpha)}$ be the expression obtained after substituting the left end α-cut of two state - elements' fuzzy probabilities in Q_2. The left end point of the α-cut of $\mathbf{Q}_L$ is then obtained by

$$\mathbf{Q}_{2L}^{(\alpha)} = \min Q_{2L}^{(\alpha)} (p_{11}, \ldots, p_{23}) \tag{6}$$

Subject to the constraints

$$p_{11} + p_{12} + p_{13} = 1 \tag{7}$$
$$p_{21} + p_{22} + p_{23} = 1 \tag{8}$$
$$C(.5,1,3.5)(p_{11})^{0.5}(p_{12})^{1.0}(p_{13})^{3.5} = \alpha \tag{9}$$
$$C(1,1,8)(p_{21})^{1}(p_{22})^{1}(p_{23})^{8} = \alpha \tag{10}$$
$$p_{ij} > 0.0; \quad i=1,2. \text{ and } j=1,2,3 \tag{11}$$

In this formulation it is assumed that only one state of the two state elements is appearing in the fault tree. If both states are appearing, then, it will introduce two constraints as is the case for multistate elements.

Let us now find α-cuts of $\mathbf{Q}_2$ at α=0.5. The left end point of α- cut of fuzzy number $\mathbf{P}_3$ corresponding to the failure state probability of resistor $\mathbf{X}_3$ is 0.075. Substituting p_{31} = 0.075 in the expression (5) for Q_2, we obtain $Q_{2L}^{(0.5)}$ as

$$Q_{2L}^{(0.5)} = p_{22}+0.075(1-p_{22})+(1-0.075)(p_{12}p_{23}+p_{11}p_{23}+p_{11}p_{21}) \tag{12}$$

Minimizing the value of this expression with respect to the constraints given in (7) to (11) (with α = 0.5), we obtain $\mathbf{Q}_{2L}^{(0.5)}$ = 0.185676. Similarly, we can obtain $\mathbf{Q}_{2R}^{(0.5)}$ = 0.723967.

Table 1 α-cuts of $\mathbf{Q}_2$

α	$\mathbf{Q}_{2L}^{(\alpha)}$	$\mathbf{Q}_{2R}^{(\alpha)}$
0.1	0.084371	0.874967
0.3	0.135255	0.790998
0.5	0.185676	0.723967
0.7	0.241888	0.656924
0.9	0.317282	0.574724
1.0	0.415000	0.480000

Table 2 provides the probability vectors of multistate elements which have given the extreme points of α-cuts of $\mathbf{Q}_2$.

4.2 Methods of Reducing Computational Requirements

If the system contains many multistate elements, the terms of the function to be optimized and the number of constraints become too large to be handled by a ordinary personal computer. This necessitates some method of reducing the computational requirements. We propose a method, which achieves the goal at the cost of some accuracy. Instead of finding the optimal value by considering the constraints induced by all multistate components, we find an optimal vector for each multistate component by optimizing the function $Z_i = p_{i1}p_{i2}...p_{in}$ and substitute it in the expression for top event probability.

Mathematically,

Min (Max) $Z_i = p_{i1}p_{i2}...p_{in}$

Subject to the constraints

$p_{i1} + p_{i2} + ... + p_{in} = 1$

$C(\beta_{i1},\beta_{i2},...,\beta_{in})(p_{i1})^{\beta i1}(p_{i2})^{\beta i2}...(p_{in})^{\beta in} = \alpha$

Table 2 Optimal Probability Vectors

α	Optimal vector $\mathbf{P}_1^{(\alpha)}$	Optimal vector $\mathbf{P}_2^{(\alpha)}$
0.1	[(0.008845, 0.021472,0.969682), (0.307176,0.447414,0.245411)]	[(0.183760,0.004856,0.811383), (0.029918,0.379732,0.590350)]
0.3	[(0.020844,0.048480,0.930676), (0.246377, 0.394264, 0.359359)]	[(0.153288,0.014233, 0.832478), (0.046937, 0.291238, 0.661826)]
0.5	[(0.033239, 0.074045, 0.892716), (0.207043, 0.352965, 0.439992)]	[(0.139699, 0.025440, 0.834861), (0.058341, 0.237305, 0.704354)]
0.7	[(0.103241, 0.047526, 0.849233), (0.173677, 0.311642, 0.514681)]	[(0.129011, 0.040054, 0.830935), (0.069916, 0.193189, 0.736895)]
0.9	[(0.070306, 0.142028, 0.787666), (0.137819, 0.260705, 0.601476)]	[(0.114451, 0.062891, 0.822658), (0.083539, 0.146833, 0.769629)]
1.0	[(0.1, 0.2, 0.7), (0.1, 0.2, 0.7)]	[(0.1, 0.1, 0.8), (0.1, 0.1, 0.8)]

Thus for each α, we find these optimal vectors for each multistate component and substitute them in the top event probability expression along with α-cuts of two state elements. If some of the state probabilities of a multistate component are not appearing in expression for fault tree, they can be combined into a single state and a reduced joint possibility distribution (in dimension) may be used. If only one state-probability of a multistate component is appearing in the expression for top event probability, we need not go for optimization as its marginal distribution offers the required α-cuts.

Example : Consider the fault tree shown in Fig. 7 of section 4.1. The top event probability expression is given by

$$Q_3 \quad = \quad p_{12}p_{21}p_{32} + p_{13}p_{21}p_{32} \quad = \quad (p_{12} + p_{13})p_{21}p_{32} \quad = \quad \bar{p}_{11}p_{21}p_{32} \qquad (13)$$

Only one state-probability of each component is appearing in the expression (13) for Q_3. Therefore, α-cuts of Q_3 can be easily obtained using α-cuts of marginal possibility distribution of $\bar{\mathbf{P}}_{11}$, $\mathbf{P}_{21}$ and $\mathbf{P}_{32}$. Assuming the same distribution as in the previous example for probability vectors and using (4), we can obtain

$$\mu(\bar{p}_{11}) = C(0.5,4.5)(\bar{p}_{11})^{4.5}(1-\bar{p}_{11})^{0.5} \qquad (14)$$

$$\mu(p_{21}) = C(1,9)\, p_{21}\, (1-p_{21})^{9} \qquad (15)$$

$$\mu = \mu(\bar{p}_{31}) = T(0.75,0.8,0.9,0.95) \qquad (16)$$

The α-cuts of $\mathbf{Q}_3$ is obtained using the expressions

$$\mathbf{Q}_{3L}^{(\alpha)} = \bar{\mathbf{P}}_{11L}^{(\alpha)}\ \bar{\mathbf{P}}_{21L}^{(\alpha)}\ \bar{\mathbf{P}}_{32L}^{(\alpha)} \qquad (17)$$

$$\mathbf{Q}_{3R}^{(\alpha)} = \bar{\mathbf{P}}_{11R}^{(\alpha)}\ \bar{\mathbf{P}}_{21R}^{(\alpha)}\ \bar{\mathbf{P}}_{32R}^{(\alpha)} \qquad (18)$$

5. CONCLUSION

Using resolution identity, extension principle, and an optimization procedure, fuzzy probability of occurrence of the top event of a fault tree with multistate elements can be evaluated. To reduce computational load, an approximate method of evaluation can be effectively used. This is evident from the illustration provided in this paper.

REFERENCES

[1] H. Tanaka, L.T. Fan, F.S. Lai and K. Toguchi, Fault *Tree Analysis by Fuzzy Probability*, IEEE **Trans. on Reliability**, Vol. 32, pp. 455- 457, 1983.

[2] K.B. Misra and G.G. Weber, *A New Method for Fuzzy Fault Tree Analysis*, **Microelectronics and Reliability**, Vol. 29, pp. 195-216, 1989.

[3] K.B. Misra and G.G. Weber, *Use of Fuzzy Set Theory for Level-I Studies in Probabilistic Risk Assessment.* **Fuzzy Set Theory and Systems**, Vol. 37, pp. 139-160, 1990.

[4] D. Dubois and H. Prade, *Operations on Fuzzy Numbers*, **International Journal of System Science**, Vol. 9, pp. 613-626, 1978.

[5] M. Mizumoto and K. Tanaka, *Some Properties of Fuzzy Numbers*, in **Advances in Fuzzy Set Theory and Applications**, North Holland, Amsterdam, pp. 53-164, 1979.

[6] A. Kaufmann and M.M. Gupta, *Introduction to Fuzzy Arithmetic*, **North Holland**, Amsterdam, Holland, pp. 153-164, 1979.

[7] K.B. Misra, *Reliability Analysis and Prediction (A Methodoloty Oriented Treatment)*, **Elsevier Science Publishers**, Amsterdam, Holland, 1992.

[8] K.B. Misra (Ed.), *New Trends in System Reliability Evaluation*, **Elsevier Science Publishers**, Amsterdam, Holland, 1993.

[9] W.E. Stein, *Fuzzy Probability Vectors*, **Fuzzy sets and Systems**, Vol. 16, pp. 263-267, 1985.

[10] L.A. Zadeh, *Fuzzy Sets as a Basis for a Theory of Possibility*, **Fuzzy Sets and Systems**, Vol. 1, pp. 3-28, 1978.

[11] A.N.S. Freeling, *Fuzzy Sets and Decision Analysis*, **IEEE Trans. on System Man and Cybernetics**, Vol. 10, pp. 341-354, 1980.

[12] D. Dubois and H. Prade, *Additions of Interactive Fuzzy Numbers*, **IEEE Trans. on Automatic Control**, Vol. 26, pp. 926-936, 1981.

[13] S. Nahmias, *Fuzzy Variables*, **Fuzzy Sets and Systems**, Vol. 1, pp. 97-110, 1978.

[14] K.P. Soman and K.B. Misra, *Fuzzy Fault tree Analysis Using Resolution Identity and Extension Principle*, **Int. Jour. of Fuzzy Mathematics**, Vol. 1, pp. 193-212, 1993.

FUZZY FAULT TREE ANALYSIS : CASE STUDIES

A.W. Deshpande and P. Khanna

National Environmental Engineering Research Institute (NEERI)
Nagpur 440 020, INDIA
FAX 91-712 - 522725

ABSTRACT: Estimation of the fuzzy probability of occurrence of an hazardous event (such as : accidental release of chemicals) taking recourse to Fuzzy Sets Theory (FST) is the topic of immediate relevance in Probabilistics Risk Assessment for Chemical Industry. The paper relates to fuzzification of fault trees of 15000 MT capacity atmospheric storage tank and nitric acid reactor. The case studies on fuzzy fault tree analysis using available interfailure statistics of process control instruments brings out its utility over the conventional probabilistic approach.

KEY WORDS : risk assessment, fuzzy fault tree, membership function, ammonia storage tank, nitric acid reactor, process control instruments, interfailure statistics, fuzzy top event probability

1 INTRODUCTION

Environmental risks are inherent in design and operation of large and complex industrial units such as the chemical plants. Any major failure or a series of minor failures in the system operation may lead to a disaster resulting in heavy toll of human life and devastating losses to ecology and property. Growing concern over the risk of major chemical accidents have led to research on new techniques for identification, assessment and evaluation of risks for devising risk mitigation measures. Probabilistic Risk Assessment (PRA) is hereto the most widely employed technique for risk management.

The steps involved in PRA are hazard identification; hazard analysis & assessment (Hazard Quantification), risk evaluation; identification of risk mitigation measures; and preparation of Disaster Management and Emergency Preparedness Plans. Fig. 1 presents the conceptual framework while component activities of PRA are depicted in Fig.2

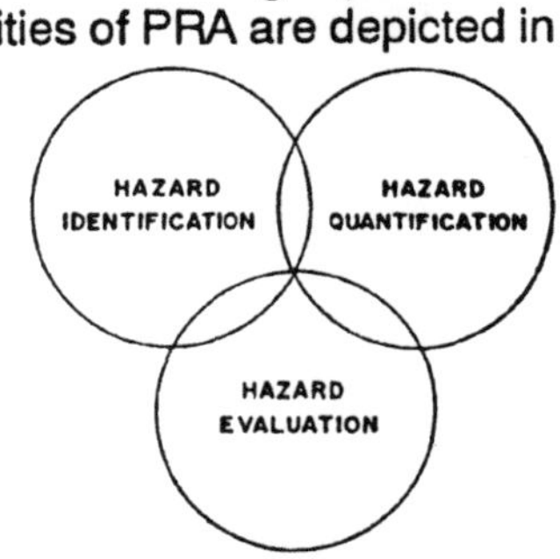

FIG. 1 : CONCEPTUAL ACTIVITIES :PROBABILISTIC RISK ASSESSMENT

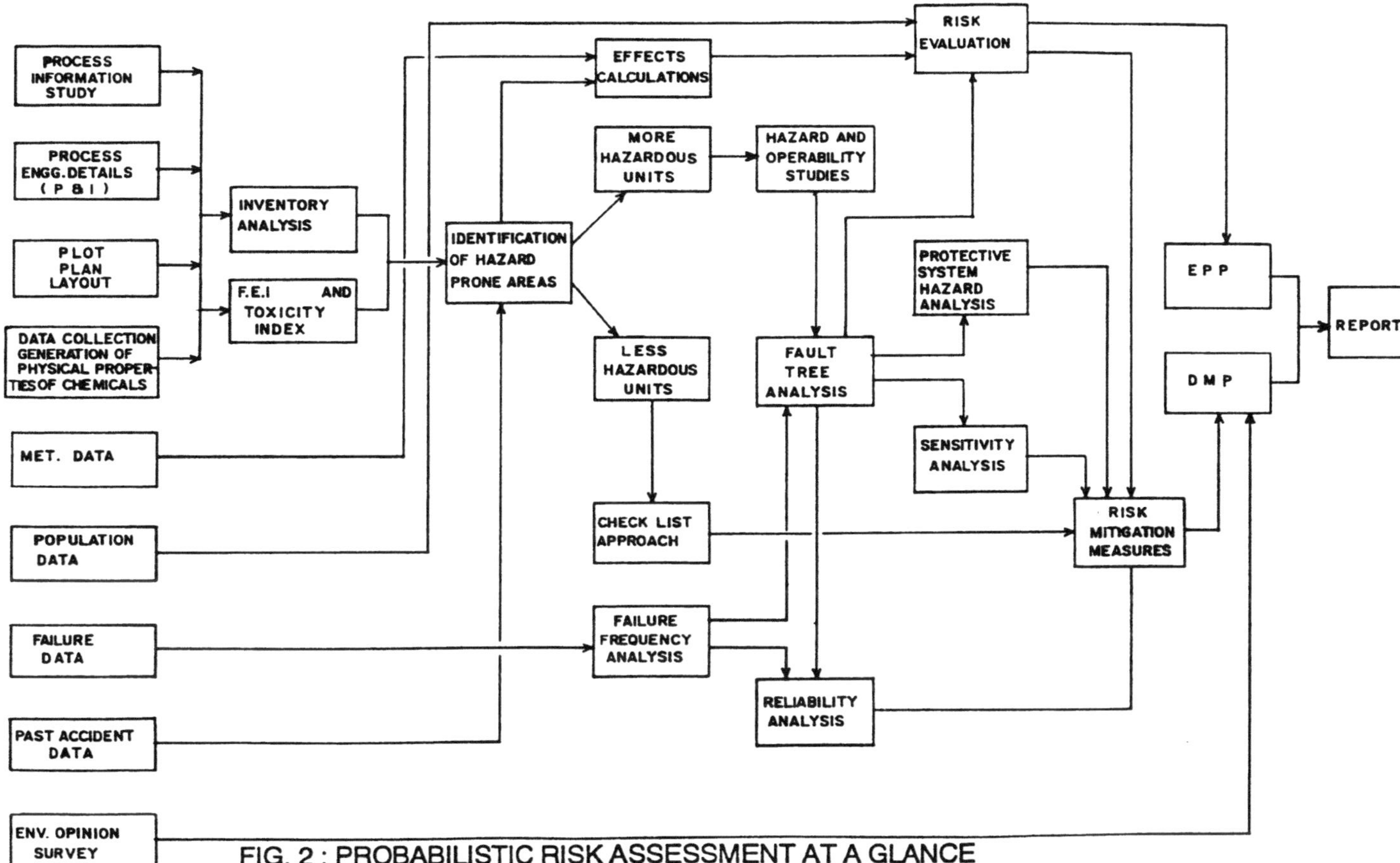

FIG. 2 : PROBABILISTIC RISK ASSESSMENT AT A GLANCE

2 DEFINITION OF RISK

The evaluation of risk, arising out of an undesired event, comprises estimation of expected frequency of undesirable events per unit time and the expected damage (1). Then, a customary definition of risk R, interms of the expected frequency of occurrence and the expected damage D, is

$$R = F.D \tag{1}$$

The other possible definitions of risk are :

$$R_k = F.D^k \tag{2}$$

where, k > 1 is a risk function which signifies the importance of events with larger damages.

Another approach to the interpretation of risk may be developed by using a mathematical notation, such as risk density $R_i(X_j,t)$, which is the frequency of occurrence of an event E_i at time t producing an ultimate damage of consequence type j between X_j and X_j+dX_j per unit dX_j. The risk is considered a density function because it has units of (consequence)$^{-1}$.

In terms of $R_i(X_j, t)$, the risk $R_i(X_j,t)$ may be defined as

$$R_i(\geq X_j,T) = \int_{X_j}^{\infty} R_i(\geq X_j,T)\, dX_j \tag{3}$$

When the operation of a system leads to risks from more than one consequence type, yet another type of risk can be defined. Aggregated risk from event E_i, $R_ik(t)$, as measured from the consequence type k, is the risk arising from event E_i due to all the possible consequences. Aggregated risk may be written as

$$R_i^{(k)}(t) = \sum_j aj^{(k)} \int_0^{\infty} R_i(X_j, t)\, dXj \tag{4}$$

Where, $a_j(k)$ are aggregation factors. Such an aggregation could involve a combination of all types of consequences.

3 UNCERTAINTIES IN PROBABILISTIC RISK ASSESSMENT (PRA)

Probabilistic Risk assessment is initiated with a qualitative phase which includes the identification of accident contributors and the consequence modelling of accidents visualising release scenarios. Many mathematical models exist for the accidental release of hazardous chemicals using toxicity and meterological data. Subjectivity in accidental release scenarios of hazardous chemicals can often eliminate the potential advantage offered by the existence of variety of these models. Koivisto et al (2) have developed a method based on fuzzy logic for the selection of a computer model to be used in consequence analysis under specific situations.

The modelling phase is then continued with the quantification of accident frequencies by using component failure and human error data. Consequence modelling, coupled with estimation of accident frequencies and the population at risk in turn, help in plotting F-X and F-N curves and risk contours. Probabilistic Risk Assessment has problems and deficiencies arising out of :

- incompleteness of accident contributors and thereby the modelling of accidents, and
- inaccuracy in the quantification of risk in terms of frequency and consequences.

Table 1 presents a summary of criticisms often levelled against

PRA studies.

Table 1. Criticism of PRA (3)

S. No.	Phase of analysis	Deficiencies
1.	Definition and description of system	Relevant subsystems or activities are excluded
2.	Hazard identification	Description of the system does not correspond to the real life situation
3.	Accident modelling	Important accident contributors or families thereofare excluded or omitted
4.	Quantification of risks	Uncertainties in component failure rate or human error data Inaccuracies in consequence modelling
5.	Documentation of results	Boundaries of - and assumption in - analyses are not described Source of quantitative data are not presented

The prerequisites to risk assessment studies include event tree development and/ or fault tree construction wherein much of the uncertainties are introduced. It is essential to mitigate the uncertainties associated with Fault Tree Analysis (FTA) to enhance the credibility of risk assessment. A probabilistic approach is used in FTA and a considerable amount of uncertainty builds up in the top event probability evaluation in the procedure in-vogue. The representation of an uncertainty band through mean and variance in probabilistic evaluation is not adequate (4). The probabilistic approach, as a whole, lacks adequate mechanism for tackling the problem of uncertainities associated with PRA. An attempt is made here to model the uncertainities with respect to top event in fault tree analysis which is further demonstrated with two case studies.

4 FAULT TREE ANALYSIS

The technique of Fault Tree Analysis (FTA) for the estimation of the frequency of occurrence of an event was formalised in 1962 at Bell Telephone Laboratories, USA. The construction of Fault Tree (FT) requires professional expertise. Checking and rechecking of FT helps in attaining the desired accuracy. In order to describe the computational procedures, it would be appropriate to consider an illustrative example.

One of the intermediate event (Flash vessel empty) from the fault tree of atmospheric storage tank depicted in Fig. 3 has been identified for describing the FTA procedure.

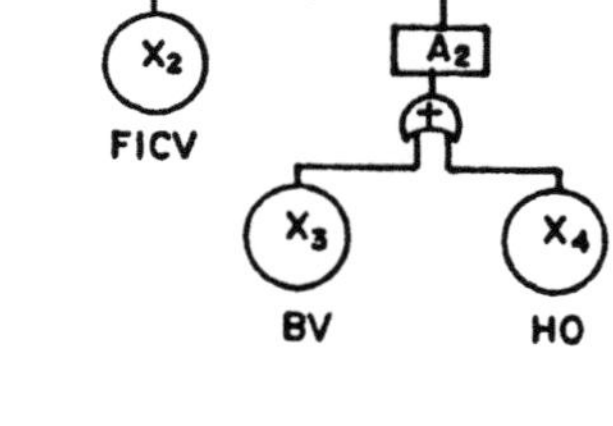

Fig.3 Intermediate event- flash vessel empty

The fault tree shown in Fig. 3 has been reduced to a logically equivalent form as follows :

Minimum cut sets $\{X_1, X_2\}, \{X_3\}, \{X_4\}$ (5)

Event Relationship $T = X_1 \cap A_1$ (6)

Where $A_1 = X_2 \cap A_2$, $A_2 = X_3 \cup X_4$ (7)

Top Event Probability (TEP)

$$P_T = 1 - \{(1 - px_1)(1 - pA_1)(1 - pA_2)\} \quad (8a)$$

Where $pA_1 = px_2 * pA_2$

and $$pA_2 = 1 - \{(1 - px_3)(1 - px_4)\} \quad (8b)$$

5 FUZZY FAULT TREE ANALYSIS

The analysts are confronted with inaccurate data on probability of failure of primary events. Occurrence of an extremely hazardous event is rare and therefore, the data is rarely available. In the absence of accurate data, it becomes necessary to work with rough estimates of probabilities that are provided by the system designers. It is evident that these experts offer their subjective assessment on human error and component failure probabilities. Under these circumstances it is considered inappropriate to use conventional FTA for computing system failure probability.

Fuzzy Set Theory (FST) offers a frame of analysis which could model imprecision in input failure probabilities used in FTA. The estimation of Top Event Probability (TEP) in fuzzy set would be hereafter termed as Fuzzy Top Event Probability (FTEP) and such analysis would be termed as Fuzzy Fault Tree Analysis (FFTA).

The concept of fuzzy probability in FT represents fuzzy number, the value of which ranges between zero and one for each primary event.

5.1 UNCERTAINTY NUMBER OR FUZZY NUMBER

An interval of confidence is one way of reducing the uncertainty of using lower and upper bounds. It is a practical and logical process for treating uncertainty with whatever information is available. This information can be objective or subjective (5). Let us relate the concept of the interval of confidence to another called the level of presumption. For example, that the failure probability of certain process control instrument (eg. PSV) has been estimated from the data to be in the interval : $500 * 10^{-4}$ - $5250 * 10^{-4}$ f/yr. This is termed as the interval of confidence. On the other hand, let us assume that in order to bring down the risk level to the acceptance level of 10^{-6} fatality/yr, the estimated failure probability is $1000 * 10^{-4}$ f/yr. Two levels of confidence could be assigned to the above two situations, **0** for ($500 * 10^{-4}$ - $5250 * 10^{-4}$ f/yr) and **1** for ($1000 * 10^{-4}$ - $1000 * 10^{-4}$ f/yr.). These two levels of confidence are infact levels of presumption and could be represented by **[0,1]**.

The level of presumption could also lie between **0** and **1** and could be represented as :

$$\forall \mu_1, \mu_2 \varepsilon [0,1]$$

$$(\mu_1 < \mu_2) \Rightarrow ([\alpha_1^{(\mu 2)}, \alpha_2^{(\mu 2)}] \subset [\alpha_1^{(\mu 1)}, \alpha_2^{(\mu 1)}]) \quad (9)$$

which means that if μ increases the interval of confidence never increases (Fig. 4). The coupling between the level of presumption and the interval of confidence at level will be a way defining the concept of an uncertain number or a fuzzy number. This association corrosponds to the natural, often implicit, mechanism of human thinking in the subjective estimation of a value for a dimension.

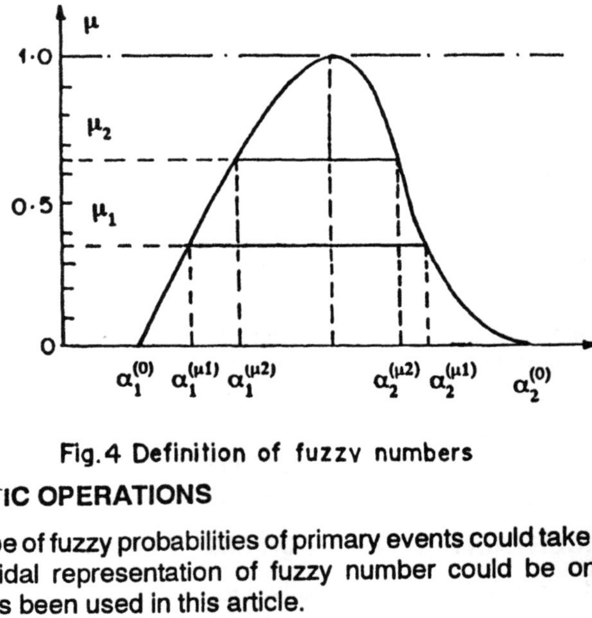

Fig.4 Definition of fuzzy numbers

5.2 ARITHMATIC OPERATIONS

The shape of fuzzy probabilities of primary events could take various forms. Trapezoidal representation of fuzzy number could be one such shape which has been used in this article.

Fuzzy probabilities of the basic events must be combined according to the rule of combination of fuzzy numbers in order to compute fuzzy probability of the top event in FFTA. Primarily, only two logical operations are required to compute FTEP, multiplication, and complimentation of fuzzy sets.

5.2.1 MULTIPLICATION

According to the extension principle (6), the computational procedure for multiplication of two fuzzy sets Px_i and Px_j is quite involved and therefore, Tanaka et al. (7) have suggested an approximation of the multiplication procedure by defining as follows :

$$P_{x_1} * P_{x_2} = \{(\alpha_{i1}\alpha_{j1}), (\alpha_{i2}\alpha_{j2}), (\beta_{i2}\beta_{j2}), (\beta_{i1}\beta_{j1})\} \quad (10)$$

Multiplying the corrosponding elements of fuzzy probabilities of event X_i and X_j respectively.

5.2.2 COMPLIMENTATION

The complimentation of any fuzzy set P_{x_i} is given as

$$\bar{P}_{x_i} = \{(1-\alpha_{i1}), (1-\alpha_{i2}), (1-\beta_{i2}), (1-\beta_{i1})\} \quad (11)$$

6 CASE STUDIES

A chemical process industrial complex can be divided broadly into three categories, viz. storages, process units and transfer lines. Storage facility for hazardous chemical is by far the most vulnerable unit in any chemical process industry.

Ammonia, an important intermediate in the production of fertilizers is either manufactured and stored in the pressurised storages or could be procured and then stored in atmospheric tanks near the fertiliser complexes while Nitric acid is an intermediate in the production of Ammonium Nitro Phosphate (ANP). The two case studies describe fuzzification of fault trees of atmospheric storage tank of ammonia and a reactor in the nitric acid plant at the fertilizer complex in Bombay.

6.1 CASE STUDY - I

No item of equipment is included in more accidents than storage tanks, probably because these are fragile and easily damaged by slight more pressure or vacuum. The major failures of the storage tanks are due to overfilling, overpressurisation, sucking in, explosions / fires, floating roof accident etc.

The case study relates to the ammonia storage tank (atmospheric storage) located at about 3 kms from the large fertilizer complex in Bombay (8). Ammonia is stored at - 33°C and at atmospheric pressure. The ammonia is transfered to a tank via a 2.2 km pipe line from Jetty. The storage has been identified as the most hazardous unit as the Toxicity Index (TI) exceeds the acceptable limit of 10 (8). The damage distance has been computed visualizing the accident scenario of release of ammonia from the tank. The main cause of rupture of the tank could be attributed to overpressurisation. Majority of the storage tanks are designed to withstand gauge pressure of only 8 inch (2 kpa or 0.3 psi) and will burst at about three times this pressure and could thus be easily damaged. Rupture of storage tank has, therefore, been assumed at a top event in the fault tree which has been depicted in Fig. 5.

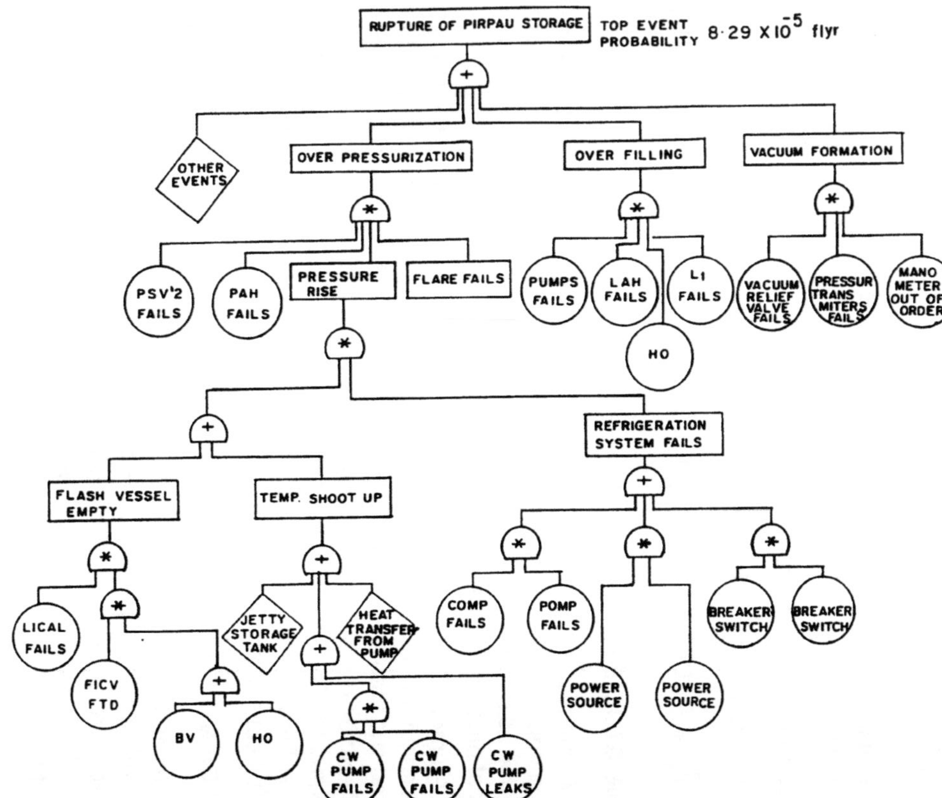

FIG. 5 : FAULT TREE OF AMMONIA STORAGE TANK

6.1.1 FAILURE DATA

To collect, store and retrieve failure information effectively for devices of many types is an enormous task and requires an organisation that is consistent with defined objectives. In order to carry out PRA studies for chemical plants a need for failure data of process control instruments was recognised.

The data required for failure probability computations was collected from the log books available at the fertilizer complex which maintains interfailure statistics of the majority of process control instruments. Table 2 shows the computed values of failure probabilities of the instruments indicating upper and lower values. Data on human error probability was not available and accordingly, international data has been used in the computation of TEP or FTEP values.

Table 2 . Failure Probabilities of Primary Events

Instrument / Event	**Failure Probability (x 10^{-4})** Abbreviation	Lower bound	Upper bound
Bypass valve	BV	1.00	20.00
Human operator	HO	1.00	20.00
Human operator (overfilling)	HO	4000.00	8000.00
Flow indicator and Controller valve	FICV	2500.00	7200.00
Level indicator, Controller and alarm	LICAL	3000.00	7500.00
Jetty storage		1.00	20.00
Cooling Water Pump - 1	CW Pump	500.00	5250.00
Cooling Water Pipe	CW Pipe	1000.00	5500.00
Heat transfer from pump		1.00	20.00
Flare fails		1.00	20.00
Compressor	Comp	8900.00	9500.00
Power Source		50.00	150.00
Breaker Switch		2500.00	4000.00
Pump		500.00	5250.00
Level Alarm	LAH	1500.00	2500.00
Level Indicator	Li	5.00	20.00
Vacuum Relief Valve		500.00	5250.00
Pressure Transmitter		3500.00	7200.00
Manometer out of order		1.00	2.00
Pressure Safety Valve	PSV 2	500.00	5250.00
Pressurealarm	PA	500.00	2500.00

6.1.2 COMPUTER CODE

The procedure for the estimation of FTEP (or possibility of the failure probability) is quite involved warranting a writing of a computer code. A computer program was developed for FTEP (μ) in various kinds of representation of fuzzy probability of the primary event. For a particular fault tree, the program takes the lower and upper bounds of the failure probabilities of the primary events as input, and calculates FTEP. In this particular case, the nature of the membership function (μ) of the primary events is not known. The only data available is the lower and upper failure probabilitiesof the primary events. The membership grade function is assumed to be a trapezium as shown in figure 6. The two known lower and upper failure probabilities represent the points $\alpha_1^{(0)}$ and $\beta_1^{(0)}$. The other two points $\alpha_2^{(1)}$and $\beta_2^{(1)}$ which lie on the $\mu = 1$ line are not known. Hence, they are generated as any two random numbers ranging between $\alpha_1^{(0)}$ and $\beta_1^{(0)}$ using standard pseudo random number generator technique. In a nutshell, the program works as under :

- Generates trapezium for each of the primary events
- Traverses the FFT bottom up computing the intermediate results according to the **AND** and **OR** fuzzy operations at the various interior nodes
- Finally the trapezium (which corresponds to the membership function curve) of the top event is generated

Certain simplifications have been incorporated taking recourse to standard switching theory techniques; taking into consideration the new expressions for the **AND** and **OR** fuzzy operations which are presented in Fig. 7 (9).

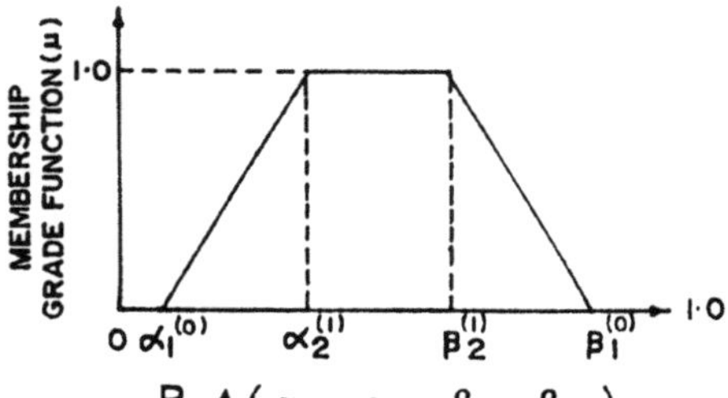

$$P_{x_i} \triangleq (\alpha_{i1}, \alpha_{i2}, \beta_{i2}, \beta_{i1})$$

Fig. 6 : Fuzzy probablity – trapizoidal representation

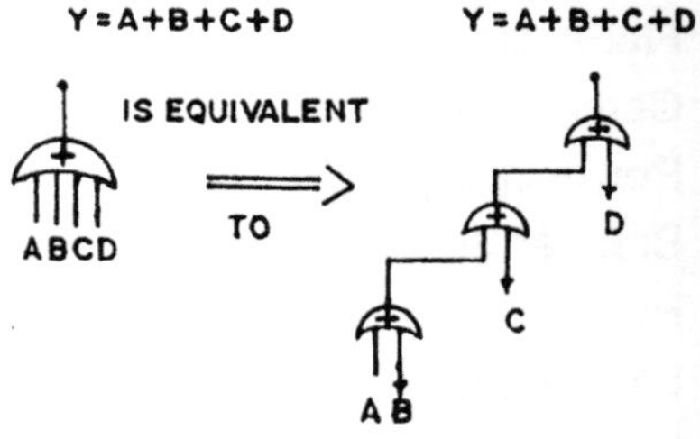

Fig. 7: Fuzzy operation – OR function

6.1.3 COMPLEXITY ANALYSIS

Both, time and memory requirements of the program are directly proportional to the number of primary events in the fuzzy fault tree. Following are the memory requirements for the storage tank FFT in the case study :

Code size - 2800 bytes;

Data size - 450 bytes;

Stack/heap size - 20 k bytes (This includes space for the system stack).

Thus any IBM compatible PC can handle FFT upto 1000 primary events. Though the program has been implemented on TURBO PASCAL version 3.0, the source code can be recompiled on any of the existing machines and executed.

6.2 CASE STUDY - II

The second study relates to a reactor in nitric acid plant (10) of the aforementioned fertilizer complex manufacturing 1250 ton/day of nitric acid (60%). The plant operates according to mono medium pressure i.e. almost same pressure is maintained in the combustion and in the absorption sections. Prior to the ammonia-combustion, the air quality required for the process is supplied by the turbo compressor. The combustion of ammonia to nitric oxide is catalytically performed on platinum gauges. The process unit is known as ammonia air reactor (RO1), which was identified as the most vulnerable using formalised FETI technique.

Reactor falls under explosive and toxic hazard category as the Toxicity Index TI is 30 exceeding the limit of 10. The unit also handles NOx gases hence it is under severe category. The extent of damage likely due to occurrence of hazardous event which is explosion in the reactor or release of toxic gases (ammonia and NO_x) was computed using various mathematical models. The scenario considered help in computing the damage distances for release of ammonia for 1% lethality limits (1330 mg/m^3). Mixture of ammonia and air enter the reactor under red hot platinum gauge in temperature of which is 850°C. The process gas on release will form a turbulent free jet. The source strength was computed for the inlet pipe rupture to the reactor which would lead to dispersion of toxic cloud and concentration contour for 1% lethality will approach to the maximum distance of 3.35 km in the wind direction.

Fault Tree was generated for reactor[10] and the most hazardous event identified was explosion in reactor due to ammonia percentage reducing explosion limit (14%, 26%) leading to ammonia gas release (Fig. 8).

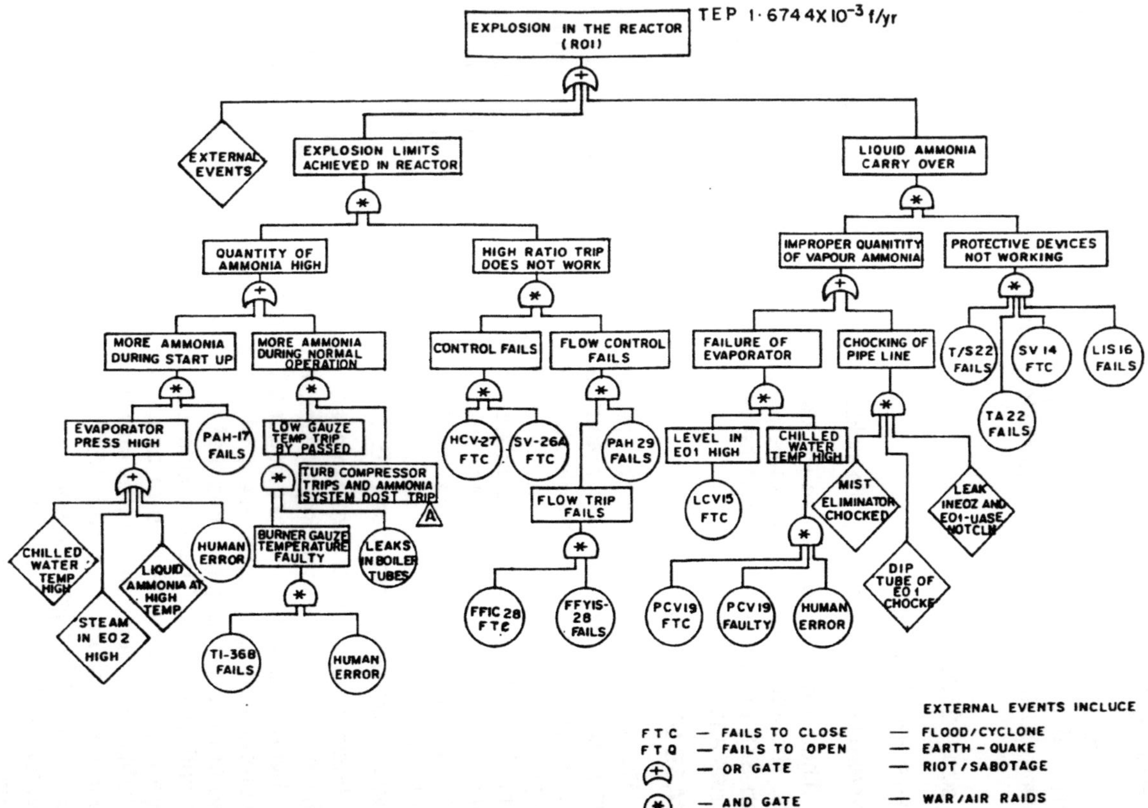

FIG. 8 : FAULT TREE FOR REACTOR (ROI)

Table 3 presents the data on failure probabilities of instruments of nitric acid plant which was collated and used for the computation of TEP using conventional approach and FTEP using fuzzy probability approach.

Table 3 . Failure Probabilities of Primary Events for Nitric Acid Reactor

Instrument / Event	Failure Probability (x 10^{-4})		
	Abbreviation	Lower bound	Upper bound
Chilled Water Temp High	CWTH	3324	7626
Steam in EO_2 High	SEH	971	9128
Liquid Ammonia at high temp	LAHT	3324	7624
Human operator	HO	1.0	20.00
PAH 17 FAILS	PAHF	1389	1576
TI - 36 B Fails	TIF	1341	1541
Leak in Boiler tubes	LBT	538	3624
Turbo compressor trips	TURCT	2322	8966
HCV - 27 FTC	HCV	971	9128
SV - 26, FTC	SV	375	2505
FFVIS 28		3822	5812
PAH - 29 FAILS	PAH 29	1389	1576
LCV - 15 FTC	LCV - 15	4696	8318
PCV - 19 FTC	PCV - 19	4843	8789
PIC - 19 Faulty	PIC - 19	1363	1608
Mist Eliminator chocked etc.	MEC	1313	2894
T/S - 22 Fails	TS - 22	2070	2773
TA - 22 Fails	TA - 22	1389	1576
SV - 14	SV - 14	2506	3750
LIS - 16 Fails	LIS - 16	128	1457

6.3 RESULTS AND DISCUSSION

The TEP of the hazardous event (release of ammonia from the storage tank) using conventional FTA approach has been estimated as 8.29 x 10^{-5} f/yr. as shown in Fig. 5. Random combination of the fuzzy probability of primary events were used as per the procedure outlined in section 6.1.2 for $\mu = 1$ and the range of FTEP was estimated which is presented in Fig. 9. It is seen that the occurrence of Fuzzy Top Event Probability (FTEP) ranges between 3.27 x 10^{-6} and 3.018 x 10^{-4} f/yr with a most likely value between 5 x 10^{-5} - 7 x 10^{-5} f/yr range with possibility (μ) of 1. Employing conventional approach, top event probability works out to 8.29 x 10^{-5} i.e. a value contained in the interval (7.0 x 10^{-5} - 3.01 x 10^{-4}) obtained through FFTA (Fig. 9).

In case of explosion of nitric acid reactor (RO1) the TEP is 1.6744 x 10^{-3} f/yr while FTEP works out to be with in band width 0.5 x 10^{-3} and 3.0 x 10^{-3} f/yr for μ = 1. (Fig. 10)

It could be stated that the hazardous event (explosion in nitric acid reactor) would occur with possibility (μ) equal to 1 with maximum fuzzy probability of 3 x 10^{-3} f/yr but the occurrence of FTEP of 5 x 10^{-3} is impossible with the fuzzy probability approach described in this article. It is, therefore, suggested that FTEP value of 3 x 10^{-3} f/yr for μ = 1 would be an improved proposition over TEP value of 1.6744 x 10^{-3} f/yr as the analysis is based on exhaustive Fuzzy Probability approach. Risk is generally described in terms of fatality probability. The authors further believe that the utilisation of the FTEP should be considered while plotting risk contours instead of TEP using conventional approach.

FIG. 9: FUZZY TOP EVENT PROBABILITY (ATMOSPHERIC STORAGE TANK)

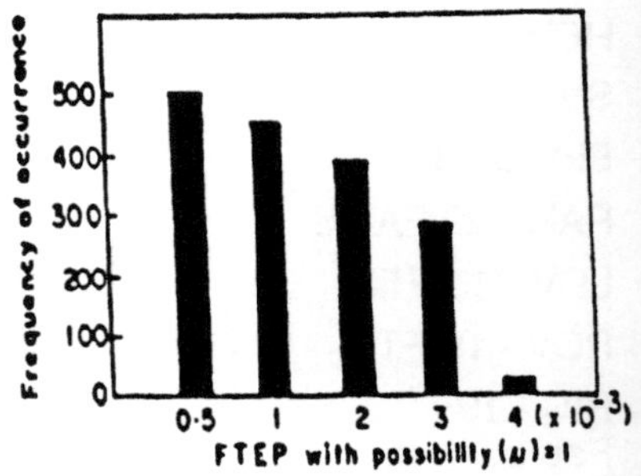

FIG. 10: FUZZY TOP EVENT PROBABILITY (μ=1) (NITRIC ACID REACTOR)

7 CONCLUSION

The case studies on FFTA using the available interfailure statistics of process control instruments brings out its utility over conventional probabilistic approach. Modelling of common cause failures, dependent failures, human error etc. are some of the issues require study in FFTA. The Fuzzy Sets Theory presents a viable alternative to improve upon the state of art in Probabilistic Risk Assessment.

ACKNOWLEDGMENTS

The authors thank Prof. T. Onisawa for his suggestions on the study. The assistance for the construction of fault tree and the development of computer software from Mr. R.S. Olaniya, Mr. U.A. Deshpande and Mrs. B. Dabir is gratefully acknowledged.

REFERENCES

1. Norman J McCormik., Reliability and Risk Analysis-Methods and Nuclear Power Applications, Academic Press New York 1981 (Book)

2. R. Koivisto, kakku,R; Dohral, Mand Jarvelainen; M Accedent Release-Fuzzy selection of adequate formal model. Journal of Loss Prevention process Ind.,1991,Vol 4, oct. pp 317-330

3. Suokas, J.and kokko, R., On the problem and future of safety and risk analysis Journal of Hazardous Materials, 21 (1989) pp 105-124

4. Misra, K.B. and Weber, G.G, Use of fussy set theory for level- I Studies in Probabilistic risk assessment, Journal of Fuzzy sets and system, 37 1990 pp 139 -160

5. A. Kaufman and M.M. Gupta., Introduction to Fuzzy Arithmatic Theory and Applications, Van Nostrand Reinhold, New York, 1984, (first edition)

6. M. Mizumoto and k. Tanaka, Some properties of Fuzzy number, in M.M. Gupta, R.K. Ragade and R.R. Yoger, Eds, Advances in Fuzzy Set Theory and Applications, North Holland Amesterdan (1979) 153-164.

7. Tanaka, H ,Fan, L.T. and Toguch, K.,Fault tree analysis by fuzzy Probability, IEE Trans . on reliability, 32 (5) (1983) pp 455 -457

8. NEERI report on Hazard Study and Quantitative Risk Assessment of RCF Complex,Chembur Bombay,Vol.1 New Ammonia Plant (Trombay) November 1990

9. Deshpande A.W., Deshpande U.A. and Khanna P., Fuzzy Fault Tree Analysis - A case study, paper presented at the 2nd International conference on Fuzzy Logic and Neural Network (IIZUKA ,92) July 17-22 Japan.

10. NEERI reports on Hazard Study and Quantitetive Risk Assessment of RCF Complex,Chembur Bombay, Vol 4, Nitric Acid Plant November 1990.

FAES – FAULT ANALYSIS EXPERT SYSTEM

M. Schneider* and A. Kandel**

*Computer Science, Florida Tech
Melbourne, FL 32901, USA

**Computer Science and Engineering
University of South Florida
Tampa, FL 33620, USA

Abstract: In the development of expert systems, the process of acquiring knowledge from the domain expert and translating it to a form which the expert system can understand, is costly and time consuming. Also, expert systems available for fault isolation, detection and assessment are not fully matured. In this paper we will describe a system that is capable of automating the knowledge acquisition process, translating that knowledge into a fully validated knowledge base, and using this knowledge base for fault detection.

Keywords: knowledge acquisition, knowledge validation, knowledge generation, fuzzy expert system tool, learning function.

1 INTRODUCTION

FAES is logically diveded into two main parts: the knowledge acquisition process and the fault isolation (detection) process. The knowledge acquisition process can be farther divided into three main modules. The first is the knowledge acquisition module. In this module, the user is requested to input the system's description. The second module has the task of validating the knowledge entered by the user. It is vital that the knowledge will be complete and non contradictory. The third module has the function of translating the captured knowledge into rule structured knowledge base. For additional details regarding the fuzzy-rule structured knowledge base and its control, see Kandel(1992), Kandel and Langholz(1992), and Kandel and Langholz(1994).

The second part of the expert system uses the knowledge base as its input to fault isolate any faulty parts in a given target system. It is composed of several main modules. First, it has a database, which has a function of tracking past behaviors of the LRU's (Lowest Replaceable Units) of the target system. This information is used by the inference engine and the learning module to improve the inference process, and the decision making of the expert system. Also, it has an explanation module which explain the user its reasoning path. It has an editor to edit the knowledge base (if needed) and the database and a learning module which learns to improve the decision making of the expert system as the system matures.

In the next two sections, we will describe in details the structure and the functions of each module.

2 THE KNOWLEDGE ACQUISITION PROCESS

In attempting to solve the problems in knowledge acquisition, several approaches to automating the acquisition process come immediately to mind. Automation of knowledge acquisition involves utilizing tools to expedite the process, make it more efficient and precise. There are several way of approaching this and there has been work done in this area in the past few years.

Previous tools and approaches for automaton have been directed at facilitating the domain experts entries and organizing the entries to help in creating the internal knowledge base. These include TEIRESIAS (McGraw and Harbison-Briggs, 1989), ETS (Davis, 1977), BDMKAT (Kitto and Boose, 1987) and TDE (Blaxon, Geesey and Reeker, 1987). They also include the development of strategies within the tools for interviewing the experts and reinforcing the acquired knowledge base, as in MORE (Kahn, Breaux, Deklerk, and Joseph, 1987).

Our approach is somewhat different. The usefulness of our approach is in its ability to allow anyone to generate the knowledge base. But not only that, we validate the knowledge and produce a knowledge base which is complete and ready to be used by the inference engine.

2.1 The Interview Session

A very important aspect of any interactive computer program is its Man-Machine Interface (MMI). Interactive programs that have poor MMI will suffer from disuse no matter how well they perform their designated functions. Our approach combines simple questions and responses with instructions and examples that guide the user. There are several steps in the interview session (Wheeler and Schneider, 1990):

1. The user has to provide the system name. Each target system is associated with a different knowledge base.
2. The user has to provide the names of the components of the target system (the LRU's names).
3. The user has to provide the name of the interfaces between varies LRUs
4. The user has to provide the interconnections between the LRUs.
5. The user has to provide the characteristics of each LRU.

Upon completion of the interview session, the system may form the relations between the LRUs and relate characteristics to its components. In any stage of the interview session, the user may edit the entries without altering the flow of the interview session. When the user has completed to enter the knowledge about the target system, the validation process has to take place.

2.2 Validation

Automatic knowledge acquisition requires a validation process to assure that the user inputs will create viable, non-ambiguous knowledge and database. For automatic knowledge acquisition, validation must be performed by a tool rather than by a knowledge engineer and therefore needs to follow a set of rules and guidelines. The goals of the validation process are (1) to provide sufficient input checking because the user is not assumed to be an expert either in the target system or in fault isolation techniques and (2) to assure that the target system model created is viable for knowledge base rule generation.

The validation process consists of five distinct steps which collectively check for any errors or conditions present in the data that may cause the creation of faulty knowledge and data bases (Wheeler and Schneider, 1991),

1. First, it is necessary to validate the LRU names for uniqueness in order to prevent ambiguity in the knowledge base. Ambiguity in the knowledge base may lead to ambiguous or even wrong results which may confuse the user and reduce his/her confidence in the system.

2. Next, it is necessary to validate the interface names for uniqueness. We want to avoid ambiguity for the same reason as describes above.

3. In this step we check that each interface is connected to some LRU which was defined. If the user were to enter LRU names that do not exist as receiving or sourcing an interface, the knowledge base would contain rules that could infer fault in a non existing LRU.

4. Here, we check that each LRU has at least one input and one output. An LRU or a set of LRUs with no input and no output has no affect on the rest of the system. Thus, it may be assumed that if such a case is encountered, the user is to blamed for faulty entry of the knowledge in the interview session.

5. The fifth and final validation step is the most complex and comprehensive. Here we examine the LRU's interconnections for any closed loops and check the whole system to assure that there exist at least one external input and one external output.

The validation process is a vital and imperative part of the automatic knowledge acquisition tool. It works interactively with the editor to allow quick and easy corrections to any errors or conditions found that may produce faulty knowledge and data bases for the expert system. The error messages are sufficient for any non expert user to know what is requires to eliminate these errors. The validation process checks for ambiguities, consistency and viability which are pertinent to the construction of a knowledge base.

2.3 Knowledge Base Generation

In order to fulfill the goal of providing a low cost fault isolation knowledge acquisition tool, there is a need to produce a set of knowledge base rules that are efficient in isolation faults and can be stored in a usable, transportable format. But most importantly, the knowledge in the knowledge base has to be complete and non ambiguous.

The rules that are generated by the expert system contain all the knowledge which was transferred from the user and validated by the validation process. The knowledge base contains knowledge about the characteristics of the LRUs, the relations among them, the input and output description of the LRUs etc. (Wheeler and Schneider, 1992).

When the process of rule generation is completed, the expert system has a complete knowledge base containing all the knowledge necessary for fault isolation.

3 THE DECISION TOOL

The decision tool of FAES incorporates an inference engine, a data base, a knowledge base, and a learning program to detect faults in a given architecture. We will describe the components of the decision tool and then describe how FAES performs its fault detection.

3.1 The Data Base

The role of the data base is to store information regarding each LRU in the target system. This include the name of the LRU, its characteristics, and its history of failures. Let *Max_fail* be a global variable which stores the largest LRU failure count, then we define the *Fail_rate* (FR) as (Elliott and Schneider, 1990a):

$$FR = \frac{failures}{Max\_fail} \tag{1}$$

Consider, for example, the following table:

Table 1: Statistics of LRU failures

LRU #	Failures	FR
1	6	0.85
2	1	0.14
3	0	0.00
4	7	1.00
5	2	0.29

As can be seen, the first column in Table 1, indicates the LRU number, the second column the number of time that LRU failed, and the third column shows the FR. Since LRU 4 has the most failures, its value of failures becomes *Max_fail*, and the FR' s are updated accordingly.

The FR s are used in the inference process and updated after each run.

3.2 The Knowledge Base

The knowledge base utilized by FAES is the knowledge base generated by the knowledge acquisition program. It consists of fuzzy production rules and each rule is associated with a rule certainty (RC), number of times the rule was tested (NT), and the number of times the rules was fired (NS). A learning factor (L) is added to each rule to prioritize the rule testing (Elliott and Schneider, 1990b). Initialy, L is set to RC and as the system matures, it is updated using the following equation:

$$L = L - 0.5(L - \frac{NS}{NT}) \tag{2}$$

The 0.5 is added to the equation to cut down the rate of learning. To fully understand the functionality of L, consider the following example:

Example 1: Assume, the knowledge base, generated by the knowledge acquisition program generated the rule:

if the power_supply is failed and the system is functioning
then the system is degraded and LRU1 is failed

Also assume that the user provide the rule certainty to be 0.9. First, FAES, will initialize L to 0.9 and set NT and NS to 0. Next, assume the inference engine fired that rule, so L is updated to

$$L = 0.9 - 0.5(0.9 - 1/1) = 0.95$$

Next, FAES is executed again, but the rule was not fired, so L is updated to

$$L = 0.95 - 0.5(0.95 - 1/2) = 0.725$$

Again, FAES was executed and the rule was not fired, then L is set to:

$$L = 0.725 - 0.5(0.725 - 1/3) = 0.529$$

The learning factor is updated after each run of FAES.

3.3 The Inference Engine

The inference engine is responsible for inferring conclusions from a given knowledge base and user/sensor entered data. The process of inferencing involves matching the premise of the rules in the knowledge base with the data stored on the blackboard and producing the proper conclusions.

Thus, we match every clause in the premise of every rule with the blackboard. If the matching factor (MF) is above a certain threshold, we fire the rule (Schneider and Kandel, 1993; Chew, Schneider and Kandel, 1994). If the rule is fired, we compute the final certainty (FC) of the conclusion:

$$FC = \frac{MF + L + RC}{3} \tag{3}$$

If more then one conclusion are generated, we compute a value (V) which is a composite of the FC and the FR. First we define certainty ranges for the FR:

and the certainty range for the FC:

When combining the two tables, we have to take into account the possible combinations of the ranks. Since each table contains three ranks, the total combined ranks is nine. In addition to the total points computed by V, we associate each combination with a class (C). The role of C is to rank each combination of ranks by its importace. That is, if two conclusions have the same V, then C detemines which rule is to be tested first. The value of C is determined by the user and can be changed via the FAES' editor (Elliott and Schneider, 1990c). Table 4 depicts the possible combinarions of the ranks from Tables 2 and 3, the combine grade (V), and the subjective class assignment.

where the class ranks the V's in the order of importance.

If two conclusion fall in the same class then a weighted average is used to determine the wining rule:

$$WA = \frac{1.1 * FC + 0.9 * FR}{2} \tag{4}$$

Example 2: Assume, the inference process produced two conclusions with the following values:
Conclusion 1: $FC = 0.90$, $FR = 0.5$
Conclusion 2: $FC = 0.75$, $FR = 0.9$

Using Tables 2, 3 and 4, we compute the corresponding V's. The value of V for both conclusions is computed to be 5. But, since conclusion 1 is in class 2 and conclusion 2 is in class 3, conclusion 1 is tried first.

Example 3: Assume, the inference process produced two conclusions with the following values:
Conclusion 1: $FC = 0.75$, $FR = 0.25$
Conclusion 2: $FC = 0.80$, $FR = 0.20$

Using Tables 2, 3 and 4, we compute the corresponding V's. The value of V for both conclusions is computed to be 3. Also, both conclusions fall in the same class. So, we apply the weighted average (4). Since the weighted average for the first conclusion is 0.525 and for the second conclusion is 0.53, the second conclusion wins and will be tried first.

4 SUMMARY

In this paper we describe the fault analysis expert system (FAES) which detects faults in a given target system. FAES is diveded into two main parts. The knowledge acquisition program, which extract knowledge from the engineer, validates it and transforms it to a set of fuzzy if-then rules. The second part of FAES utilizes a fuzzy inference engine and some learning function to generate the proper recommendations for the user.

Table 2: *FR*'s certainty assignments

Range	Rank	points
0.7 - 1.0	High	3
0.3 - 0.7	Med.	2
0.0 - 0.3	Low	1

Table 3: FC's certainty assignments

Range	Rank	points
0.9 - 1.0	High	3
0.7 - 0.9	Med.	2
0.6 - 0.7	Low	1

Table 4: *V*'s certainty assignments

FC	*FR*	*V*	Class
High	High	6	1
High	med.	5	2
High	low	4	5
Med.	High	5	3
Med.	Med.	4	4
Med.	Low	3	7
Low	High	4	6
Low	Med.	3	8
Low	Low	2	9

References

[1] Blaxon, T., R. Geesey, and L. Reeker (1987). A Knowledge Acquisition Tool for Process Applications. Proceedings of the Fifth Intelligence Community AI Symposium.

[2] Chew, G., M. Schneider, and A. Kandel (1994). Designing Fuzzy Inference Procedures. The second World Congress on Expert Systems, Lisbon.

[3] Davis, R. (1977). Interactive Transfer of Expertise: Acquisition of New Inference Rules. Proceedings of the Fifth International Joint Conference on AI, pp. 321 - 328.

[4] Elliott, W., and M. Schneider (1990a). Fault Finder: A PC Based Expert System. The 1990 ACM Symposium on Personal and Small Computers, Arlington, Virginia, March 28 - 30, pp. 13 - 23.

[5] Elliott, W., and M. Schneider (1990b). The Fault Finder Expert System. The 1990 IEEE Southeastcon, New Orleans, LA, April 1-4, pp. 764 - 769.

[6] Elliott, W., and M. Schneider (1990c). Learning Aspects of the Fault Finder Expert System. 20th International Symposium on Multiple-Valued Logic, Charlotte, NC, May 23-25, pp. 378 - 385.

[7] Kahn, G. S., E. Breaux, P. Deklerk, and R. Joseph (1987). A Mixed-Initiative Workbench for Knowledge Acquisition. International Journal of Man-Machine Studies, No. 27, pp. 167-179.

[8] Kandel, A (Ed.)(1992). Fuzzy Expert Systems.CRC Press.

[9] Kandel, A. and G. Langholz (Ed.)(1992). Hybrid Architectures for Intelligent Systems. CRC Press.

[10] Kandel, A. and G. Langholz (Ed.)(1994). Fuzzy Control Systems. CRC Press.

[11] Kitto, C., and J. Boose (1987). Selecting Knowledge Acquisition Tools and Strategies Based on Application Characteristics. Proceedings of the Second AAAI Knowledge Acquisition for Knowledge-Based Systems Workshop.

[12] McGraw, K. L., and K. Harbison-Briggs (1989). Knowledge Acquisition Principles and Guidelines. Prentice-Hall.

[13] Schneider, M., and A. Kandel (1993). On Uncertainty Management in Fuzzy Inference Procedures. Information Sciences, Soft-Computing (To appear).

[14] Wheeler, M., and M. Schneider (1990). Automatic Knowledge Acquisition for Expert Systems. 1990 Symposium on Applied Computing, Fayetteville, Arkansas, April 5 - 6, pp. 96 - 101.

[15] Wheeler, M., and M. Schneider (1991). Input Validation For AUTOKNAQ: Knowledge Acquisition System. International Conference on Systems, Man, and Cybernetics, Charlottesville, Virginia, October 13 - 16, pp. 1883 - 1888.

[16] Wheeler, M., and M. Schneider (1992). AUTOKNAQ: Automatic Knowledge Acquisition for Fault Isolation Expert Systems. 1992 Symposium on Applied Computing, Kansas City, Kansas, March 1 - 3, pp. 387 - 395.

4

LIFE TIME ANALYSIS AND FUZZY SETS

Reliability Estimation based on Fuzzy Life Time Data

R. Viertl and W. Gurker
Institut für Statistik und Wahrscheinlichkeitstheorie
Technische Universität Wien
Wiedner Hauptstrasse 8–10/107
A–1040 Wien

Abstract

Real life time data are usually more or less fuzzy. Therefore it is necessary to generalize statistical estimation of the reliability characteristics. For life time data in form of fuzzy numbers x^* with characterizing function $\xi(.)$ this is possible and explained in the paper. Especially classical parameter estimations for stochastic life time models and the concept of the empirical reliability function as well as Bayesian reliability estimation procedures are generalized. Moreover the important technique of accelerated life testing is revisited in face of non–precise observed life times.

1 Introduction

For the estimation of the reliability of an element or a system observed life times are necessary. There are different kinds of uncertainty connected with life times.

The first kind of uncertainty is statistical variation of observed life times. This kind of uncertainty is modelled by a stochastic model using probability distributions.

Another kind of uncertainty are errors in observations. This kind of uncertainty is modelled using statistical error models. Usually in these models it is assumed that observations are non–negativ real numbers.

But looking at real observations there is another kind of uncertainty of every single observation namely vagueness in the sense that an observation is not a real number but is more or less fuzzy, i.e. the real numbers are not a realistic model for real life times. This kind of uncertainty is mostly neglected. But this can produce non–justified fictious precision of the results of estimation procedures for the reliability.

In order to obtain realistic results of estimation procedures for life time distributions the fuzziness of data has to be modelled also.

2 Fuzzy life times

In order to model real observed life times a generalization of real numbers is necessary. This generalization must contain as special cases the situation of interval data and precise data. An appropriate model for such generalized numbers are special fuzzy subsets of $\mathbb{R}$ called *fuzzy numbers* x^*.

These fuzzy numbers are characterized by so called *characterizing functions* $\xi(.)$ of a fuzzy number x^* which obey the following conditions.

(1) $\xi : \mathbb{R} \to [0,1]$ is Borel–measurable.

(2) $\exists\, x_0 \in \mathbb{R} : \ \xi(x_0) = 1$.

(3) The so–called *α–cuts* $(0 < \alpha \leq 1)$

$$B_\alpha(x^*) = B_\alpha(\xi(.)) = \{x \in \mathbb{R} : \ \xi(x) \geq \alpha\}$$

are all closed intervals, i.e. $B_\alpha(x^*) = [a_\alpha, b_\alpha] \ \ \forall\, \alpha \in (0,1]$.

For precise real observations $x_0 \in \mathbb{R}$ the corresponding characterizing function is $\xi(.) = I_{\{x_0\}}(.)$ the indicator function of the set $\{x_0\}$. For interval data $[c, d]$ the corresponding characterizing function is the indicator function $I_{[c,d]}(.)$.

Remark: Characterizing functions $\xi(.)$ describe the fuzziness of *one* observed life time. The area under $\xi(.)$ is a measure for the fuzziness of the observation.

For n precise observations $x_1, \ldots, x_n$ of a life time X with observation space M the estimation procedures for the reliability function and for parameters depend on the combination $(x_1, \ldots, x_n)$ which is an element of the sample space M^n.

For fuzzy life times $x_1^*, \ldots, x_n^*$ with corresponding characterizing functions $\xi_1(.), \ldots, \xi_n(.)$ these fuzzy observations have to be combined to a fuzzy element $\underline{x}^*$ of the sample space M^n in order to be able to generalize statistical estimation procedures. This *fuzzy combined sample element* $\underline{x}^*$ is different from the vector $(x_1^*, \ldots, x_n^*)$ of fuzzy observations.

There are different possibilities to obtain the characterizing function $\xi(.)$ of the fuzzy combined sample element $\underline{x}^*$. Motivated by fuzzy set theory one possibility is, using the notation $\underline{x} = (x_1, \ldots, x_n)$,

$$\xi(\underline{x}) = \min_{i=1(1)n} \xi_i(x_i) \quad \text{for all} \quad \underline{x} \in M^n.$$

This combination is suitable for computer calculations connnected with estimation procedures.

Another possibility would be

$$\xi(\underline{x}) = \prod_{i=1}^{n} \xi_i(x_i) \quad \text{for all} \quad \underline{x} \in M^n.$$

The different combination rules are also motivated by properties of α–cuts of the fuzzy combined sample element $\underline{x}^*$. For details compare the comments in Frühwirth–Schnatter (1992).

3 Classical parametric estimation procedures for fuzzy data

Let $X \sim f(.|\theta)$, $\theta \in \Theta$ be a parametric life time model with density $f(.|\theta)$ and parameter space Θ. For n fuzzy observed life times $x_1^*, \ldots, x_n^*$ with corresponding characterizing functions $\xi_1(.), \ldots, \xi_n(.)$ the generlizations of estimation procedures are based on the fuzzy combined sample element $\underline{x}^*$ with characterizing function $\xi(., \ldots, .)$.

For an estimator $\hat{\theta} = \vartheta(x_1, \ldots, x_n)$ for the true parameter θ_0 the generalized fuzzy estimator $\hat{\theta}^*$ based on fuzzy observed life times is a fuzzy element of the parameter space with characterizing function $\psi(.)$. The characterizing function $\psi(.)$ is given by its values

$$\psi(\theta) := \sup\{\xi(x_1, \ldots, x_n) : \ \vartheta(x_1, \ldots, x_n) = \theta\},$$

where $(x_1, \ldots, x_n)$ is varying in the sample space M^n.

Example 1: In figure 1 an example of ten fuzzy life times from an exponential distribution Ex_θ,

Figure 1: Fuzzy sample of an exponential distribution and fuzzy estimation of the parameter

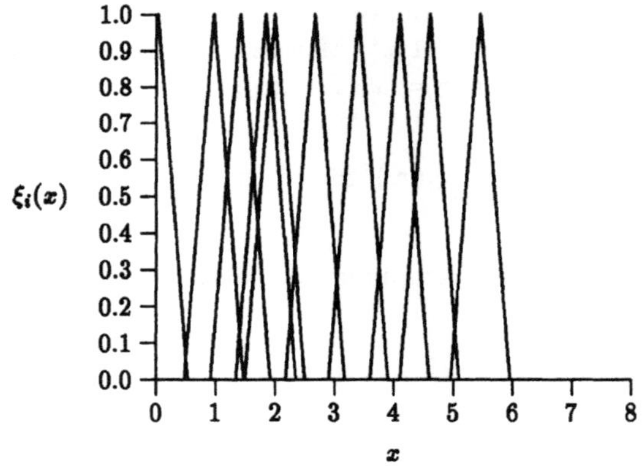

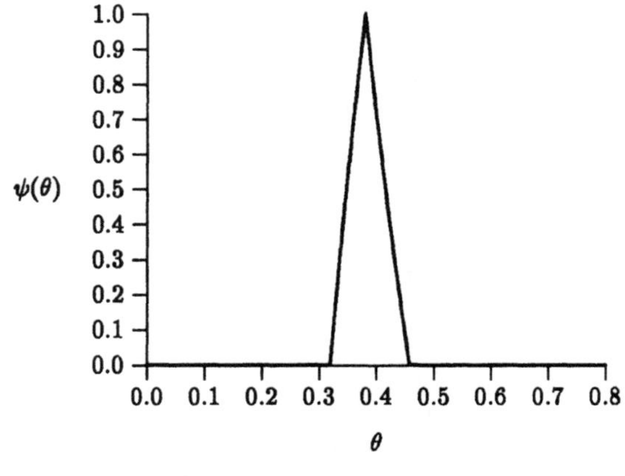

$$f(x|\theta) = \theta e^{-\theta x}, \ x > 0,$$

and the characterizing function of the estimated parameter $\hat{\theta}^*$ is given. In fact, the peaks of the triangle–type fuzzy observations were chosen from an exponential distribution with $\theta = \frac{1}{3}$ and for $\vartheta(x_1, \ldots, x_n)$ the maximum–likelihood–estimate is used.

$$\hat{\theta} = \frac{n}{\sum_{i=1}^{n} x_i}$$

In analogous way the concept of confidence regions for parameters can be generalized.

Figure 2: Fuzzy confidence interval for the parameter θ of the exponential distribution

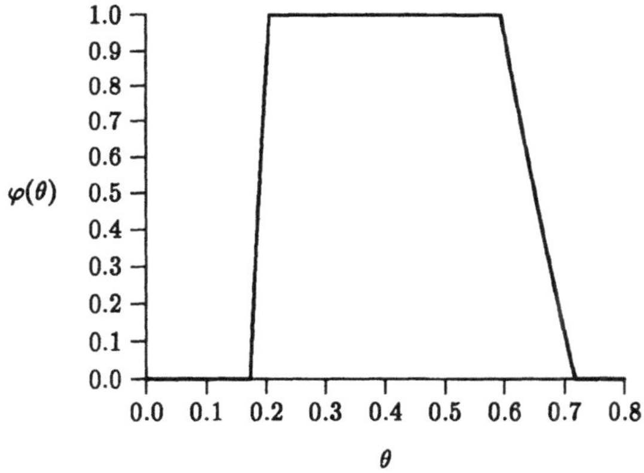

$\varphi(.)$ characterizing function of the fuzzy confidence interval for θ

Let $\kappa(x_1, \ldots, x_n)$ be a classical confidence region for the parameter θ of a stochastic model $X \sim f(.|\theta)$, $\theta \in \Theta$. In a natural way for fuzzy data a generalization has to yield a fuzzy subset of the parameter space Θ.

The characterizing function $\varphi(.)$ of this fuzzy subset – also called membership function – is given by its values

$$\varphi(\theta) = \sup\{\xi(x_1, \ldots, x_n) : \theta \in \kappa(x_1, \ldots, x_n)\},$$

where $(x_1, \ldots, x_n)$ is varying in the sample space M^n.

Example 2: In figure 2 the characterizing function of a fuzzy confidence interval for the parameter θ of the exponential distribution based on the fuzzy data from figure 1 is depicted. The classical confidence region was chosen as

$$\kappa(x_1, \ldots, x_n) = \left[\frac{\chi^2_{2n; \frac{1-\gamma}{2}}}{2\sum\limits_{i=1}^{n} x_i}, \frac{\chi^2_{2n; \frac{1+\gamma}{2}}}{2\sum\limits_{i=1}^{n} x_i} \right],$$

with coverage probability $\gamma = 0.95$.

4 Classical non–parametric estimation of the reliability function based on fuzzy life time data

In order to generalize the empirical reliability function $\hat{R}_n(.)$ based on precise life time data $x_1, \ldots, x_n$, i.e.

$$\hat{R}_n(x) = \frac{1}{n} \sum_{i=1}^{n} I_{(x,\infty)}(x_i),$$

there are different possibilities.

For n fuzzy observed life times $x_1^*, \ldots, x_n^*$ with characterizing functions $\xi_1(.), \ldots, \xi_n(.)$ a natural generalization of the empirical reliability function is given by

$$R_n^*(x) = \frac{1}{n} \sum_{i=1}^{n} \frac{\int_x^\infty \xi_i(t)\, dt}{\int_0^\infty \xi_i(t)\, dt} \quad \forall\, x \geq 0.$$

Another generalization is motivated by interval data of life times, i.e. the characterizing functions are of the form

$$\xi_i(.) = I_{[a_i, b_i]}(.), \; i = 1, \ldots, n.$$

In this situation an interval–valued generalization of the empirical reliability function is given by two functions $R_u(.)$ and $R_l(.)$ which define the limits of these intervals.

Example 3: In figure 3 a sample of four interval data and the corresponding interval–type generalized empirical reliability function is depicted. The shaded area between $R_u(x)$ and $R_l(x)$ covers all possible empirical reliability functions $\hat{R}_n(.)$ for exact observations x_i in the interval $[a_i, b_i]$.

For general fuzzy observations with characterizing functions $\xi_1(.), \ldots, \xi_n(.)$ the two functions $R_u(.)$ and $R_l(.)$ are defined using the ordered sample. If the characterizing functions $\xi_i(.)$ have pairwise disjoint supports then $\xi_{(i)}(.)$ denotes the characterizing function of the i–th largest observed fuzzy life time. The functions $R_u(.)$ and $R_l(.)$ are given in the interval $\text{supp}(\xi_{(i)}(.))$ by

Figure 3: Interval life time data

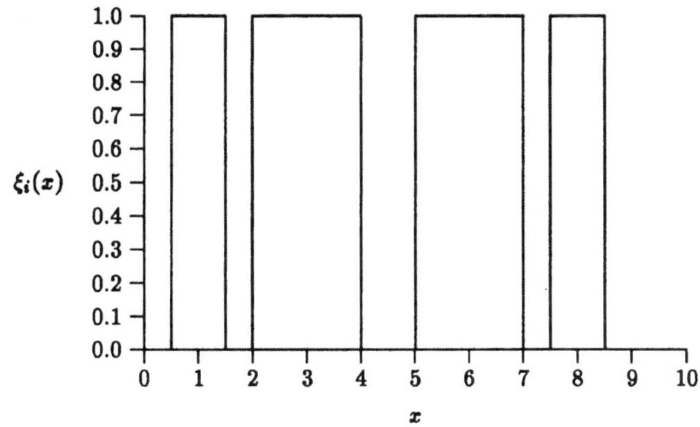

Generalized empirical reliability function

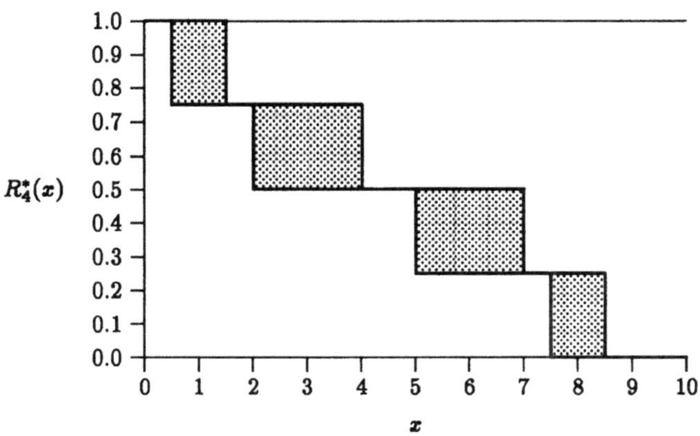

$$R_u(x) = \begin{cases} \dfrac{n-i+1}{n} & \text{for } x \text{ with } \xi_{(i)}(x) \text{ increasing or } \xi_{(i)}(x) = 1 \\[2ex] \dfrac{n-i+\xi_{(i)}(x)}{n} & \text{for } x \text{ with } \xi_{(i)}(x) \text{ decreasing} \end{cases}$$

and

$$R_l(x) = \begin{cases} \dfrac{n-i+1-\xi_{(i)}(x)}{n} & \text{for } x \text{ with } \xi_{(i)}(x) \text{ increasing} \\[2ex] \dfrac{n-i}{n} & \text{for } x \text{ with } \xi_{(i)}(x) \text{ decreasing or } \xi_{(i)}(x) = 1. \end{cases}$$

Figure 4: Fuzzy life times and general interval–valued empirical reliability function

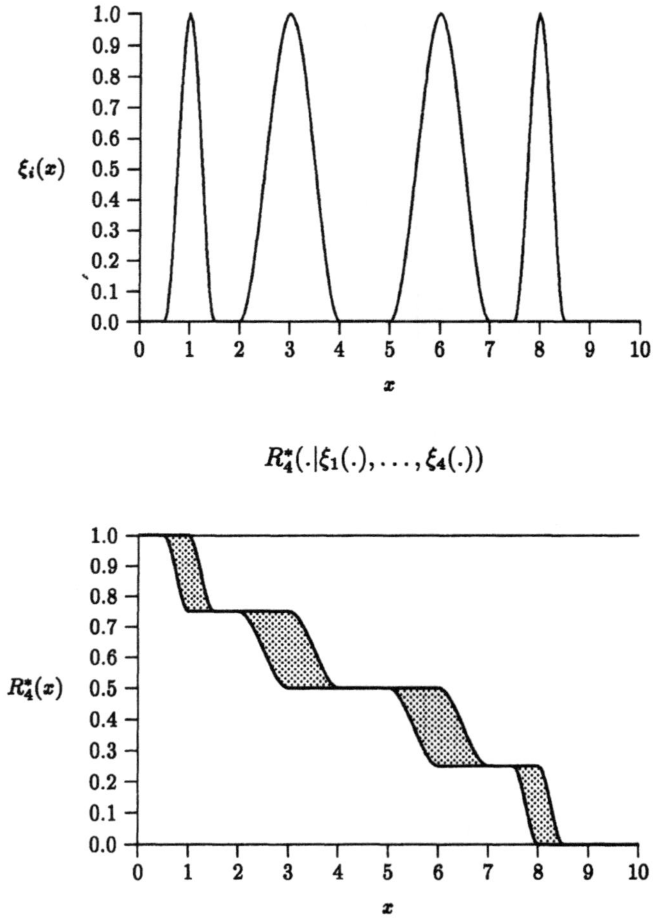

Example 4: In figure 4 the characterizing functions of four fuzzy life time observations and the corresponding generalized interval–valued empirical reliability function is depicted. $R_n^*(.|\xi_1(.),\ldots,\xi_n(.))$ covers the shaded area between the two functions $R_l(.)$ and $R_u(.)$.

There are also other generalizations of the empirical reliability function. They are explained in the book Viertl (to appear) for which a list of contents is available from the author.

5 Bayesian reliability estimation for fuzzy life time data

For a parametric stochastic life time model $X \sim f(.|\theta)$, $\theta \in \Theta$ and a-priori-distribution of the parameter with density $\pi(.)$ Bayes' theorem for precise data $x_1, \ldots, x_n$ of X,

$$\pi(\theta|x_1, \ldots, x_n) \propto \pi(\theta) \cdot l(\theta; x_1, \ldots, x_n),$$

has to be generalized to the real situation of fuzzy data $x_1^*, \ldots, x_n^*$. This was done in Viertl and Hule (1991) obtaining a fuzzy a-posteriori-density denoted by $\pi^*(.|x_1^*, \ldots, x_n^*)$ which is a family of fuzzy values of the a-posteriori-density, i.e.

$$\pi^*(.|x_1^*, \ldots, x_n^*) = \Big(\psi_\theta(.);\ \theta \in \Theta\Big),$$

where $\psi_\theta(.)$ denotes the characterizing function of the fuzzy value of the a-posteriori-density $\pi(\theta|x_1^*, \ldots, x_n^*)$ for argument θ.

More details cannot be given here by space restrictions but these can be found in the references or in Viertl (to appear).

Using the fuzzy a-posteriori-density generalized confidence regions, which are fuzzy subsets of the parameter space, can be constructed. Based on the fuzzy combined sample element $\xi(., \ldots, .)$ for fuzzy data $D^* = (x_1^*, \ldots, x_n^*)$ a *fuzzy Bayesian confidence region* for the parameter θ is the fuzzy subset of Θ with characterizing function (= membership function) $\phi(.)$ given by

$$\phi(\theta) := \sup_{\underline{x} \in [0,\infty)^n} \Big\{\xi(\underline{x}) : \theta \in \Theta_{\underline{x}}\Big\},$$

where $\Theta_{\underline{x}}$ denotes a Bayesian confidence region for θ based on precise observations $\underline{x} = (x_1, \ldots, x_n)$ with coverage probability γ, i.e. a subset of Θ fulfilling

$$Pr_{\pi(\theta|\underline{x})}(\Theta_{\underline{x}}) = \int_{\Theta_{\underline{x}}} \pi(\theta|\underline{x})\, d\theta = \gamma.$$

It is also possible to generalize the concept of HPD-regions. This was done in Viertl and Hule (1991) and is also explained in detail in Viertl (to appear).

Figure 5: α–level–lines of the fuzzy predictive reliability function

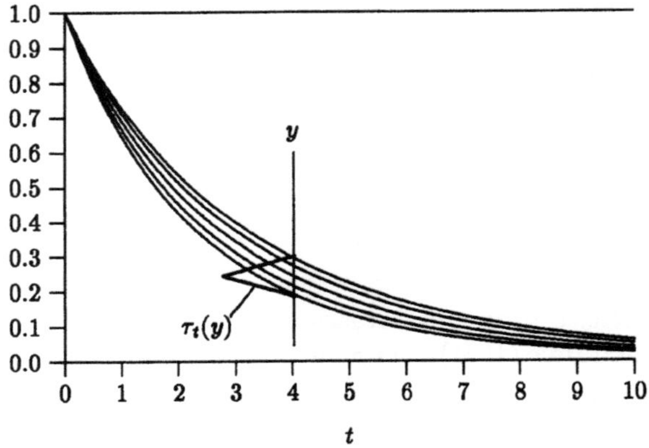

The most important concept for reliability analysis is probably the reliability function. Therefore a generalization of the predictive reliability function $R(.|D)$ based on exact data $D = (x_1, \dots, x_n)$, defined by its values

$$R(t|D) = \int_t^\infty \int_\Theta f(x|\theta)\, \pi(\theta|D)\, d\theta\, dx,$$

is necessary. This generalization for fuzzy data $D^* = (x_1^*, \dots, x_n^*)$ with fuzzy combined sample element $\underline{x}^*$ having characterizing function $\xi(., \dots, .)$ of n real variables $\underline{x} = (x_1, \dots, x_n)$ is the family of fuzzy values $R^*(t|D^*)$, $t \geq 0$, defined by its characterizing functions $\tau_t(.)$ whose values are given by

$$\tau_t(y) = \sup_{\underline{x} \in [0,\infty)^n} \{\xi(\underline{x}) : R(t|\underline{x}) = y\}.$$

The fuzzy predictive reliability function $R^*(.|D^*)$ based on fuzzy life time data D^* is the following family of fuzzy values of the reliability function

$$R^*(.|D^*) = \left(\tau_t(.);\ t \geq 0\right).$$

Example 5: In figure 5 an example of a fuzzy predictive reliability function is given by some α–level–curves based on the fuzzy sample of example 1 and a non–informative a–priori–distribution for θ of the form

$$\pi(\theta) \propto \frac{1}{\theta},\ \theta > 0.$$

The two outer curves correspond to $\alpha = 0.1$, the two inner curves to $\alpha = 0.5$; the curve in the middle corresponds to $\alpha = 1$ (in this case the predictive reliability function for "exact" observations).

6 Accelerated life testing and fuzzy life time data

In accelerated life testing usually observations of the life time under different stress levels $S_1, \ldots, S_m$ with $S_1 > S_2 > \cdots > S_m > S_0$ are taken and the goal is to estimate the reliability function $R(.|S_0)$ of the life time X under use–stress S_0. Many parametric and nonparametric (classical and Bayesian) models are considered in the literature (see Viertl (1988)). Here only a special nonparametric model is considered in more detail based on so called (linear) time transformations $\alpha \cdot x$ for the relationship between the reliability function of the life time under different stress levels,

$$R(x|S) = R(\alpha(S) \cdot x|S_0) \quad \forall\, x \geq 0,$$

where $R(.|S)$ denotes the reliability function of the life time under stress S. The stress–life time relationship between two accelerated stress levels $S_i < S_j$ is given by

$$R(x|S_j) = R(\alpha(S_i, S_j) \cdot x|S_i) \quad \forall\, x \geq 0,$$

where $\alpha(S_i, S_j)$ is now the relative acceleration factor.

If one has estimated the reliability function $R(.|S)$ under an accelerated stress S based on n fuzzy observed life times ${}^S x_1^*, \ldots, {}^S x_n^*$ with characterizing functions $\xi_1^S(.), \ldots, \xi_n^S(.)$ using the interval–valued generalization of the empirical reliability function (compare section 4) the following relationships for the upper and lower bounderies are obtained,

$$R_u(x|S_0) = R_u\left(\frac{x}{\alpha(S)}\middle|S\right) \quad \forall\, x \geq 0$$

and

$$R_l(x|S_0) = R_l\left(\frac{x}{\alpha(S)}\middle|S\right) \quad \forall\, x \geq 0,$$

Figure 6: Interval-valued empirical reliability function for an accelerated stress level S and for the use-stress level S_0 based on interval-type fuzzy observations

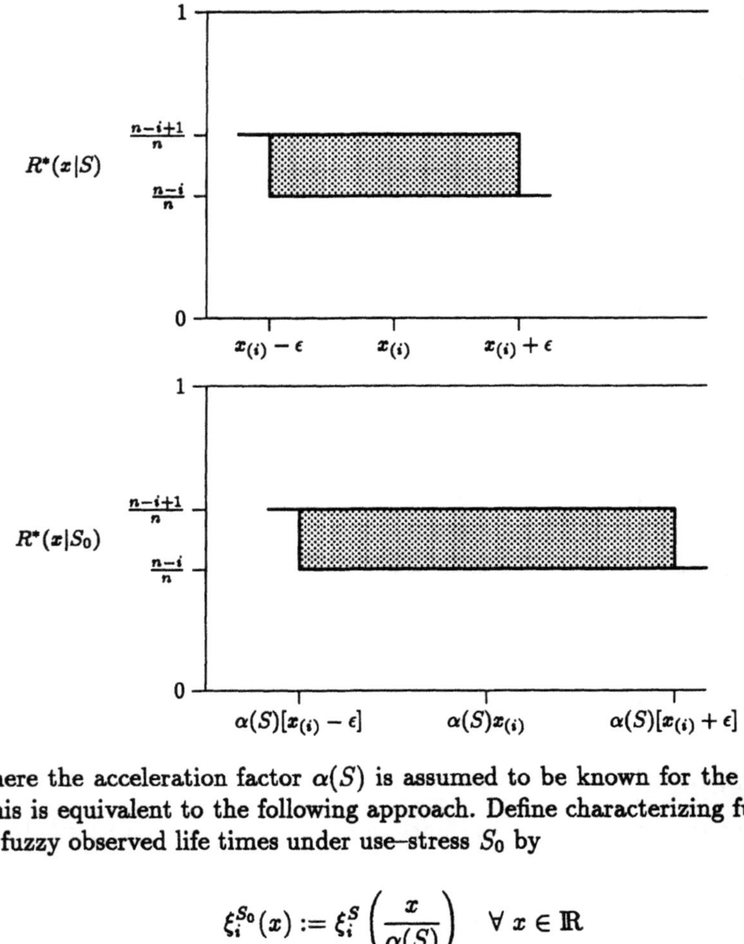

where the acceleration factor $\alpha(S)$ is assumed to be known for the present. This is equivalent to the following approach. Define characterizing functions of fuzzy observed life times under use-stress S_0 by

$$\xi_i^{S_0}(x) := \xi_i^S\left(\frac{x}{\alpha(S)}\right) \quad \forall\, x \in \mathbb{R}$$

and calculate the interval-valued empirical reliability function based on these fuzzy life times. This corresponds to an increase of imprecision as illustrated by the following simple example.

Example 6: In figure 6 a typical section of the fuzzy empirical reliability function for an accelerated stress-level S, based on interval-type observations of the form

$$\xi_i^S(.) = I_{[x_{(i)}-\epsilon,\ x_{(i)}+\epsilon]}(.),\ i = 1, \ldots, n,$$

and for the use–stress S_0 is depicted.

Note that in accelerated life testing the acceleration factor $\alpha(S)$ is greater than one, $\alpha(S) > 1$, for all stress levels $S > S_0$.

Yet another approach is possible. If one does not use the interval–valued generalization of the empirical reliability function, but uses instead the extension principle pointwise for every $t > 0$, a family of fuzzy estimates of the reliability function is obtained,

$$R^*(.|S) \hat{=} \left(\eta_t^S(.);\ t > 0\right)$$

with

$$\eta_t^S(y) = \sup_{\underline{x}\in[0,\infty)^n} \left\{\xi(\underline{x}) : \hat{R}_n(t|S) = y\right\},$$

where $\hat{R}_n(.|S)$ is the empirical reliability function for exact observations $\underline{x} = (x_1, \ldots, x_n)$ and $\xi(., \ldots, .)$ is the characterizing function of the fuzzy combined sample element under stress level S. Now the time transformation to use–stress S_0 is characterized by

$$R^*(.|S_0) \hat{=} \left(\eta_t^{S_0}(.);\ t > 0\right)$$

with

$$\eta_t^{S_0}(.) = \eta^S_{\frac{t}{\alpha(S)}}(.).$$

Dropping the assumption of a known acceleration factor $\alpha(S)$, it has to be estimated too. Omitting the details (see Viertl (1988)) $\alpha(S)$ can be expressed by a so called *infinitesimal characterization*

$$\alpha(S) = \exp\left\{\ln\left[\alpha(S_i, S_j)\right] \frac{H(S) - H(S_0)}{H(S_j) - H(S_i)}\right\},$$

where $H(.)$ is a known function. For example, for $H(S) = \ln(S)$ the *power rule model* and for $H(S) = \frac{1}{S}$ the *Arrhenius reaction rate model* is obtained. It suffices therefore to estimate the relative acceleration factor $\alpha(S_i, S_j)$ for two accelerated stress levels $S_i < S_j$. In the case of fuzzy observed life times on both accelerated stress levels S_i and S_j the characterizing function of a fuzzy estimator of $\ln \alpha(S_i, S_j)$ is given by

$$\psi_{\ln\alpha(S_i,S_j)^\bullet}(b) = \sup_{(\underline{x}^{(i)},\underline{x}^{(j)})\in(0,\infty)^{2n}} \left\{ \xi(\underline{x}^{(i)},\underline{x}^{(j)}) : \frac{1}{n}\sum_{l=1}^{n} \ln\frac{x_l^{(i)}}{x_l^{(j)}} = b \right\},$$

with $\xi(\underline{x}^{(i)},\underline{x}^{(j)})$ the characterizing function of the fuzzy combined sample element of the fuzzy observations on the two accelerated stress levels. The characterizing function of a fuzzy estimator of $\alpha(S)$ is now given by

$$\psi_{\alpha(S)^\bullet}(a) = \sup_b \left\{ \psi_{\ln\alpha(S_i,S_j)^\bullet}(b) : \exp\left\{ b\cdot\frac{H(S)-H(S_0)}{H(S_j)-H(S_i)} \right\} = a \right\}.$$

Predicting to use–stress S_0 raises the problem of how to combine this estimate with the already fuzzy characterization of the reliability function of the life time under use–stress. A sort of weighing may be a very natural approach (other possibilities are feasible),

$$\hat{\xi}_i^{S_0}(x) := \sup_{a\geq 1} \left\{ \xi_i^S\left(\frac{x}{a}\right) \psi_{\alpha(S)^\bullet}(a) \right\}$$

or

$$\tilde{\xi}_i^{S_0}(x) := \frac{\displaystyle\int_1^\infty \xi_i^S\left(\frac{x}{a}\right) \psi_{\alpha(S)^\bullet}(a)\,da}{\displaystyle\int_0^\infty\int_1^\infty \xi_i^S\left(\frac{x}{a}\right) \psi_{\alpha(S)^\bullet}(a)\,da\,dx},$$

assuming integrability of the characterizing functions. Both approaches lead again to characterizing functions of fuzzy numbers in the sense of section 2. Note that in contrast to the first case in the second case an extra normalization has to be carried out to guarantee the existence of a $x_0 \in \mathbb{R}$ with $\tilde{\xi}_i^{S_0}(x) \leq \tilde{\xi}_i^{S_0}(x_0) = 1$ for $x \in \mathbb{R}$.

Example 7: Figure 7 illustrates the two weighing approaches for a triangle–type fuzzy observation on stress–level S

$$\xi^S(x) = xI_{[0,1)}(x) + (2-x)I_{[1,2]}(x)$$

and a (artificial) triangle–type characterizing function for the acceleration factor $\alpha(S)$

$$\psi_{\alpha(S)^\bullet}(a) = 2\,(a-1.5)I_{[1.5,2)}(a) + 2\,(2.5-a)I_{[2,2.5]}(a).$$

Figure 7: $\hat{\xi}^{S_0}(.)$ and $\tilde{\xi}^{S_0}(.)$ for a triangel–type fuzzy observation x^* on stress level S and a triangle–type characterizing function $\psi_{\alpha(S)^*}(.)$ for the acceleration factor $\alpha(S)$

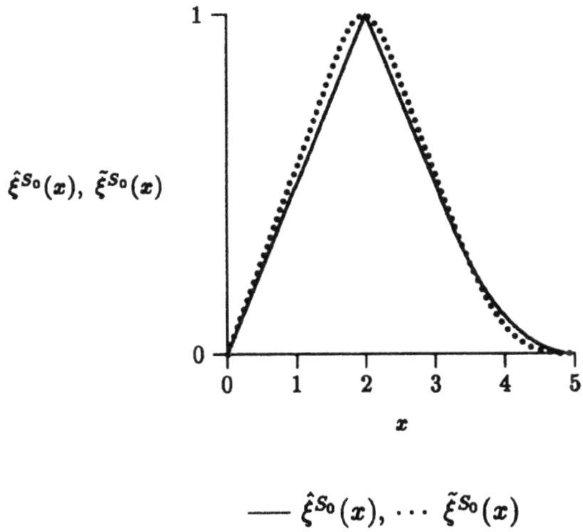

On the basis of $\hat{\xi}_i^{S_0}(.)$, $i = 1,\ldots,n$ or $\tilde{\xi}_i^{S_0}(.)$, $i = 1,\ldots,n$ the interval–valued reliability function can be obtained. If one relies on the family $R^*(.|S_0)$, $t > 0$ discussed above, the corresponding time–transformation is carried out through

$$\hat{\eta}_t^{S_0}(x) := \sup_{a \geq 1} \left\{ \eta_{\frac{t}{a}}^{S}(x)\, \psi_{\alpha(S)^*}(a) \right\}$$

or

$$\tilde{\eta}_t^{S_0}(x) \;\propto\; \int_1^\infty \eta_{\frac{t}{a}}^{S}(x)\, \psi_{\alpha(S)^*}(a)\, da.$$

The reliability function for use–stress is now characterized by

$$\hat{R}^*(.|S_0) \triangleq \left(\hat{\eta}_t^{S_0}(.);\ t > 0\right)$$

or by

$$\tilde{R}^*(.|S_0) \triangleq \left(\tilde{\eta}_t^{S_0}(.);\ t > 0\right).$$

References

[1] Frühwirth–Schnatter S. (1992) On statistical inference for fuzzy data with applications to descriptive statistics, *Fuzzy Sets and Systems* 50.

[2] Gurker W. (1989) On distribution–free filtering methods in accelerated life testing using infinitesimal characteristics, in *Proceedings of the 47th Session of the International Statistical Institute*, Book 1, Paris.

[3] Meeker W.Q. and Escobar L.A. (1993) A review of recent research and current issues in accelerated testing, *International Statistical Review* 61.

[4] Unwin S.D. (1986) A fuzzy set theoretic foundation for vagueness in uncertainty analysis, *Risk Analysis* 6.

[5] Viertl R. (1988) *Statistical Methods in Accelerated Life Testing*, Vandenhoeck & Ruprecht, Göttingen.

[6] Viertl R. (1989a) Estimation of the reliability function using fuzzy life time data, in P.K. Bose, S.P. Mukherjee, K.G. Ramamurthy (Eds.): *Quality for Progress and Developement*, Wiley Eastern, New Delhi.

[7] Viertl R. (1989b) Modelling for fuzzy measurements in reliability estimation, in V. Colombari (Ed.): *Reliability Data Collection and Use in Risk and Availability Assessment*, Springer–Verlag, Berlin.

[8] Viertl R. (1991) On descriptive statistics for non–precise data, in *Bulletin of the 48th Session of the International Statistical Institute*, Contributed Papers, Book 2, Cairo.

[9] Viertl R. (1993) On statistical inference based on non–precise data, in H. Bandemer (Ed.): *Modelling Uncertain Data*, Akademie Verlag, Berlin.

[10] Viertl R. (to appear) *Statistics with Non–precise Data*, CRC–Press, Florida.

[11] Viertl R. and Gurker W. (1988) On filtering methods in accelerated life testing using infinitesimal characteristics, in *Proceedings Relectronic 88*, Vol. 1, Budapest.

[12] Viertl R. and Hule H. (1991) On Bayes' theorem for fuzzy data, *Statistical Papers* 32.

LIFETIME TESTS FOR IMPRECISE DATA AND FUZZY RELIABILITY REQUIREMENTS

Olgierd Hryniewicz
Systems Research Institute
Newelska 6, 01-447 Warsaw, Poland
e-mail: hryniewi@ibspan.waw.pl

Abstract: The paper deals with the problem of the estimation of the average lifetime when observed survival are described by fuzzy numbers. The fuzzy estimator of the average lifetime is given. A simple method for the verification of reliability requirement is proposed.
Keywords: average lifetime estimation, fuzzy reliability data.

1. INTRODUCTION

Statistical methods for the analysis of reliability data were developed mainly for data coming from industrial tests. In this case there exist clearly stated requirements for the parameters of tested items, and life-times of tested items can, in principle, be determined exactly. It works perfectly in the case of simple objects described by few parameters with obvious definition of a failure (e.g., simple electronic elements). However, in the case of complex technical objects the definition of a failure is usually not so easy. It is widely accepted that in such cases a simple two-state (failed or non-failed) description of the tested items is hardly sufficient. To cope with this problem many mathematical models for multi-state objects have been developed. Practical application of those models seems, however, to be limited to relatively simple three-state models. Mathematical complexity of multi-state reliability models is not the only obstacle in their practical applications. With the increased number of analyzed reliability states the number of the parameters that have to be estimated increases rapidly. It causes another problem that is extremely difficult to overcome when the number of tested items is small, and that is the case when we test complex objects. In such cases classical statistical methods do not offer practically acceptable solutions.

Industrial tests are not the only tests that are used for the evaluation of reliability of complex objects. Usually it is more efficient to collect data from normal users to estimate such parameters as the average lifetime in field conditions. In such a case we can hardly expect the normal user to deliver data based on precise measurements, and on the comparison of those measurements with technical specifications. Usually the user is even not able to define precisely the character of the failure. Failures are often described verbally without clear correspondence to the technical specification. Also the time when the problems began cannot be defined precisely. In all these cases we face reliability data that are not only random but also imprecise. To describe formally such data we propose the usage of the formalism of fuzzy sets. We assume that the observed lifetimes are described by fuzzy random variables of a certain type. In the paper we propose relatively simple algorithms for solving such basic reliability problems as the estimation of the average lifetime or the comparison of the test results with certain requirements.

Before we proceed to the main sections of the paper it seems necessary to give an answer to a possible question: "Why to use a fuzzy description for a lack of precision in such cases?". We argue that in the case of field data there exists an inherent lack of precision coming from the subjective character of the user's definition of a failure. In the case of a precise definition of a failure and imprecise measurements we can use probabilistic models. However, when the definition of a failure is not precise (e.g., the engine consumes too much gasoline) the application of pure probabilistic methods is insufficient. We believe that pure probabilistic models can be applied only in those instances when the lack of precision comes from imprecise measurements, but not from imprecise notions and definitions.

2.MATHEMATICAL MODEL OF RELIABILITY DATA

Assume that the results of life tests of n elements are available. For all tested elements the test time hasn't exceeded T units of time, i.e. these elements either failed before the test time T or the test was stopped at that time. Suppose that all the tested elements can be divided into three groups:

a) elements that totally failed before the time T,

b) elements which were partially failed at the time T,

c) elements which didn't revealed any signs of failure at the time T.

Let the number of the elements from the first group be equal to $m, (0 \leq m \leq n)$. For those elements we assume that two times have been recorded: the signalization time $x_{m,i}$, $i = 1,\ldots m$, when the first sign of a failure was noticed, and the failure time $x_{M,i}, i = 1,\ldots m$, when the element i completely failed $(x_{m,i} \leq x_{M,i} \leq T)$. Now assume that the number of the elements from the second group is equal to $d, (0 \leq d \leq n)$. For those elements we observe their signalization times $z_{m,i}, i = 1,\ldots d$, and a certain measure $\mu_i, 0 < \mu_i < 1, i = 1,\ldots,d$ that describes the grade to which the element i is considered sound at the time T. The remaining $k = n - m - d$ elements have survived the test time T.

Now we describe the results of the reliability test using notions of fuzzy sets. For the elements belonging to the first group we assume that their lifetimes are described by fuzzy sets with the following piecewise linear membership function

$$\mu(x_i) = \begin{cases} -\dfrac{1}{x_{M,i} - x_{m,i}}(x_i - x_{M,i}) & , \quad \text{for } x_{m,i} \leq x_i \leq x_{M,i} \\ 0 & , \quad \text{otherwise} \end{cases} \quad , i = 1,\ldots,m \tag{1}$$

For the elements from the second group each observation can be described as a fuzzy number with the following membership function

$$(z_i) = \begin{cases} -\dfrac{1}{T - z_{m,i}}[(1-\mu_i)z_i - (T - \mu_i z_{m,i})] & , \quad \text{for } z_{m,i} \leq z_i \leq T \\ 0 & , \quad \text{otherwise} \end{cases} \quad , i = 1,..,d \tag{2}$$

and a number ν_i where $\nu_i = 1 - \mu_i$, which describes the grade of a failure at the time T. Further, we assume that the elements from the second group are numbered in such a way that $\mu_i \leq \mu_{i+1}, i = 1,\ldots,d-1$. For the elements from the third group it is sufficient to describe their survival times using a crisp number T.

3. ESTIMATION OF THE AVERAGE LIFETIME

Assume now that the observed lifetimes are exponentially distributed fuzzy random variables. In the case of non-fuzzy exponentially distributed random variables the maximum likelihood estimator of the average lifetime is given as the quotient of the total survival time (the sum of survival times of all the elements from the sample) and the number of observed failures. Using the extension principle we find that for fuzzy random variables in the considered case the average lifetime can be calculated from the same formula. However, in this case the total survival time and the number of failures are fuzzy. Therefore, the maximum likelihood estimator of the average lifetime is also a certain fuzzy number. In order to calculate its membership function we have to find the membership function of the total survival time S.

The total survival time S is the sum of k crisp and $m+d$ fuzzy numbers described by linear membership functions. Therefore, using the extension principle, we find that the membership function of S is piecewise linear, and is given by the following formula

$$\mu(S) = \begin{cases} 0 & , \text{ for } S < S_0 \\ -\dfrac{\mu_{d-i+1} - \mu_{d-i}}{S_{i+1} - S_i} S + \dfrac{\mu_{d-i+1} S_{i+1} - \mu_{d-i} S_i}{S_{i+1} - S_i} & , \text{ for } S_i \le S \le S_{i+1} \quad , \text{ i} = 0,\ldots,\text{d} \\ 0 & , \text{ for } S \ge S_{d+1} \end{cases}$$

$$\mu_0 = 0, \qquad \mu_{d+1} = 1$$

(3)

Parameters $S_i, i = 0,\ldots,d$ in (3) are calculated recursively from the following formulae

$$S_M = \sum_{i=1}^{m} x_{M,i} + kT \tag{4}$$

$$S_m = \sum_{i=1}^{m} x_{m,i} + kT \tag{5}$$

$$S_1^0 = S_M \tag{6}$$

$$S_0^0 = S_m \tag{7}$$

$$S_0^i = S_m + \sum_{j=1}^{i} z_{m,j} \quad ,\ldots i = 1,\ldots,d \tag{8}$$

$$S_1^i = \frac{(S_1^{i-1} - \mu_{i-1} S_0^{i-1}) - \mu_i (S_1^{i-1} - S_0^{i-1})}{1-\mu_{i-1}} + T \quad , \quad i = 1,\dots,d \tag{9}$$

$$S_j = S_1^{d-j+1} + (j-1)T \quad , \; j = 1,\dots d \tag{10}$$

$$S_0 = S_m + \sum_{i=1}^{d} z_{m,i} \tag{11}$$

$$S_{d+1} = S_M + dT \tag{12}$$

Interpretation of these parameters can be deduced from the proof of (3) in the Appendix.

Let $r = m + d$ be the fuzzy number of observed failures. Assume that the membership function of r is as follows

$$\mu(r) = \begin{cases} 1 & , \quad r = m \\ v_i & , \quad r = m + i \quad , i = 1,\dots,d \end{cases} \tag{13}$$

The average lifetime can be estimated from the formula

$$b = \frac{S}{r} \tag{14}$$

The membership function of b can be calculated using the extension principle. To do this we introduce for $j = 1,\dots,d$ a partial membership function for the average lifetime representing the membership function of the average lifetime for a given partial value of the total number of failures r.

$$\mu(\hat{b}_j) = \inf(v_j, \mu(\frac{S}{m+j})) \tag{15}$$

Hence, the membership function of the average lifetime is given by

$$\mu(\hat{b}) = \sup \mu(\hat{b}_j) \tag{16}$$

The membership functions of both the total survival time and the number of observed failures are monotoneous. The membership function of the average lifetime in the majority of cases is convex with its maximum at S_0 / m . However, in certain cases, generally - when a number of failures is small, it can be not convex. In such cases the interpretation of the final result may be not straightforward.

4. TESTING FUZZY HYPOTHESES ABOUT THE AVERAGE LIFETIME

Estimation of the average life time is only one statistical problem in reliability analysis. In many practical cases it is of interest to verify the requirement that this value is greater than a certain fixed value b_0. When the failure times are determined precisely we say that such a requirement is fulfilled at the significance level α when the following inequality holds

$$S > \frac{b_0}{2}\chi^2_{2(r+1),1-\alpha} \tag{17}$$

where $\chi^2_{2(r+1),1-\alpha}$ is the quantile of the chi-square distribution, S is the observed total survival time, and r is the number of observed failures. In the considered case of fuzzy observations both the total survival time S, and the number of observed failures are fuzzy. Therefore, we need to compare two fuzzy numbers: the total survival time, and a function of the fuzzified quantile of the chi-square distribution. For a given significance level α this quantile is given in the following fuzzy form

$$\tilde{\chi}^2(r,j,1-\alpha) = \chi^2_{2(r+1),1-\alpha}|1 + \chi^2_{2(r+2),1-\alpha}|\, v_1 + \cdots + \chi^2_{2(r+r+1),1-\alpha}|\, v_r \tag{18}$$

Hence, for the verification of a reliability requirement it is necessary to compare the fuzzy total survival time described by the membership function (3), and a fuzzy criterion

$$\tilde{C}(r) = \frac{b_0}{2}\tilde{\chi}^2(m,d,1-\alpha) \tag{19}$$

The membership function of $\tilde{C}(r)$ is given by

$$\nu(\tilde{C}(r)) = v_{r-m}, \quad r = m,\ldots m+d \tag{20}$$

There exist many methods for the comparison of fuzzy numbers. Generally they can be divided into two groups: those which are based on averaging of membership functions (like Yager's F1), and the others. Simulation experiments described in Hryniewicz(1992) have shown that in the case of statistical tests methods based on averaging are not satisfactory. From those experiments we found that the comparison based on NSD (Necessity of Strict Dominance) index proposed by Dubois and Prade(1983), or the equivalent index proposed by Tsukamoto et al.(1981) should be preferred.

For two fuzzy numbers, X and Y, described by the membership functions $\mu(x)$ and $\nu(y)$, respectively, the NSD index is calculated from the following formula

$$\text{NSD} = \text{Nec}(X > Y) = 1 - \sup_{x \le y} \min[\mu(x), \nu(y)] \tag{21}$$

In the considered case we can use the following algorithm to find the NSD index:

a) if $C(m+d) \le S_0$ then NSD=1

otherwise

b) find r^* such that r^* is the minimal value of r for which inequality

$S_0 \leq C(r)$ still holds,

c) calculate

$$\text{NSD} = 1 - \min[\nu(C(r^*)), \mu(C(r^*))] \tag{22}$$

Therefore, the verification of reliability requirements using the NSD index is much easier than the estimation of the average lifetime, and doesn't require any special computation.

It is also possible to soften the requirement by fuzzifying the value of b_0. Suppose that we require the average lifetime to be greater than a certain value $b_{0,m}$. We could also set greater values for b_0, but with smaller preference. Hence, we can describe b_0 as a fuzzy number $\tilde{b}_0$ with a nonincreasing membership function $\tilde{\pi}(b_0)$. The criterion function in this case may not be nondecreasing. Generally, in this case the values of the NSD index are smaller than in the previous case. It is also possible to fuzzify the value of the significance level α. The membership function in that case is more complicated, but the final values of the NSD index remain nearly the same.

5. A NUMERICAL EXAMPLE

To evaluate reliability of their diesel engines $n=20$ buses were observed in field conditions. The test time was fixed to $T=50000$ km. One engine failed abruptly after 18000 km. Another revealed first signs of a failure after 34000 km, and failed definitely after 40000 km. Third engine began to fail after 42000 km, and at the moment of test completion was considerd as failed in 90%. Fourth revealed first signs of a failure after 46000 km, and after 50000 km was considered as failed in 20%. The purpose of the test was to verify whether the average lifetime of the tested engines was greater than 150000 km.

Using the notation adopted in the paper we have:

$m = 2$, ($x_{m,1} = 18000$, $x_{M,1} = 18000$), ($x_{m,2} = 34000$, $x_{M,2} = 40000$),

$k = 16$, $d = 2$, ($z_{m,1} = 42000$, $\mu_1 = 0.1$), ($z_{m,2} = 46000$, $\mu_2 = 0.8$)

From (4)-(12) we have

$S_m = 852000$, $S_M = 860000$, $S_0^0 = 852000$, $S_1^0 = 860000$,

$S_1^1 = 907400$, $S_0^1 = 894000$, $S_1^2 = 852000$,

$S_0 = 940000$, $S_1 = 947000$, $S_2 = 957400$, $S_3 = 958000$

Hence, the membership function of S calculated from (3) is given by

$$\mu(S)=\begin{cases}0 & , \quad S<940000\\ \frac{1}{7000}(-200S+195000) & , \quad 940000\le S\le 947000\\ \frac{1}{10400}(-700S+671220) & , \quad 947000\le S\le 957400\\ \frac{1}{600}(-100S+95800) & , \quad 957400\le S\le 964000\\ 0 & , \quad S\ge 964000\end{cases}$$

From (13) we find that the number of observed failures r is given by the following fuzzy number

$$2|1 \;+\; 3|0.9 \;+\; 4|0.2$$

Hence, from (18)-(19) we obtain the fuzzy criterion $\tilde{C}(r)$

$945000|1 \;+\; 1162500|0.9 \;+\; 1372500|0.2$

To find the NSD index we apply the algorithm proposed in section 3. From step b) of the algorithm we find that $r^{*}=2$, and from (22) we find

$$\mathrm{NSD}=1-\min(1,0.83)=0.17$$

The interpretation of the above result can be the follwing. If we take into account only full failures the total survival time equals $\overline{S}=958000$, and is slightly greater than $C(2)=945000$. However, when we take into account all failures the minimal value of the total survival time equals $\underline{S}=940000$, and is significantly smaller than $C(4)=1372500$. Hence, the hypothesis that the average lifetime of the tested engines is greater than 150000 holds with rather small grade of necessity.

REFERENCES

Dubois,D.,Prade,H.(1983), Ranking fuzzy numbers in the setting of possibility theory, Information Sciences, v.30, 183-244.

Hryniewicz,O.(1992), Statistical quality control with imprecise information from a sample and fuzzy quality requirements. Reports of IBS PAN, Warsaw (in Polish).

Tsukamoto,Y.,Nikiforuk,P.N.,Gupta,M.M.(1981), On the comparison of fuzzy sets using fuzzy chopping. Proceedings of Eight Triennial IFSA World Congress, v.5, 46-52.

APPENDIX

For the elements from the first and the third groups their total survival time can be calculated as the sum of m fuzzy numbers, each with a linear membership function, and a crisp number kT. Denote this sum by S^0. The membership function of this sum is given as follows

$$\mu^0(S^0)=\begin{cases}-\dfrac{1}{S_M-S_m}(S^0-S_M) & , \text{ for } S_m\le S^0\le S_M\\ 0 & , \text{ otherwise}\end{cases} \tag{A1}$$

Let $S^1=S^0+Z_1$, where Z_1 is a fuzzy number with a membership function given by (2). For a given value of S^1 the maximum value of a membership function can be obtained for Z_1 fulfilling the following equation

$$-\frac{1}{S_M-S_m}[(S^1-Z_1)-S_M]=-\frac{1}{T-z_{m,1}}[Z_1(1-\mu_1)-(T-\mu_1 z_{m,1})] \tag{A2}$$

Denote this value by $z_1^0(S^1)$. From (A2) we find

$$z_1^0(S^1)=\frac{(T-z_{m,1})(S^1-S_M)+(S_M-S_m)(T-\mu_1 z_{m,1})}{(1-\mu_1)(S_M-S_m)+(T-z_{m,1})} \tag{A3}$$

From the condition

$$z_1^0(S^1)\le T \tag{A4}$$

we find that

$$S^1\le S_1^1=S_M+T-\mu_1(S_M-S_m) \tag{A5}$$

Denote by $\mu^1(S^1)$ the membership function of S^1. For $S^1\le S_1^1$ the membership function of S^1 is linear, and such that

$$\mu^1(S^1)=\nu(z_1^0(S^1))=\begin{cases}0 & , \text{ for } S^1<S_0^1\\ -\dfrac{1-\mu_1}{S_1^1-S_0^1}S^1+\dfrac{S_1^1-\mu_1 S_0^1}{S_1^1-S_0^1} & , \text{ for } S_0^1\le S^1\le S_1^1\end{cases} \tag{A6}$$

where

$$S_0^1=S_m+z_{m,1} \tag{A7}$$

For $S^1\ge S$ function $\mu^1(S^1)$ is also linear, and such that

$$\mu^1(S^1)=\mu^0(S^1-T)=\begin{cases}-\dfrac{\mu_1}{S_2^1-S_1^1}+\dfrac{\mu_1 S_2^1}{S_2^1-S_1^1} & , \text{ for } S_1^1\le S^1\le S_2^1\\ 0 & , \text{ for } S^1\ge S_2^1\end{cases} \tag{A8}$$

where

$$S_2^1 = S_M + T \tag{A9}$$

Now, let us notice that S_1^1 is a breakpoint of $\mu^1(S^1)$, and $S_1^1 = S_1^* + T$, where

$$S_1^* = S_M - \mu_1(S_M - S_m) \tag{A10}$$

is the solution of the equation

$$\mu^0(S^0) = \mu_1 \tag{A11}$$

Define now $S^2 = S^1 + Z_2$, and denote the membership function of S^2 by $\mu^2(S^2)$. Analogous considerations lead us to the conclusion that $\mu^2(S^2)$ is also piecewise linear with two breakpoints S_1^2 and S_2^2. The leftmost break point S_1^2 is calculated from the formula

$$S_1^2 = S_2^* + T \tag{A12}$$

where S_2^* is the solution of the equation

$$\mu^1(S^1) = \mu_2 \tag{A13}$$

The second breakpoint S_2^2 is given by

$$S_2^2 = S_1^1 + T \tag{A14}$$

Hence, the membership function of S_2 is the following

$$\mu^2(S^2) = \begin{cases} -\dfrac{1-\mu_2}{S_1^2 - S_0^2} + \dfrac{S_1^2 - \mu_2 S_0^2}{S_1^2 - S_0^2} & , \; S_0^2 \le S^2 \le S_1^2 \\ -\dfrac{\mu_2 - \mu_1}{S_2^2 - S_1^2} + \dfrac{\mu_2 S_2^2 - \mu_1 S_1^2}{S_2^2 - S_1^2}, & S_1^2 \le S^2 \le S_2^2 \\ -\dfrac{\mu_1}{S_3^2 - S_2^2} + \dfrac{\mu_1 S_3^2}{S_3^2 - S_2^2} & , \; S_2^2 \le S^2 \le S_3^2 \\ 0 & , \; S^2 < S_0^2, S^2 \ge S_3^2 \end{cases} \tag{A15}$$

where

$$S_0^2 = S_m + z_{m,1} + z_{m,2} \tag{A16}$$

$$S_3^2 = S_M + 2T \tag{A17}$$

In a general case when $S^i = S^{i-1} + Z_i$, $i = 1,\dots d$, the leftmost breakpoint is calculated from the formula

$$S_1^i = S_i^* + T \tag{A18}$$

where S_i^* as the solution of the following equation

$$\mu^{i-1}(S^{i-1}) = \mu_i \tag{A19}$$

is given as follows

$$S_i^* = \frac{(S_1^{i-1} - \mu_{i-1} S_0^{i-1}) - \mu_i (S_1^{i-1} - S_0^{i-1})}{1 - \mu_{i-1}} \tag{A20}$$

The remaining $i-1$ breakpoints can be found by shifting all breakpoints of $\mu^{i-1}(S^{i-1})$ to the right by T

$$S_j^i = S_{j-1}^{i-1} + T \quad , \quad j = 2,\dots,i \tag{A21}$$

Hence, the membership function of S^i is given by the following expresion

$$\mu^i(S^i) = \begin{cases} 0 & , \quad S^i < S_0^i \\ -\dfrac{\mu_{i-j+1} - \mu_{i-j}}{S_{j+1}^i - S_j^i} S^i + \dfrac{\mu_{i-j+1} S_{j+1}^i - \mu_{i-j} S_j^i}{S_{j+1}^i - S_j^i} & , \quad S_j^i \le S^i \le S_{j+1}^i \quad , j = 0,\dots,i \\ 0 & , \quad S^i \ge S_{i+1}^i \end{cases} \tag{A22}$$

where

$$S_0^i = S_m + \sum_{j=1}^{i} z_{m,j} \tag{A23}$$

$$S_{i+1}^i = S_M + iT \tag{A24}$$

and $\mu_0 = 0$, $\mu_{i+1} = 1$.

5

RELIABILITY AND QUALITY CONTROL IN ENGINEERING SYSTEMS

RELIABILITY BEHAVIOR OF COMBINED HARDWARE-SOFTWARE SYSTEMS

*CAI Kai-Yuan**
WEN Chuan-Yuan
ZHANG Ming-Lian

Department of Automatic Control
Beijing University of Aeronautics and Astronautics
Beijing 100083, China

ABSTRACT With the recognization that software reliability behavior is fuzzy or possibilistic, rather than probabilistic , in nature, and the *assumption* that hardware reliability behavior is probabilistic in nature, in this paper we show how to integrate fuzzy methodology and probabilistic methodology to characterize combined hardware-software reliability behavior. In concrete, the notions of canonical computer, possibilistic survival (PS-) function, and possibilistic vulnerary (PV-) function are introduced, and the combined hardware-software reliability behavior is demonstrated, respectively, for canonical computers, series computer systems and parallel computer systems. Thus this paper provides basic concepts and fundamental analysis and modeling methodology to characterize combined hardware-software reliability behavior of computer systems.

Keywords Hardware-software system, Canonical computer, Possibilistic survival (PS-) function, Possibilistic vulnerary (PV-) function, Reliability, Fuzzy varaible

1 INTRODUCTION

As computer plays a more and more important role in technological systems and society, computer reliability becomes a crucial problem. This is particularly the case when life critical systems, e.g., nuclear reaction control devices, manned spacecraft, are put into operation. In order to measure computer reliability, in the past enormous efforts have been seperately taken in hardware relaibility research and software reliability research. Consequently, a (conventional) theory of hardware reliability has reached a nearly mature status and various well developed hardware reliability techniques have been widely and successfully applied, and great advances have been made in software reliability modeling, software defect avoidance, software fault-tolerance, and software defect removal (testing). However this doesn't prevent people from the desire to model hardware reliability and software reliability in a comprehensive way and to reveal their inherent

---

* Present Address: **Centre for Software Reliability, City University, Northampton Square, London EC1V 0HB, U.K.**

interrelevances. As a result, a logical step is to integrate hardware reliability and software reliability and to develop proper models for combined hardware-software reliability behavior. To this end, a number of efforts were taken in the last decade [1-3]. Undoubtedly, there efforts help us grasp a preliminary understanding on combined hardware-software reliability behavior. If we take a critical and deep insight into these efforts, however, we have the following observations.

1. All the efforts are based on probabilistic methodology. That is, both hardware reliability behavior and software relaibility behavior are postulated to be probabilistic in nature.
2. All the efforts are involved in software debugging phase. Various software debudding-related measures are computed, e.g., the distributions of the time until the first system failure and the time required for completely eliminating all software errors, the distribution of the number of software failures that occur before complete debugging.
3. The resultant models or results are very complex and thus hard to apply to real-world computer systems or to provide necessary and helpful information. This is because various forms of stochastic processes, e.g., Markov processes, are employed to characterize combined hardware-software reliability behavior and the exhaustive number of system states may be very great.

To be confronted with the proceding three points, we have to take some comments and point out how to characterize combined hardware-software reliability behavior. First, software reliability behavior is fuzzy or possibilistic in nature, rather than probabilistic [4,5]. This is because software is unique. Various unsuccesses of Probabilistic Software Reliability Models (PSRMs) make us face the unpleasant fact that more and more PSRMs are proposed and none of them has been proved to be universally applicable and mathematically tractable. In this sense we say PSRMs are unsuccessful on the whole and probabilistic methodology is not the most appropriate approach to characterizing software reliability behavior. In contrast, it has been shown that fuzzy methodology may provide a powerful tool to characterizing software reliability behavior [5]. This implies combined hardware-software reliability behavior cannot be properly characterized if only probabilistic methodology is employed.

Second, although software reliability models are generally software debugging-related and software debugging (failure) data are used to predict the software reliability behavior in the operational phase, this doesn't imply reliability models for combined hardware-software systems should be software debugging phase-oriented. During debugging phase, hardware failures generally don't make serious or fatal consequences. Further, the hardware devices used to debug or test software are put into operation in an intermittent rather than continuous way. In a word, during debugging phase, the topic of combined hardware-software reliability behavior is of little interest. The most important is to demonstrate the predicted combined hardware-software reliability behavior in the operational phase, since any failure, software-induced or hardware-induced, in the operational phase may result in fatal consequences. In sum, reliability models for combined hardware-software systems should be operational phase oriented, rather than debugging phase oriented.

Third, to make useful proposed measures or models for combined hardware-software reliability behavior, mathematical tractability is neccessary.

In response to the proceding three comments, this paper aims to show how to model combined hardware-software reliability behavior with the recognization that software reliability behavior is fuzzy or possibilistic in nature and the *assumption* that hardware reliability behavior is probabilistic in nature. In Section 2 we introduce Nahmias' concept of fuzzy variable [6]. In Section 3 we propose the notion of canonical computer and show how to model the combined hardware-software reliability behavior for a canonical computer. The combined hardware-software reliability behavior of series systems is demonstrated in Section 4, whereas the combined hardware-software reliability behavior of parallel systems is formulated in Section 5. Finally, concluding remarks are contained in Section 6.

Here a note may be helpful. Conventional reliability theory and practice has shown that the assumption that hardware reliability behavior is probabilistic in nature is acceptable and successful in a large variety of circumstances. This doesn't mean that the probability assumption is valid in every case. In fact, in an extensive scope of situations the probability assumption or the basis of conventional reliability theory should be relaxed or even discarded and thus fuzzy reliability theories can be developed [7,8]. However in this paper we confine ourselves to the probability assumption for hardware reliability behavior.

2 MATHEMATICAL PRELIMINARIES

In this section we introduce the concept of fuzzy variable [6] and prove a theorem as a preliminary to the subsequent sections.

DEFINITION 2.1 Suppose that $\wp$ is the discrete topology of Γ(that is, the class of all subsets of Γ). A scale, σ, will be defined as special type of Choquet capacity on $\wp$ which satisfies the following properties:

(1). $\sigma(\phi)=0$, and $\sigma(\Gamma=1)$

(2). For any arbitary collection of sets A_α of $\wp$ (finite, countable or uncountable),

$$\sigma\left(\bigcup_\alpha A_\alpha\right)=\sup_\alpha \sigma(A_\alpha)$$

Then the triplet $(\Gamma,\wp,\sigma)$ is referred to as the pattern space.

DEFINITION 2.2 A fuzzy variable, X, is a real valued function defined on a pattern space $(\Gamma,\wp,\sigma)$.

DEFINITION 2.3 The membership function of a fuzzy varaible X, denoted by μ_X, is a mapping from $\Re$ (the real number line) to the unit interval $[0,1]$ and is given by

$$\mu_X(x)=\sigma(\gamma{:}X(\gamma)=x) \tag{2.1}$$

for all $x\in\Re$. Note that

$$\sup_x \mu_X(x)=\sigma\left\{\bigcup_x(\gamma{:}X(\gamma)=x)\right\}=\sigma(\Gamma)=1 \tag{2.2}$$

In general,. we use the briefer notation $\{X=x\}$ to denote the subset $\{\gamma{:}X(\gamma)=x\}$ of $\wp$.

DEFINITION 2.4 Given a pattern space $(\Gamma,\wp,\sigma)$, the sets $A_1, A_2, \ldots, A_n \subset \wp$ are said to mutually unrelated if for any permutation of the set $\{1,2,\ldots,n\}$, denoted by $i_1, i_2, \ldots, i_k$ for $1 \le k \le n$,

$$\sigma\left(A_{i_1} \cap A_{i_2} \cap \ldots \cap A_{i_k}\right) = \min\left(\sigma\left(A_{i_1}\right), \sigma\left(A_{i_2}\right), \ldots, \sigma\left(A_{i_k}\right)\right) \tag{2.3}$$

DEFINITION 2.5 Given a pattern space $(\Gamma,\wp,\sigma)$, the fuzzy variables $X_1, X_2, \ldots, X_n$ are said to be (mutually) unrelated if for any permutation of the set $\{1,2,\ldots,n\}$, denoted by $i_1, i_2, \ldots, i_k$ for $1 \le k \le n$, the sets $\{X_{i_1} = x_1\}, \{X_{i_2} = x_2\}, \ldots, \{X_{i_k} = x_k\}$ are unrelated for all $x_1, x_2, \ldots, x_k \in \Re$.

THEOREM 2.1 Suppose $f: A \to \Re, A$ is some set, and C_α is some constant irrelevant to A. Then there always holds

$$\sup_{\beta \in A} \min(C_\alpha, f(\beta)) = \min\left(C_\alpha, \sup_{\beta \in A} f(\beta)\right) \tag{2.4}$$

PROOF: Let

$$\sup_{\beta \in A} \min(C_\alpha, f(\beta)) = C$$

Then for each positive integer n, there exists some $\beta_n \in A$ such that

$$C - \frac{1}{n} \prec \min\left(C_\alpha, f(\beta_n)\right) \le C \tag{2.5}$$

First, if $C_\alpha \le C$, then we have

$$C - \frac{1}{n} \prec C_\alpha \le C \qquad \text{for all } n$$

This implies $C_\alpha = C$. Further, from equality (2.5) we have

$$C - \frac{1}{n} \prec f(\beta_n) \le \sup_{\beta \in A} f(\beta)$$

or

$$C_\alpha = C \le \sup_{\beta \in A} f(\beta)$$

With the above equality we immediately arrive at equality (2.4). Otherwise we assume $C_\alpha \succ C$, then from equality (2.5)

$$C - \frac{1}{n} \prec f(\beta_n) \le C$$

So

$$\sup_{\beta \in A} f(\beta) = C \prec C_\alpha$$

With the equality we also arrive at equality (2.4).

3 COMBINED HARDWARE-SOFTWARE RELIABILITY BEHAVIOR OF CANONICAL COMPUTERS

In this section we aim to define the combined hardware-software reliability for canonical computers. First we introduce the notion of canonical computer.

DEFINITION 3.1 From the reliability point of view, a canonical computer consists of one hardware unit (component) and one software unit (component), each of which only exhibits two states: functioning and failed. These units operate independently and unrelatedly, and constitute a series sytem. The independence is interpreted in the probabilistic sense and the unrelatedness is in Nahmias' sense.

Let X_{sw} be the software lifetime. Since software reliability behavior is fuzzy or possibilistic in nature, it is certain that X_{sw} is a fuzzy variable

$$X_{sw} = x:\mu_{sw}(x) \tag{3.1}$$

Typically, $\mu_{sw}(x)$ may be of the form

$$\mu_{sw}(x) = e^{-(x-a)^2} \tag{3.2}$$

where a is some positive constant.

DEFINITION 3.2 The software reliability is defined as

$$R_{sw}(t) = \sigma(X_{sw} \succ t) = \sup_{x \succ t} \mu_{sw}(x) \tag{3.3}$$

and the software unreliability is defined as

$$\overline{R_{sw}}(t) = \sigma(X_{sw} \le t) = \sup_{x \le t} \mu_{sw}(x) \tag{3.4}$$

THEOREM 3.1 At any time instant t, there holds

$$\max\left(\overline{R}_{sw}(t), R_{sw}(t)\right) = 1 \tag{3.5}$$

PROOF: This is obvious from equalities (3.3) and (3.4).

EXAMPLE 3.1 Suppose $\mu_{sw}(x)$ is governed by equality (3.2). Then we have

$$R_{sw}(t) = \begin{cases} 1 \ldots\ldots\ldots\ldots if \ldots t \le a \\ e^{-(t-a)^2} \ldots\ldots if \ldots t \succ a \end{cases} \tag{3.6}$$

and

$$\overline{R_{sw}(t)} = \begin{cases} e^{-(t-a)^2} \ldots\ldots if \ldots t \le a \\ 1 \ldots\ldots\ldots\ldots if \ldots t \succ a \end{cases} \tag{3.7}$$

Let X_{hw} be the hardware lifetime. Since in this paper hardware reliability behavior is postulated to be probabilistic in nature, X_{hw} can be interpreted as a random variable governed by some probability measure P. In this way the hardware reliability is

$$R_{hw}(t) = P\{X_{hw} \succ t\} \tag{3.8}$$

and the hardware unreliability is

$$\overline{R}_{sw}(t) = P\{X_{hw} \le t\} \tag{3.9}$$

However, from another point of view, X_{hw} can be taken as a fuzzy variable whose membership function is just a random variable. That is,

$$X_{hw} = x:\mu_{hw}(x) \tag{3.10}$$

where

$$\mu_{hw}(x)=\delta(x_0)=\begin{cases}1......if...x=x_0\\0......if...x\neq x_0\end{cases} \tag{3.11}$$

where x_0 is just a random variable governed by the probability measure P. Then we denote

$$Q_{hw}(t)=\sigma(X_{hw}\succ t)=\begin{cases}1......if...t\prec x_0\\0......if...t\geq x_0\end{cases} \tag{3.12}$$

and

$$\overline{Q_{hw}}(t)=\sigma(X_{hw}\leq t)=\begin{cases}0......if...t\prec x_0\\1......if...t\geq x_0\end{cases} \tag{3.13}$$

DEFINITION 3.3 The hardware reliability is defined as

$$R_{hw}(t)=E.Q_{hw}(t)=P\{t\prec x_0\} \tag{3.14}$$

and the hardware unreliability is

$$\overline{R_{hw}}(t)=E.\overline{Q_{hw}}(t)=P\{t\geq x_0\}$$

where E is the probabilistic expectation operation.

The validity of this definition is justified by the fact that equalities (3.14) and (3.15) coincide with equalities (3.8) and (3.9) respectively.

THEOREM 3.2 At any time instant there hold

$$\max\left(Q_{hw}(t),\overline{Q_{hw}(t)}\right)=1 \tag{3.16}$$

and

$$R_{hw}(t)+\overline{R_{hw}(t)}=1 \tag{3.17}$$

PROOF: This is obvious from Definition 3.3.

Let X be the system lifetime. Then for a canonical computer,

$$X=\min(X_{hw},X_{sw}) \tag{3.18}$$

Viewing X as a fuzzy varaible, we have

$$X=x:\mu_X(x) \tag{3.19}$$

THEOREM 3.3 For a canonical computer, there holds

$$\mu_X(x)=\begin{cases}\mu_{sw}(x).........if...x\prec x_0\\R_{sw}(x).........if...x=x_0\\0..............if...x\succ x_0\end{cases} \tag{3.20}$$

PROOF: From Definition 2.3

$$\begin{aligned}\mu_X(x)&=\sigma(X=x)=\sigma(\min(X_{hw},X_{sw})=x)\\&=\sigma(\{X_{hw}=x,X_{sw}\geq x\}\cup\{X_{hw}\geq x,X_{sw}=x\})\\&=\max(\sigma(X_{hw}=x,X_{sw}\geq x),\sigma(X_{hw}\geq x,X_{sw}=x))\end{aligned}$$

However

$$\begin{aligned}\sigma(X_{hw}=x,X_{sw}\geq x)&=\sigma\left(\sup_{\alpha\geq x}\{X_{hw}=x,X_{sw}=\alpha\}\right)\\&=\sup_{\alpha\geq x}\sigma(X_{hw}=x,X_{sw}=\alpha)=\sup_{\alpha\geq x}\min(\mu_{hw}(x),\mu_{sw}(\alpha))\end{aligned}$$

Further, from Theorem 2.1, we arrive at

$$\sigma(X_{hw}=x,X_{sw}\geq x)=\min\left(\mu_{hw}(x),\sup_{\alpha\geq x}\mu_{sw}(\alpha)\right)$$

$$=\min(\delta(x_0),\sigma(X_{sw}\geq x))=\begin{cases}R_{sw}(x)\ldots\ldots if\ldots x=x_0\\ 0\ldots\ldots\ldots\ldots if\ldots x\neq x_0\end{cases}$$

Similarly,

$$\sigma(X_{hw}\geq x,X_{sw}=x)=\min(\sigma(X_{hw\geq x}),\sigma(X_{sw}=x))$$

$$=\min\left(\sup_{\alpha\geq x}\delta(x_0),\mu_{sw}(x)\right)=\begin{cases}\mu_{sw}(x)\ldots\ldots if\ldots x\leq x_0\\ 0\ldots\ldots\ldots\ldots if\ldots x\succ x_0\end{cases}$$

Then equality (3.20) is obtained.

DEFINITION 3.4 Let

$$Q(t)=\sigma(X\succ t) \tag{3.21}$$

and

$$\overline{Q(t)}=\sigma(X\leq t) \tag{3.22}$$

Then $Q(t)$ and $\overline{Q(t)}$ are referred to as possibilistic survival (PS-) function and possibilistic vulnerary (PV-) function, respectively.

THEOREM 3.4 For a canonical computer, the PS-function and the PV-function are given, respectively, by

$$Q(t)=\begin{cases}R_{sw}(t)\ldots\ldots if\ldots t\prec x_0\\ 0\ldots\ldots\ldots\ldots if\ldots t\geq x_0\end{cases} \tag{3.23}$$

and

$$\overline{Q(t)}=\begin{cases}\overline{R}_{sw}(t)\ldots\ldots if\ldots t\prec x_0\\ 1\ldots\ldots\ldots\ldots if\ldots t\geq x_0\end{cases} \tag{3.24}$$

PROOF: First we note

$$Q(t)=\sigma(X\succ t)=\sup_{x\succ t}\mu_X(x)$$

Then from equality (3.20), for $t\geq x_0$, we immediately arrive at

$$Q(t)=\sup_{x\succ t\geq x_0}\mu_X(x)=0$$

For $t\prec x_0$,

$$Q(t)=\max\left(\sup_{t\prec x\prec x_0}\mu_X(x),\mu_X(x_0),\sup_{x\succ x_0}\mu_X(x)\right)$$

$$=\max\left(\sup_{t\prec x\prec x_0}\mu_X(x),\sigma(X_{sw}\geq x_0)\right)=\sup_{t\prec x}\mu_{sw}(x)$$

Thus equality (3.23) is obtained. Equality (3.24) can be obtained in a similar way.

Here we must note that $Q(t)$ (and $\overline{Q(t)}$ can be obtained in a direct way. This is because X_{hw} itself can be directly viewed as a random variable. That is,

$$Q(t)=\sigma(X\succ t)=\sigma(\min(X_{hw},X_{sw})\succ t)$$

Then for $X_{hw} \succ t, \min(X_{hw}, X_{sw}) \succ t$ implies $X_{sw} \succ t$. So

$$Q(t) = \sigma(X_{sw} \succ t)$$

For $X_{hw} \le t, \min(X_{hw}, X_{sw}) \succ t$ is just a null set. Thus

$$Q(t) = \begin{cases} R_{sw}(t)........if...X_{hw} \succ t \\ 0..................if...X_{hw} \le t \end{cases} \tag{3.25}$$

Similarly,

$$\overline{Q(t)} = \begin{cases} \overline{R}_{sw}(t)........if...X_{hw} \succ t \\ 1..................if...X_{hw} \le t \end{cases} \tag{3.26}$$

For the sake of convenience, in the sequel we view X_{hw} directly as a random varaible.

COROLLARY 3.1 At any time instant t, there holds

$$\max\left(Q(t), \overline{Q(t)}\right) = 1 \tag{3.27}$$

PROOF: This is obvious from equalities (3.25) and (3.26).

DEFINITION 3.5 For a canonical computer, the system reliability is

$$R(t) = E.Q(t) \tag{3.28}$$

and the system unreliability is

$$\overline{R(t)} = E.\overline{Q(t)} \tag{3.29}$$

THEOREM 3.5 For a canonical computer, the system reliability and the systen unreliability are determined, respectively, by

$$R(t) = R_{hw}(t).R_{sw}(t) \tag{3.30}$$

and

$$\overline{R(t)} = \overline{R}_{hw}(t) + \overline{R}_{sw}(t) - \overline{R}_{hw}(t).\overline{R}_{sw}(t) \tag{3.31}$$

PROOF: Equality (3.30) can be directly obtained from equality (3.25), and equality (3.31) is just due to equality (3.26).

4 COMBINED HARDWARE-SOFTWARE RELIABILITY BEHAVIOR OF SERIES SYSTEMS

Consider a series system comprising two independent and unrelated canonical computers. Let X_{hw_i} be the hardware lifetime of computer i and X_{sw_i} be the software lifetime of computer i. Let X_i be the lifetime of computer i. Let X be the system lifetime. Then

$$X = \min(X_1, X_2) \tag{4.1}$$

Denote

$$Q_i(t) = \begin{cases} R_{sw_i}(t)........if...X_{hw_i} \succ t \\ 0..................if...X_{hw_i} \le t \end{cases} \tag{4.2}$$

$$\overline{Q}_i(t) = \begin{cases} \overline{R}_{sw_i}(t)........if...X_{hw_i} \succ t \\ 1..................if...X_{hw_i} \le t \end{cases} \tag{4.3}$$

where

$$R_{sw_i}(t) = \sigma\left(X_{sw_i} \succ t\right) \tag{4.4}$$

$$\overline{R}_{sw_i}(t) = \sigma\left(X_{sw_i} \leq t\right) \tag{4.5}$$

Further,

$$R_{hw_i}(t) = P\left(X_{hw_i} \succ t\right) \tag{4.6}$$

$$\overline{R}_{hw_i}(t) = P\left(X_{hw_i} \leq t\right) \tag{4.7}$$

$$Q(t) = \sigma(X \succ t) \tag{4.8}$$

$$\overline{Q(t)} = \sigma(X \leq t) \tag{4.9}$$

Then we have the following theorem.

THEOREM 4.1 For a series system comprising two independent and unrelated canonical computers, there holds

$$Q(t) = \min(Q_1(t), Q_2(t)) \tag{4.10}$$

or

$$Q(t) = \begin{cases} \min\left(R_{sw_1}(t), R_{sw_2}(t)\right) \ldots\ldots\ldots if \ldots X_{hw_1} \succ t, X_{hw_2} \succ t \\ 0 \ldots\ldots\ldots\ldots\ldots\ldots\ldots\ldots\ldots\ldots\ldots otherwise \end{cases} \tag{4.11}$$

PROOF: First we have

$$Q(t) = \sigma(X \succ t) = \sigma\left(\min(X_1, X_2) \succ t\right)$$

$$= \sigma\left(X_1 \succ t, X_2 \succ t\right) = \sigma\left(\bigcup_{\alpha \succ t}\{X_1 = \alpha, X_2 \succ t\}\right)$$

$$= \sup_{\alpha \succ t} \sigma(X_1 = \alpha, X_2 \succ t) = \sup_{\alpha \succ t} \sigma\left(\bigcup_{\beta \succ t}\{X_1 = \alpha, X_2 = \beta\}\right)$$

$$= \sup_{\alpha \succ t} \sup_{\beta \succ t} \min\left(\sigma(X_1 = \alpha), \sigma(X_2 = \beta)\right)$$

With the help of Theorem 2.1, we arrive at

$$Q(t) = \sup_{\alpha \succ t}\left(\min\left(\sigma(X_1 = \alpha), \sup_{\beta \succ t} \sigma(X_2 = \beta)\right)\right)$$

$$= \min\left(\sup_{\alpha \succ t} \sigma(X_1 = \alpha), \sup_{\beta \succ t} \sigma(X_2 = \beta)\right)$$

Thus equality (4.10) is obtained. By use of equality (4.2), then equality (4.11) immediately becomes true.

THEOREM 4.2 For a sereis system comprising n independent and unrelated canonical computer, the system reliability is given by

$$\begin{aligned} R(t) &= E.Q(t) \\ &= R_{hw_1}(t).R_{hw_2}(t).\min\left(R_{sw_1}(t), R_{sw_2}(t)\right) \end{aligned} \tag{4.12}$$

PROOF: This is obvious from equality (4.11).

THEOREM 4.3 For a series system comprising n independent and unrelated canonical computer, the PS-function is given by

$$Q(t)=\min\left(Q_1(t),Q_2(t),\ldots,Q_n(t)\right) \tag{4.13}$$

or

$$Q(t)=\begin{cases}\min\left(R_{sw_1}(t),\ldots,R_{sw_n}(t)\right)\ldots\ldots if\ldots X_{hw_1}\succ t,\ldots,X_{hw_n}\succ t\\ 0\ldots\ldots\ldots\ldots\ldots\ldots\ldots\ldots\ldots otherwise\end{cases} \tag{4.14}$$

and the system reliability is expressed as

$$R(t)=E.Q(t)=R^e_{hw}(t).R^e_{sw}(t) \tag{4.15}$$

where

$$R^e_{hw}(t)=R_{hw_1}(t)\ldots R_{hw_n}(t) \tag{4.16}$$

and

$$R^e_{sw}(t)=\min\left(R_{sw_1}(t),\ldots,R_{sw_n}(t)\right) \tag{4.17}$$

PROOF: Obvious.

THEOREM 4.4 For a series system comprising two independent and unrelated canonical computer, the PV-function is given by

$$\overline{Q(t)}=\sigma(X\le t)$$

$$=\begin{cases}\max\left(\min\left(\overline{R}_{sw_1}(t),\overline{R}_{sw_2}(t)\right),\min\left(\overline{R}_{sw_1}(t),R_{sw2}(t)\right),\min\left(R_{sw_1}(t),\overline{R}_{sw_2}(t)\right)\right)\\ \ldots\ldots\ldots if\ldots X_{hw_1}\succ t,X_{hw_2}\succ t\\ \max\left(\overline{R}_{sw_1}(t),R_{sw_2}(t)\right)\ldots\ldots\ldots if\ldots X_{hw_1}\succ t,X_{hw_2}\le t\\ \max\left(\overline{R}_{sw_2}(t),R_{sw_2}(t)\right)\ldots\ldots\ldots if\ldots X_{hw_1}\le t,X_{hw_2}\succ t\\ 1\ldots\ldots\ldots\ldots\ldots\ldots\ldots\ldots\ldots if\ldots X_{hw_1}\le t,X_{hw_2}\le t\end{cases} \tag{4.18}$$

and the system unreliability is expressed as

$$\overline{R(t)}=E.\overline{Q(t)}$$

$$=R_{hw_1}(t).R_{hw_2}(t)\left[\max\begin{pmatrix}\min\left(\overline{R}_{sw_1}(t),\overline{R}_{sw_2}(t)\right),\\ \min\left(\overline{R}_{sw_1}(t),R_{sw_2}(t)\right),\\ \min\left(R_{sw_1}(t),\overline{R}_{sw_2}(t)\right)\end{pmatrix}\right] \tag{4.19}$$

$$+R_{hw_2}(t).\overline{R}_{hw_2}(t).\max\left(\overline{R}_{sw_1}(t),R_{sw_1}(t)\right)$$

$$+\overline{R}_{hw_1}(t).R_{hw_2}(t).\max\left(\overline{R}_{sw_2}(t),R_{sw_2}(t)\right)+\overline{R}_{hw_1}(t).\overline{R}_{hw_2}(t)$$

PROOF: To prove equality (4.18), it is sufficient to note

$$\overline{Q(t)} = \sigma(X \le t) = \sigma(\min(X_1, X_2) \le t)$$
$$= \sigma(\{X_1 \le t, X_2 \le t\} \cup \{X_1 \le t, X_2 \succ t\} \cup \{X_1 \succ t, X_2 \le t\})$$
$$= \max\begin{pmatrix} \min(\sigma(X_1 \le t), \sigma(X_2 \le t)), \min(\sigma(X_1 \le t), \sigma(X_2 \succ t)), \\ \min(\sigma(X_1 \succ t), \sigma(X_2 \le t)) \end{pmatrix}$$
$$= \max(\min(\overline{Q}_1(t), \overline{Q}_2(t)), \min(\overline{Q}_1(t), Q_2(t)), \min(Q_1(t), \overline{Q}_2(t)))$$

Using equalities (4.2) and (4.3), we immediately arrive at equality (4.18). Further we note equality (4.19) is just a direct result of equality (4.18).

5 COMBINED HARDWARE-SOFTWARE RELIABILITY BEHAVIOR OF PARALLEL SYSTEMS

Consider a parallel system comprising two independent and unrelated canonical computers. Adopting the various notations used in the last section, we have

$$X = \max(X_1, X_2) \tag{5.1}$$

THEOREM 5.1 For a parallel system comprising two independent and unrelated canonical computer, the PS-function is given by

$$Q(t) = \sigma(X \succ t)$$
$$= \begin{cases} \max(\min(R_{sw_1}(t), R_{sw_2}(t)), \min(R_{sw_1}(t), \overline{R}_{sw_2}(t)), \min(\overline{R}_{sw_1}(t), R_{sw_2}(t))) \\ \dots\dots\dots\dots\dots\dots if \dots X_{hw_1} \succ t, X_{hw_2} \succ t \\ R_{sw_1}(t) \dots\dots\dots if \dots X_{hw_2} \succ t, X_{hw_2} \le t \\ R_{sw_2}(t) \dots\dots\dots if \dots X_{hw_1} \le t, X_{hw_2} \succ t \\ 0 \dots\dots\dots\dots\dots if \dots X_{hw_1} \le t, X_{hw_2} \le t \end{cases} \tag{5.2}$$

and the system reliability is expressed as

$$R(t) = E.Q(t)$$
$$= R_{hw_1}(t).R_{hw_2}(t).\max\begin{pmatrix} \min(R_{sw_1}(t), R_{sw_2}(t)), \\ \min(R_{sw_1}(t), \overline{R}_{sw_2}(t)), \\ \min(\overline{R}_{sw_1}(t), R_{sw_2}(t)) \end{pmatrix} \tag{5.3}$$
$$+R_{hw_1}(t).\overline{R}_{hw_2}(t).R_{sw_1}(t) + \overline{R}_{hw_1}(t).R_{hw_2}(t).R_{sw_2}(t)$$

PROOF: In order to prove equality (5.2), it is sufficient to note

$$Q(t) = \sigma(X \succ t) = \sigma(\max(X_1, X_2) \succ t)$$
$$= \sigma(\{X_1 \succ t, X_2 \succ t\} \cup \{X_1 \succ t, X_2 \le t\} \cup \{X_1 \le t, X_2 \succ t\})$$
$$= \max(\min(Q_1(t), Q_2(t)), \min(Q_1(t), \overline{Q}_2(t)), \min(\overline{Q}_1(t), Q_2(t)))$$

Then by use of equalities (4.2) and (4.3), equality (5.2) is obtained. Further equality (5.3) is just a just result of equality (5.2).

THEOREM 5.2 For a parallel system comprising two independent and unrelated canonical computers, the PV-function is given by

$$\overline{Q(t)} = \sigma(X \le t)$$

$$= \begin{cases} \min\left(\overline{R}_{sw_1}(t), R_{sw_2}(t)\right)......if...X_{hw_1} \succ t, X_{hw_2} \succ t \\ \overline{R}_{sw_1}(t)....................if...X_{hw_1} \succ t, X_{hw_2} \le t \\ \overline{R}_{sw_2}(t)....................if...X_{hw_1} \le t, X_{hw_2} \succ t \\ 1..........................if...X_{hw_1} \le t, X_{hw_2} \le t \end{cases} \tag{5.4}$$

and the system unreliability is expressed as

$$\overline{R}(t) = E.\overline{Q(t)}$$

$$= R_{hw_1}(t).R_{hw_2}(t).\min\left(\overline{R}_{sw_1}(t), \overline{R}_{sw_2}(t)\right) + R_{hw_1}(t).\overline{R}_{hw_2}(t).\overline{R}_{sw_1}(t) \tag{5.5}$$

$$+\overline{R}_{hw_1}(t).R_{hw_2}(t).\overline{R}_{sw_2}(t) + \overline{R}_{hw_1}(t).\overline{R}_{hw_2}(t)$$

PROOF: First we note

$$\overline{Q(t)} = \sigma(X \le t) = \sigma\left(\max(X_1, X_2) \le t\right)$$

$$= \sigma\left(X_1 \le t, X_2 \le t\right) = \min\left(\overline{Q}_1(t), \overline{Q}_2(t)\right)$$

With the help of equality (4.3), we immediately arrive at equality (5.4). Further, equality (5.5) is just a direct result of equality (5.4).

6 CONCLUDING REMARKS

In response to the daily increasing importance of the topic of computer reliability, a number of efforts were made in an attempt to characterize combined hardware-software reliability behavior in the last decade. All these efforts are based on probabilistic methodology. Unfortunately, various unsuccesses of PSRMs (Probabilistic Software Reliability Models) urge us to recognize that software reliability behavior is fuzzy or possibilistic, rather than probabilistic, in nature. In the proceding sections we have showed how to integrate fuzzy methodology and probabilistic methodology to characterize combined hardware-software reliability behavior. In concrete, we introduce the notions of canocial computer, possibilistic survival (PS-) function and possibilistic vulnerary (PV-) function and demonstrate the combined hardware-software reliability behavior for canonical computers, series computer systems and parallel computer systems. The corresponding results can be used to further characterize the combined hardware-software reliability behavior for more complex systems. Thus we say this paper provides basic concepts and fundamental analysis and modeling methodology to characterize combined hardware-software reliability behavior for computer systems. However we should note that to apply the results given in this paper to real-world computer systems, efforts on the coordination between the time index for software reliability and that for hardware reliability have to be taken. This is because that the most appropriate time index to base software reliability may be the software CPU execution time [9], whereas the time index used for hardware reliability is usually chosen to be the calendary time.

REFERENCES

1. U.Sumita, Y.Masuda, "Analysis of Software Availability/Reliability Under the Influence of Hardware Failures", *IEEE Transaction on Software Engineering*, Vol.SE-12, No.1, 1986, pp32-41.
2. A.L.Goel, J.Soenjoto, "Models for Hardware-Software System Operational-Performance Evaluation", *IEEE Transaction on Reliability*, Vol.R-31, No.3, 1981, pp232-239.
3. J.E.Angus, L.E.James, "Combined Hardware/Software Reliability Models", *Proc. Annual Reliability and Maintainability Symposium*, 1982, pp176-181.
4. K.Y.Cai, C.Y.Wen, M.L.Zhang, "A Critical Review on Software Reliability Modeling", *Reliability Engineering and System Safety*, Vol.32, 1991, pp357-371.
5. K.Y.Cai, C.Y.Wen, M.L.Zhang, "A Novel Approach to Software Reliability Modeling", *Microelectronics and Reliability*, Vol.33, No.15, 1993, pp2265-2267.
6. S.Nahmias, "Fuzzy Variables", *Fuzzy Sets and Systems*, Vol.1, 1978, pp97-110.
7. K.Y.Cai, C.Y.Wen, M.L.Zhang, "Fuzzy States as a Basis for a Theory of Fuzzy Reliability", *Microelectronics and Reliability*, Vol.33, No.15, 1993, pp2253-2263.
8. K.Y.Cai, C.Y.Wen, M.L.Zhang, "Fuzzy Variables as a Basis for a Theory of Fuzzy Reliability in the Possibility Context", *Fuzzy Sets and Systems*, Vol.42, 1991, pp145-172.
9. J.D.Musa, A.Iannino, K.Okumoto, *Software Reliability: Measurement, Prediction, Application*, McGraw-Hill, 1987.

AN APPLICATION OF FUZZY SET THEORY TO RELIABILITY ANALYSIS

– RELIABILITY PREDICTION FOR EQUIPMENT USING FUZZY INFERENCE –

Tadashi MURATA

Faculty of Engineering
University of the Ryukyus
Nishihara, Okinawa 903-01, Japan

Abstract - Fuzzy estimation is made of the reliability of an actual piece of electronic equipment, as measured by the mean time between failures (MTBF). This is then compared with actual MTBF field data.

Keywords - fuzzy inference, reliability analysis, electronic equipment, MTBF

1. INTRODUCTION

When electronic equipment is used in the field, there is usually a large difference between the design reliability measured by mean time between failures (MTBF) and the actual MTBF. This problem mainly occurs as toothing problems with new equipment, but we know that the problem may continue for a considerable period. The design engineers and production department involved will then endeavor to improve the equipment reliability, which mainly depends on workmanship, part reliability and the equipment design itself. Here we deal with a piece of airborne electronic equipment (Air Data Computer). We compare the design and actual MTBFs with several predictions for the MTBF at various stages of implementation derived by the fuzzy inference method.

2. EQUIPMENT DATA

The design and actual reliability of the equipment is shown in Fig.1 as a function of time. The actual MTBF in Fig.1 is computed monthly using equation (1)[1].

$$MTBF = T / r \tag{1}$$

where T is the flight-time in hours and r the equipment failure count per month. It can be seen from Fig.1 that although the monthly MTBF fluctuates considerably, a consistent improvement in the yearly MTBF is achieved. Even so, it only reaches about 50% of the design MTBF.

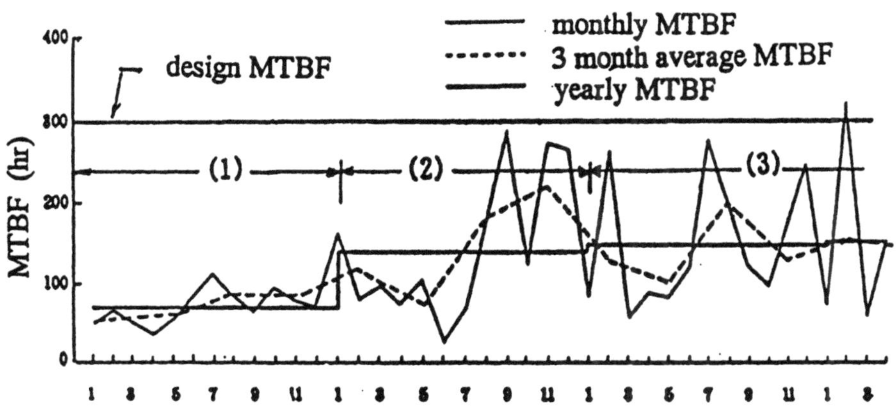

(1) an initial period, (2) reliability improving period,
(3) a later period

Fig.1 Design MTBF (reliability) and actual MTBF of equipment

3. MTBF PREDICTION WITH FUZZY INFERENCE

3.1 Fuzzy IF-THEN Rules

The MTBF y_i of the equipment for each period i in the field is predicted using equation (2).

$$y_i = k_i \cdot y_0 \qquad (2)$$

where y_0 is the design MTBF and k_i the fuzzy - derived environmental parameter for the i-th period. In order to predict the factor k_i for the i-th period, the fuzzy IF-THEN rules shown in Table 1 are used. M_1 represents the man (operational experience of factory worker), M_2 the method (contents

Table 1 Fuzzy IF-THEN rules determing the factor K_i (example)

(M_3=Z)

K_i		M_2		
		N	Z	P
M_1	N	—	N	—
	Z	N	Z	P
	P	—	P	—

of repair manuals) and M_3 the material (parts and / or raw materials). P, Z and N stand for positive, zero and negative states of these factors (refer to Table 2).

Table 2 Contents of labels (IF-part)

element		contents of lable		
		N	Z	P
M_1	MAN (worker)	poor	normal	good
M_2	METHOD (mamual)	poor	normal	good
M_3	MATERIAL (parts)	poor	normal	good

3.2 FMEA METHOD FOR FUZZY IF-THEN RULES

We can create these fuzzy IF-THEN rules from our experience and knowledge, which reliability engineers typically formulate in the form of FMEA (failure mode and effects analysis) table (see example, Table 3). The construction method [(2)] is;

(1) list up possible failure phenomena for a component part.

(2) list up the main cause of the part's failure based on an engineer's knowledge and/or experience.

(3) describe a functional defect of the equipment as a whole based on the part's failure.

(4) estimate the part's failure rate.

(5) estimate the degree of equipment function defect due to the part failure.

(6) if (4) and/or (5) are large, the design or the part should be changed in order to improve the design reliability.

In the example below, the information in (b) ~(e) relate to the IF-part, (f) and (g) to the THEN-part of the fuzzy IF-THEN rule.

Table 3 Equipment FMEA table (example)

system:		FMEA table (failure mode and effects analysis)				data: prepared by:	
(a) No.	(b) item	(c) failure function	(d) failure mode	(e) estimated cause	(f) effect to system	(g) system damage degree	(h) notes
3-1	potentiometer	noise voltage exceed	wear out	part-life reached	unstable output	[B] [3] (notes)	change to more reliable part

Notes: [B] indicates comparative failure rate and [3] comparative criticality, on some scale.

4. MEMBERSHIP FUNCTION SHAPES

Shapes of membership functions for P, Z and N in the condition-part (IF-part) and an operation-part (THEN-part) of the fuzzy inference are

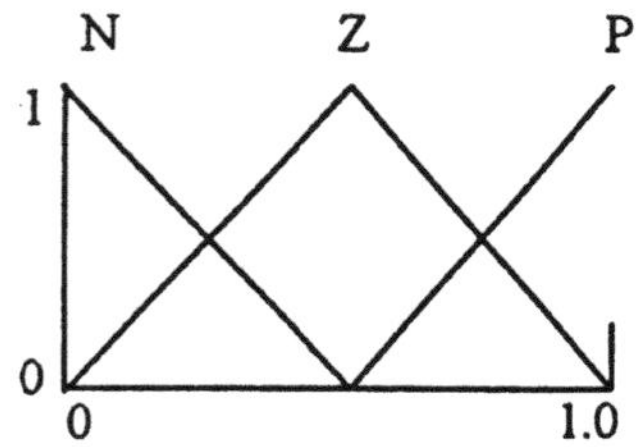

Fig.2 Function shapes for IF-part

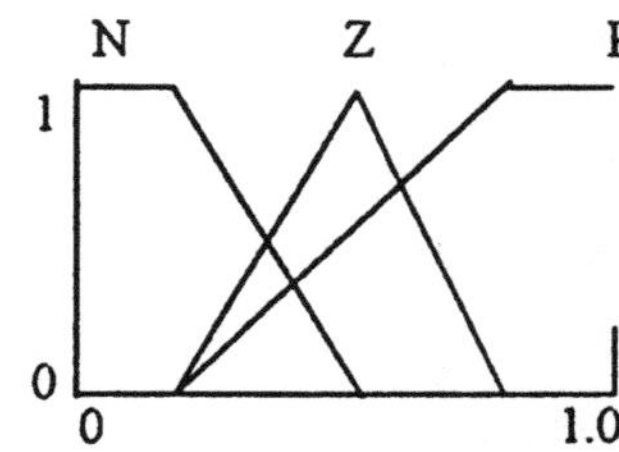

Fig.3 Function shapes for THEN-part

shown separately in Fig.2 and Fig.3. In order to calculate a defuzzy value, a center of gravity method is applied with a "bounded sum" in the THEN-part.

Membership function shape of the IF-part are reasonably standard, but those for the THEN-part are selected through trial-and-error to make the predicted MTBF coincide as closely as possible with the actual MTBF.

It is means that some elements of "P" may effect on side of "Z" and "N", and label of "P", "Z" and "N" are shown in Table 4. This is possible eventually by the use of the bounded-sum technique.

Table 4 Contents of lables (THEN-part)

lable	contents
N (negative)	small
Z (zero)	medium
P (positive)	large

5. DISCUSSION OF RESULTS

The MTBF predicted [(3)(4)] using fuzzy inference is compared with the actual MTBF for each period in Table 5. We can see again from this table that the real MTBF is very much smaller than the design MTBF of the equipment, and even in the latest period, the real MTBF only reached about 60 % of the design MTBF. This problem has been treated in a previous paper by the author [5],

It is significant that the predicted MTBF was within about 10% of the actual values (see Table 5, (b) / (a)). The abundance of technical and experiential information is skilfully handled by the breakdown of the factors causing equipment failure into those due to man, method and material, and application of the orderly FMEA method to make the fuzzy IF-THEN rules.

Table 5 A comparison of actual and predicted MTBF values.

period (see Fig.1)	(a)	(b)	(b)/(a)
(1) initial period	0.27	0.28	1.04
(2) improvement period	0.57	0.50	0.88
(3) later stage	0.62	0.68	1.10

Notes ; (a) : actual MTBF / design MTBF
(b) : predicted MTBF / design MTBF
(b)/(a) : predicted MTBF / actual MTBF, (prediction error)

6. CONCLUDING REMARKS

In this paper, the MTBF of a piece of airborne electronic equipment is predicted using fuzzy inference. The predicted MTBF in different periods following initial use is compared with field data for MTBF. We conclude that:

(1) The FMEA method developed as a reliability design technique is very useful for making the necessary fuzzy IF-THEN rules, because it facilitates the collection and organization of much technical information and personal experience on equipment failure.

(2) It is possible to produce a prediction for MTBF which agrees to within about 10% with actual values found in practice.

(3) In order to improve the consistency of MTBFs, it is essential to select the appropriate membership function shapes, and to derive a satisfactory value for the overall environmental factor k_i.

(4) Improved fuzzy estimations of MTBF should lead to more efficient repair procedures and management of spare parts.

REFERENCES

(1) T.Murata, " Reliability case history of an airborne Air Data Computer", IEEE Trans. reliability, Vol.R-24, No.2, pp98-102, 1975.

(2) T.Murata, E.Shimabukuro, " On the FMEA chart using fuzzy inference technique", EIC, R91-25, pp.23-28, 1991. (in Japanese)

(3) T.Murata, "Reliability evaluation on electronic equipment with fuzzy set theory", journal of SOFT (Japan), Vol.5, No.5, pp.981-990, 1993.

(4) T.Terano, K.Asai, M.Sugeno, " Fuzzy system theory and its applications", Ohm Co., 1987 (in Japanese)

(5) T.Murata," A study on a predicted MTBF and a real MTBF in electronic equipment", IECE, R73-24, 1973. (in Japanese)

Application of Fuzzy Relational Modelling to Industrial Product Quality Control

Y. Qian[1] *, P. J. C. Tessier[2], G. A. Dumont[2]

[1] Department of Chemical Engineering, Tsinghua University, Beijing, 100084, P.R. China

[2] Chemical Engineering Department, University of British Columbia, Vancouver, Canada

Abstract: Industrial product quality control is a typical mathematical programming and optimization problem. Unfortunately, complete and precise models are not always available for many industrial processes. A fuzzy relation modelling approach is proposed to describe approximate relationships among system variables and reconcile empirical equations. Symmetric fuzzy decision-making is transformed into a non-linear function maximization problem. The approach is applied to optimization of a wood chip refining process. It is used either to improve pulp quality within the refiner operation range, or to reduce operation costs while maintaining an acceptable pulp quality.

Keywords: product quality control, fuzzy optimization, mathematical programming, fuzzy relational model, wood chip refining process.

1. Introduction

Mathematical programming and optimization are important mathematical tools in industrial production management and product quality control. An optimization problem can be constructed by choosing a set of system variables X, an objective functions, $g(X)$, and system constraints represented as a set of equations $f_i(X)$ and inequalities $C_j(X)$:

$$
\begin{array}{lll}
\text{maximize} & g(X) & \\
\text{such that} & f_i(X) = 0, & i = 1, \ldots, N \\
 & C_j(X) < 0, & j = 1, \ldots, M \\
 & X \geq 0 &
\end{array}
\tag{1}
$$

* Author to whom correspondance should be addressed.

Depending on the mathematical character of the objective function, $g(X)$, and the constraints, $f_i(X)$ and $C_i(X)$, many types of mathematical programming algorithms exist, such as linear programming, quadratic programming, fractional programming, convex programming, etc. (Wisniewski, 1990). That can be used for searching optimal solution or decision in a deterministic environment, where models, constraints, and objectives are well defined and formalized.

Industrial product quality control is in nature an optimization and decision-making in a fuzzy environment. Many decisions are subjective, imprecise and not easily quantifiable. The situation is even worse in some processing industries, such as in pulp and paper-making, oil refining, petrochemical processing, where measurements from on-line sensors may be noisy and lack accuracy, and for which complete and accurate data and models may not be available. Knowledge of these processes are sometimes represented in empirical equations and linguistic descriptions. It is always a trade-off between a better product quality and lower production costs, within a number of system constraints of operating feasibility, product specifications, safety and environmental issues.

In this paper, we propose a fuzzy-logic-based modelling and optimization approach and its application to industrial product quality control. As a case study, we present how a fuzzy modelling and optimization approach can be used in control of a wood chip refining process and improvement of pulp quality.

2. Optimization in a Fuzzy Environment

For optimization in a fuzzy environment, most published work focuses on different forms of fuzzy linear programming (FLP) problems (Fedrizzi, Kacprzyk and Verdegay, 1991; Zimmermann, 1992). We have, however, concentrated our efforts on the following aspects of the problem:

1. In most industrial processes, linearity can not always be guaranteed. To further avoid distortion of decision-making in a fuzzy environment, non-linearity has to be properly handled.

2. In Equation (1) we distinguished two types of system constraints: equality relationships, $f_i(X)$, and inequality constraints, $C_i(X)$. There is a distinct difference in fuzziness between these two types of constraints. Inequality constraints address imprecise and non-crisp boundaries, around which a slight violation would be tolerated. On the other hand, fuzzy equations indicate fuzziness of poorly-understood relationships among system variables. In the following sections, a fuzzy relation modelling approach is described in detail, and the corresponding optimization methodology is developed.

2.1. Fuzzy Relational Modelling

In many industrial processes, interactions between operating conditions and product quality are complex. Most of them may not have been quantitatively studied and modelled in a systematical way. However, experience and knowledge may have been accumulated through years of industrial practice. When no mechanistic models is available, empirical equations can be used instead. Because of its empirical nature, this knowledge is generally valid under conditions close to those for which it was derived. Although it will be less accurate under different circumstances, it can still be useful when properly used .

Empirical knowledge can be described as a set of approximate equations:

$$f_i(X) \approx 0 \qquad i = 1, \dots, N \tag{2}$$

where the symbol $\approx$ indicates that the equalities are approximate; $X = \{X_1, \dots, X_n\}$ is the set of process variables, which consists of both operating variables and product quality variables. The number of equations is not necessarily the same as the number of variables. There are usually more equations than variables, namely $N > n$, which indicates redundancy of information.

We define a membership function for each equation as follows:

$$\mu_i = \exp\left[-\omega_i f_i^2(X)\right] \qquad i = 1, \dots, N \tag{3}$$

where ω_i is the weight of the i–th equation, denoting relative importance of each equation in the system. Membership μ_i is in the range of 0 to 1. When $f_i(X) = 0$, meaning the equation satisfactorily describes a relationship between the system variables, the membership $\mu_i = 1$. On the other hand, a low membership indicates that the equation is not compatible with the other equations of the set. In order for the model to consistently represents the true process, equations with extremely low memberships should be removed from the set.

Now we define a fuzzy model, F, which is the set of all screened equations. The degree of fuzziness of the model, μ_F, can be calculated as follows:

$$\mu_F = \min_i \mu_i = \min_i \left[\exp\left(-\omega_i f_i^2(X)\right)\right], \quad i = 1, \dots, N'; \ \left(N' \leq N\right) \tag{4}$$

μ_F indicates the consistence of the whole set of empirical equations. When $\mu_F = 1$, the model is viewed as mathematically consistent. A model with a low μ_F is said to be conflicting and imprecise.

For a mathematically precise model, the solution is a set of variables X which can completely satisfy the equation set. For a fuzzy model, however, a set of variable, X^*, is defined as the solution when it minimizes inconsistency among the empirical equations, or equivalently, maximizes membership of the fuzzy model:

$$\mu_F(X^*) = \max_X \mu_F(X) \tag{5}$$

Equation (5) shall be used recursively to get the set of optimal variables X^* when evaluating individual membership μ_i. Through the equations from (2) to (5), we have mapped a modelling problem into a functional maximization problem (Qian, Tessier, and Dumont, 1993a). That is in principle much easier to solve. As shown later in this paper, this also facilitates the merging of the modelling phase into an integrated optimization algorithm.

2.2. Non-linear Membership Functions for Fuzzy Constraints

There are different types of constraints, such as operating condition limitations and product quality specifications. In conventional approach a constraint is represented as a crisp step function:

$$\mu_C(x) = \begin{cases} 0 & x < x_0 \\ 1 & x \geq x_0 \end{cases} \tag{6}$$

where $\mu_c = 1$ stands for a feasible region, while $\mu_c = 0$ stands for a forbidden region (Figure 1). When a variable x is too close to the boundary, a slight disturbance can result in totally different consequents. Process models operated on this kind of constraints would behave sensitively in those boundary regions.

The constraints in an industrial system might be vague in several ways. In the fuzzy-logic-based approach, constraints are represented as "soft" constraints. When x moves away from the inside to the outside of a constrained region, the membership of the fuzzy set changes gradually from 1.0 to 0.0, meaning from fully allowable to completely forbidden.

Linear membership functions could be defined by fixing two points, the upper and lower bounds of the tolerance interval. One approach to handle nonlinear membership functions is to approximate them by piecewise linear functions. We suggest to use sigmoid functions:

$$\mu_C(x) = \frac{1}{1 + \exp\left[-(x - x_0)/\tau\right]} \tag{7}$$

where τ determines the weight or importance of the constraint: The smaller the τ,

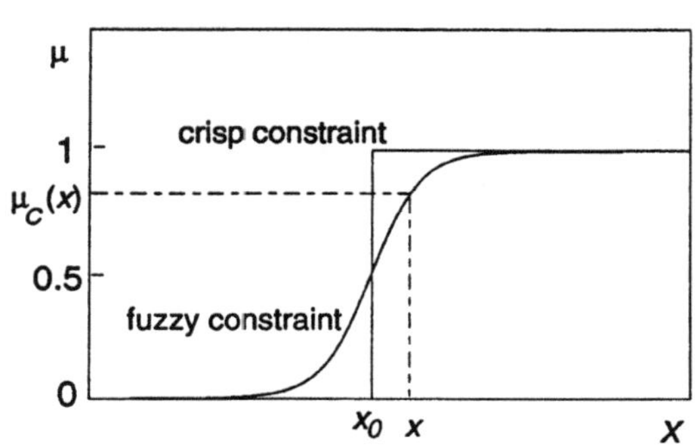

Figure 1. Crisp constraint and fuzzy constraint.

the higher the importance of the constraint. When $\tau \to 0$, the constraint becomes crisp, i.e. no violation is allowed. Comparison between a crisp constraint and a fuzzy constraint is shown in Figure 1. The sigmoid function has the following properties:

1. $\mu_c(x)$ is monotonously increasing and the derivative of $\mu_c(x)$ is continuous;
2. $\mu_c(x) = 1/2$ when $x = x_0$;
3. $\mu_c(x)$ is strictly convex on $(-\infty, x_0)$ and strictly concave on (x_0, ∞).
4. $\forall x \in \mathbf{R}$, $0 < \mu_c(x) < 1$ and $\mu_c(x)$ approaches asymptotically $f(x) = 0$ and $f(x) = 1$, respectively.

These properties facilitate the solution of the resultant optimization problem. The membership function for an upper bound constraint can be defined accordingly. The decisive constraint depends upon the toughest constraint in the system:

$$\mu_C(\mathbf{x}) = \min\{\mu_{C_1}(\mathbf{x}), \ldots, \mu_{C_k}(\mathbf{x})\} = \min_i \{\mu_{C_i}(\mathbf{x})\} \tag{8}$$

2.3. Maximizing Set of the Objective Function

Let g: $X \to \mathbf{R}$ be the objective function, R is the fuzzy feasible region, $S(R)$ is the support of R: $S(R) = \{ X \mid \mu_R(X) > 0 \}$. R_λ is the λ-level cut of R for $\mu_R(X) > \lambda$, R_1 is the kernel of R, namely the λ-level cut of R when $\lambda=1$.

Given the solution space R, the membership of the objective function $g(X)$ is defined as:

$$\mu_G(X) = \begin{cases} 0 & g(X) \le \sup_{R_1} g \\ \left(g(X) - \sup_{R_1} g\right) \Big/ \left(\sup_{S(R)} g - \sup_{R_1} g\right) & \sup_{R_1} g < g(X) < \sup_{S(R)} g \\ 1 & \sup_{S(R)} g \le g(X) \end{cases} \tag{9}$$

This is a linear monotonic function.

2.4. Symmetric Fuzzy Decision-Making Optimization

Engineers and mill managers make decisions in the presence of uncertainty and imprecise information. Decision-making objectives can be represented as a fuzzy set D. Decision is made upon the compromising solution from the fuzzy model $\mu_F(X)$, the fuzzy constraints $\mu_C(X)$, and the fuzzy objective function $\mu_G(X)$.

Bellman and Zadeh (1970) proposed that in a fuzzy environment, goals and constraints have the same nature and can be represented by fuzzy sets on X. This has been done in our approach for all three sub-sets. A fuzzy decision D is obtained from the intersection of the fuzzy sub-sets G, F and C:

$$D = (\cap C_i) \cap (\cap G_j) \cap (\cap F_k), \tag{10}$$

that is,

$$\forall x \in X, \ \ \mu_D(x) = \min\left[\min_{i=1,M} \mu_{C_i}(x), \ \min_{j=1,L} \mu_{G_j}(x), \ \min_{k=1,N} \mu_{F_k}(x)\right] \tag{11}$$

This model is symmetric. It does not distinguish between constraints and objectives. When criteria and constraints have unequal importance, membership functions can be weighted by *x*-dependent coefficients α_i, β_j and γ_k such that

$$\forall x \in X, \qquad \sum_{i=1}^{m} \alpha_i(x) + \sum_{j=1}^{n} \beta_j(x) + \sum_{k=1}^{l} \gamma_k(x) \ = \ 1\,, \tag{12}$$

and we have similar to Bellman and Zadeh (1970):

$$\mu_D(x) = \sum_{i=1}^{m} \alpha_i(x)\, \mu_{C_i}(x) + \sum_{j=1}^{n} \beta_j(x)\, \mu_{G_j}(x) + \sum_{k=1}^{l} \gamma_k(x)\, \mu_{f_k}(x)\,. \tag{13}$$

Note that *D* satisfies the property of *the convex combination of fuzzy sets* (Zadeh, 1965):

$$(\cap C_i) \cap (\cap G_j) \cap (\cap F_k) \ \subseteq D \subseteq (\cup C_i) \cup (\cup G_j) \cup (\cup F_k) \tag{14}$$

Process optimization and product quality control can be reached at a maximal decision set M_f, which maximizes the membership of the general fuzzy set *D(X)*:

$$M_f = \{\, X^*,\ \mu_D(X^*) \geq \mu_D(X),\ \forall x \in X \,\} \tag{15}$$

Or in a simpler representation form,

$$\mu_D\,(X^*) \ = \ \max_X\ \mu_D\,(X) \tag{16}$$

The work described above constitutes a systematic approach to modelling and optimizing poorly-understood industrial processes. The original mathematical programming form (Equation 1) is now transformed into a search for a maximum in a continuous variable space $\mathbf{R}^n$. Due to the non-monotonic and local maximal properties of the membership functions in fuzzy relation model *F*, and the existence of inconsistency among fuzzy equations, the search space for the maximization problem is poorly-structured and may present local maxima. Conventional derivative-based optimization algorithms turned out to be incapable of finding the global maximal. To solve this fuzzy optimization problem, we use a genetic algorithm (Goldberg, 1989) as the search strategy. Genetic algorithms

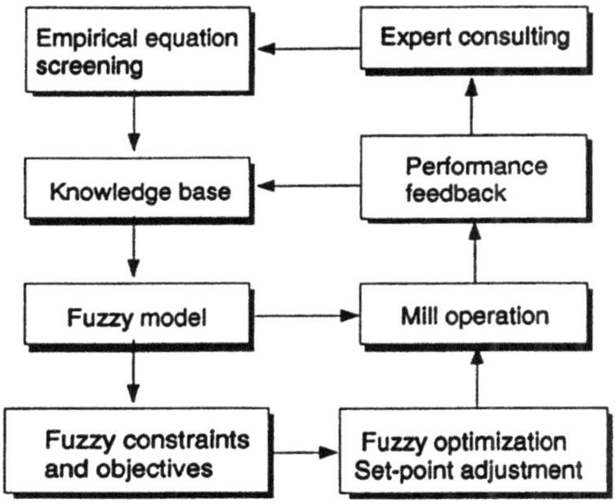

Figure 2. Schematic of fuzzy-logic-based modelling and optimization.

have been shown to behave well in multimodal function optimization (Qian et al., 1993b). For a poorly-structured system with 20 variables and 42 equations as described later in this paper, it takes around 25 minutes for the genetic algorithm to find the global optimal on a Sun SPARC Station 2.

The proposed fuzzy modelling and optimization methodology for an industrial process is schematically shown in Figure 2. The application of this new approach to a wood chip refining process is described in detail in the next section.

3. Refining Process and Pulp Quality Optimization

The approach described above has been applied to a fuzzy model of an industrial wood chip refining system. A 175 ton/day thermo-mechanical pulping line in a western Canadian newsprint mill was chosen as a case study. Shown in Figure 3

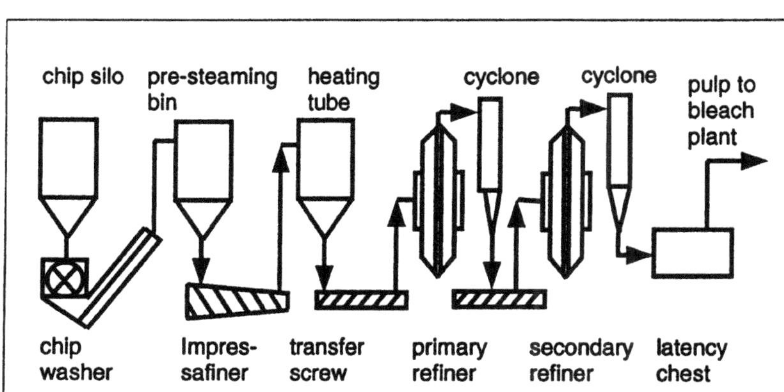

Figure 3. Schematic of a two-stage chemi-thermo-mechanical chip refining plant.

is a schematic representation of the process. The line uses hemlock and balsam wood chips. The wood chips are delivered from the chip silos to the chip washer, where sand, heavy materials and contaminants are removed. A metering screw feeder feeds the washed chips to the presteaming bin, where they are heated with steam during approximately 20 minutes. The chips are then compressed into an impressafiner and impregnated with a solution of sodium sulphite. In the heating tube, the chips are heated to a higher temperature under pressure. From there they fall into a surge tube and are sent to the first refining stage by means of a transfer screw feeder. After the primary stage, the wood pulp goes to a cyclone where pulp and steam are separated. The pulp then goes through a second refiner. The steam is separated from the pulp in another cyclone and the pulp is finally stored into a storage tank with a retention time of approximately 20 minutes for latency removal before it is sent to the paper machine.

3.1. Relationships and Interactions among Refiner Variables

The interactions between the process variables are extremely complex. The pulp quality, most of the time expressed in terms of freeness (which is a measurement of the drainability of the pulp), fibre fractions, and hand-sheet properties (tear, burst, breaking length, scattering coefficient, etc.), depends upon the raw fibres

Table 1. Variables in the chip refining process

1. Chip quality	2. Fibre characteristics
Wood species	Fibre strength
Chip density (CD)	Average fibre length (FL)
Chip size distribution	Fibre coarseness (FC)
Chip moisture content	Fibre length distribution
	Shive content
3. Operating condition	**4. Pulp and paper quality**
Production rate (PR)	Canadian standard freeness (CSF)
Screw feeder speed	Tear index (TI)
Specific energy (SE)	Tensile index (TN)
Motor load (ML)	Burst index (BU)
Plate gap (PG)	Breaking length (BL)
Hydraulic pressure	Sheet density (SD)
Plate age (PA)	Brightness (BR)
Refining consistency (CN)	Scattering coefficient (SC)
Dilution water flow-rate	Opacity (OP)
Steaming temperature	

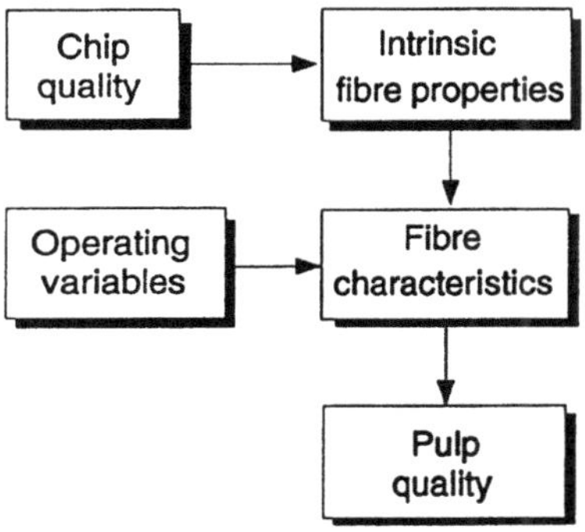

Figure 4. Interactions between the different groups of variables.

intrinsic properties such as fibre strength, length, coarseness, specific surface. Fibre properties are modified during the plant operation and the thermal, chemical, and mechanical treatment of the wood. Table 1 lists the most commonly used variables to represent wood chip refiner operation.

Mill operators generally use screw feeder speed, dilution water flow-rate, and motor load as the three main operating variables to control the refiners, and thus the pulp quality. The screw feeder speed is used to set the production rate, the dilution water is used to control the refining consistency of the pulp between the plates of the refiners, while the motor load is used to change the specific energy which characterizes the mechanical treatment received by the fibres. Plate age is also an important independent variable which will significantly affect the refiner operation, but has a more slower effect. Besides these variables, a number of other process variables also affect the refining but to a lesser degree. To start with a simple case, only production rate (PR), specific energy (SE), and refining consistency (CN) are viewed as manipulated variables in the model, while other variables are assumed constant.

For the pulp quality, we use a set of inequalities to indicate the upper and lower bounds. Pulp quality specifications depend upon the end-products (newsprint, tissue, ...). We selected fibre strength, length, coarseness, the Bauer McNett fibre fractions and the shive content to characterize the fibres (Table 1).

After analyzing and simplifying the system, we kept twenty four variables for inclusion into our model. The interactions between these different groups of variables are shown in Figure 4. A mathematical model describing the input-output relationships among the process variables would be extremely useful for pulp quality optimization and control. Much research has focused on the subject, but unfortunately, such a sophisticated model has yet to be developed.

Our main objectives in this application were the construction of a fuzzy model to describe the interactions between the different variables and the optimization of the process operating conditions and pulp quality. A total of forty five pieces

Table 2. Some of the empirical equations describing the chip refining process

$\Delta PG = -.78\Delta SE -.184\Delta PR -4e\text{-}5\Delta PA +.01\Delta CN$
$\Delta CSF = -250\ \Delta SE - 104\ \Delta PR - 0.05\ \Delta PA$
$\Delta CSF = -200\ \Delta SE - 32\ \Delta PR$
$CSF = 1130\ \exp[-.2 - .033FL(1+(2 -.11CN)SE)]$
$TI = 0.103\ CD +11.5\ FL - 57.9$
$\Delta BU = 1.2\ \Delta SE - 0.00015\ \Delta PA$
$\Delta BL = 1.8\Delta SE -0.11\Delta PR - .0023\ \Delta PA$
$BL=0.11\ SD +5.8\ FL - .99FL^2 -7.01$
$\Delta SD = -40.2\Delta FL -0.58\Delta CD$
$\Delta SC = -0.56\ \Delta SD - 16.8\ \Delta FC$

(Definition of the abbreviations are given in Table 1)

of process-related knowledge sources were collected from the literature, technical reports, and mill data analysis. They were then represented as a set of approximate equations. Since the empirical equations were derived under different operating conditions, it is not surprising that some of them might be imprecise or even in contradiction with others.

We tested the memberships of the 45 empirical equations in our model. There were three equations with memberships lower than 0.45, which were removed from the model. The 42 remaining empirical equations with memberships between 0.7 and 1.0 form the fuzzy model of the wood chip refining process. This fuzzy model constitutes the basis on which the simulation and optimization are developed. Some of the empirical equations used in the fuzzy model are listed in Table 2.

Two different types of optimization were performed. The first one tries to minimize the energy consumption while maintaining the same pulp quality. The second one tries to improve pulp quality while keeping the operating conditions within certain constraints.

3.2. Energy Consumption Minimization

The first optimization simulation for the chip refining process is shown in Table 3. In Test 1, we looked at the possibility of reducing energy consumption by adjusting other operating variables, while maintaining a constant breaking length. This can be achieved by slightly increasing the production rate. However, reduction in tear index was also observed. Typically for the same breaking length a 4.5% increase in production rate resulted in a 8.2% decrease in tear index.

In test 2, we tried to minimize the energy consumption while maintaining the pulp quality within the following window:

$$\begin{aligned} 100\,\text{ml} \le CSF &\le 120\,\text{ml} \\ TI &\ge 9.10\,\text{mN}\,\text{m}^2/\text{g} \\ BU &\ge 1.70\,\text{kPa}\,\text{m}^2/\text{g} \end{aligned} \quad (17)$$

Table 3 summarizes the results of the optimization. the Canadian standard freeness, tear index, and burst remain in the specified window, while the production rate increased by 6% and the energy consumption was reduced by 3.6%. These results were achieved by increasing the inlet consistency from 25% to 26.7%. Assuming the chip moisture was 52% before the refiner, the increase in consistency can be interpreted as a 13% decrease in the dilution water flow rate. As for the other properties (sheet density, brightness, and scattering coefficient), the changes were within 2%. However, the price to pay for the 3.6% decrease in specific energy is a 6.9% decrease in the breaking length.

Table 3. Energy consumption minimization

Test 1

variable	initial	goal or constraints	optimal
Production rate, odmt/d	175	$170 \lesssim PR \lesssim 190$	183
Specific energy, Mwhr/odmt	1.825	minimization	1.78
Consistency, %	25	fixed	25
Canadian standard freeness, ml	110		107
Tear index, $mN.m^2/g$	9.6		8.8
Burst index, $kPa.m^2/g$	1.9		1.9
Breaking length, km	3.1	fixed	3.1

Test 2

Variable	Initial	Goal or constraints	Optimal
Production rate, odmt/d	175	$170 \lesssim PR \lesssim 190$	186
Specific energy, Mwhr/odmt	1.825	minimization	1.76
Consistency, %	25	$22 \lesssim CN \lesssim 28$	26.7
Canadian standard freeness, ml	110	$CSF \gtrsim 102$	118
Tear index, $mN.m^2/g$	9.6	$TI \gtrsim 9.1$	9.3
Burst index, $kPa.m^2/g$	1.9	$BU \gtrsim 1.7$	1.84
Breaking length, km	3.1	$2.8 \lesssim BL \lesssim 3.4$	2.9

In classical optimization problems, quite often a solution is found at the boundary of the constraint conditions. In an imprecise system, however, the risk of violating the constraints is high. In our approach, as described in the previous section, constraints are viewed as soft restrictions, which are represented as descending or ascending functions. When the system approaches the boundary too closely, the constraint functions get low fitness values, and the optimization procedure forces the system away from the constraint boundary. In this example, constraints are $CSF < 120$, $TI > 9.1$, and $BU> 1.7$. While the optimal solution for the three variables, as shown in Table 3, are $CSF = 118$, $TI = 9.3$, and $BU = 1.84$, respectively. It gives the process some degree of tolerance to cope with stochastic fluctuation of process variables.

3.3. Pulp Quality Optimization

Once again, our objective is to maximize pulp quality while maintaining the manipulated operation variables within a desired range. However it is well known

Table 4. Burst index Maximization subject to freeness lower bound constraint

Scheme 1

Variable	Initial	Goal or constraints	Optimal
Production rate, odmt/d	175	fixed	175
Specific energy, Mwhr/odmt	1.825	$1.70 \lesssim SE \lesssim 1.95$	1.85
Consistency, %	25	fixed	25
Canadian standard freeness, ml	110	$CSF \gtrsim 100$	101
Tear index, $mN.m^2/g$	9.6	$TI \gtrsim 9.4$	9.93
Burst index, $kPa.m^2/g$	1.9	maximization	2.05
Breaking length, km	3.1	$BL \gtrsim 3.1$	3.28

Scheme 2

Variable	Initial	Goal or constraints	Optimal
Production rate, odmt/d	175	fixed	175
Specific energy, Mwhr/odmt	1.825	$1.70 \lesssim SE \lesssim 1.95$	1.86
Consistency, %	25	$22 \lesssim CN \lesssim 28$	26.3
Canadian standard freeness, ml	110	$CSF \gtrsim 100$	102
Tear index, $mN.m^2/g$	9.6	$TI \gtrsim 9.4$	9.49
Burst index, $kPa.m^2/g$	1.9	maximization	2.07
Breaking length, km	3.1	$BL \gtrsim 3.1$	3.19

that all properties cannot be improved at the same time. Compromises must be made and should depend upon the customer's requirements.

As an example, we tried to achieve a higher burst index while keeping freeness not below 110 ml (90% of its initial value). The fuzzy optimization approach shows that this can be achieved by changing the specific energy alone (scheme 1), or by changing both the specific energy and the consistency (scheme 2). The results of the optimization are given in Table 4. In both cases, the burst index can be increased by 8–9%, and the drop in freeness was less than 10% .

4. Conclusions

A wood chip refining model based on fuzzy relation equations has been developed. This facilitates the maximum use of available empirical, imprecise, even conflicting information and data from different sources. The model can be used as a simulation and optimization tool for a refiner process, either to minimize the refiner energy consumption while maintaining a stable pulp quality window, or to optimize pulp quality within a reasonable operating range. The model can be implemented in a fuzzy controller and incorporated into a conventional control system in complex processes. Operation experience, more reliable sensor readings, and operator's observations can all be useful for a better control decision-making system. It should not be used as a substitute to conventional control systems but be seen as a complement. It expands the use of all available information from quantitative measurement to qualitative information. Correspondingly, non-structured system knowledge, mostly represented as heuristic rules, could also be used by a fuzzy controller to achieve a better performance. An integrated control system is shown in Figure 5.

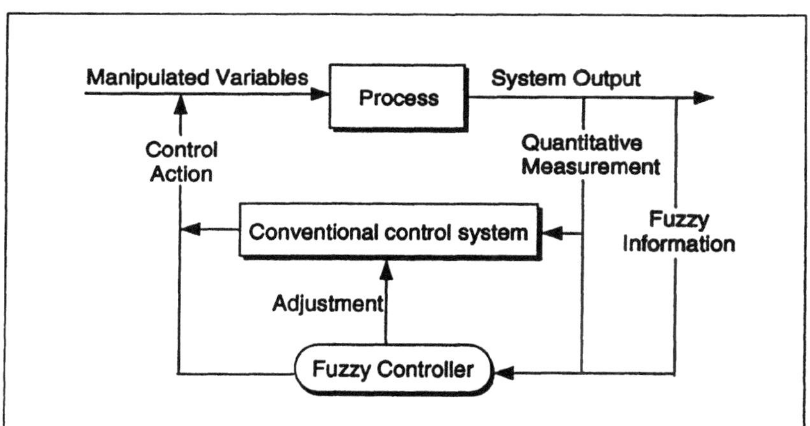

Figure 5. An integrated industrial control and optimization system.

The current model software is written in C language. When it is implemented in a specific mill, the knowledge base and the process model need to be modified and screened according to mill data and practice. A user interface can be developed for operator's convenience. When process data are provided, the model can predict the process performance. When process objectives or product specifications change, the model can make suggestions on how setpoints should be adjusted. An interface to the existing mill DCS would also be needed for the model to quickly respond to process disturbance or process changes.

Acknowledgment: The authors would like to acknowledge the financial support of the Mechanical and Chemimechanical Wood-Pulps Networks of Centres of Excellence, Canada.

References

1. Bellman, R.E. and L. A. Zadeh (1970) "Decision-making in a Fuzzy Environment", *Management Science*, **17B**, 141–164.
2. Dumont, G.A., et al. (1982) "Computer Control of a TMP Plant", *Pulp & Paper Canada*, *83*(8), 54-59.
3. Fedrizzi, M., J. Kacprzyk, and J. L. Verdegay (1991) " A survey of Fuzzy Optimization and Mathematical Programming". In M. Fedrizzi (Eds.): *Interactive Fuzzy Optimization*, Berlin: Spinger-Verlag.
4. Goldberg, D.E. (1989) *Genetic Algorithms*. Massachusetts: Addison-Wesley.
5. Pearson, A.J.(1983) "Towards a Unified Theory of Mechanical Pulping and Refining", International mechanical pulping conference, Washington D.C., 131-138.
6. Qian, Y., P.J.C. Tessier, and G.A. Dumont (1993a) "Process Modelling and Optimization of Systems with Imprecise and Conflicting Equations", *Engineering Application of Artificial Intelligence*, 6(1), 39-47.
7. Qian, Y., P.J.C. Tessier, and G. A. Dumont (1993b) "Fuzzy Logic Based Optimization and its Application". In M. Delgado and J. Kacprzyk (Eds.): *Fuzzy Optimization: Recent Advances*, Heidelberg: Physica-Verlag.
8. Wisniewski, Mik (1990) *Mathematical Programming: Optimization Models for Business and Management Decision-Making*, New York: McGraw-Hill.
9. Zadeh, L. A. (1965) "Fuzzy Sets", *Information and Control*, **8**, 94-102.
10. Zimmermann, H.-J. (1992) "Methods and Applications of Fuzzy Mathematical Programming". In R. R. Yage and L.A. Zadeh (Eds.): *An Introduction to Fuzzy Logic Applications in Intelligent Systems*. Boston: Kluwer Academic Publishers.

AN APPLICATION OF FUZZY STRUCTURAL MODELING TO RELATION DIAGRAM METHOD IN QUALITY CONTROL

Shin'ya NAGASAWA
Faculty of Business Administration, Asia University
Musashino, Tokyo 180, JAPAN

ABSTRACT: Fuzzy sets theory is relatively little known to the sphere of quality control(QC)/quality management (QM). In the clerical, sales and research and development departments, however, objective numerical data are sometimes hard to gather, and the need to handle subjective data arises quite often.

The Relation Diagram Method, one of "the Seven Management Tools for QC", is employed to solve a complex problem, such as out-of-contol caused by human factors in production process, by means of arrangement of cause and effect relationships of a problem. In applying the Relation Diagram Method to complex relationships among many elements of causes and/or results in the problem to be solved, it is often desirable and sometimes essential to create hierarchies. However, the process of rearranging elements in a hierarchy is not dealt with in the existing Relation Diagram Method.

In order to deal with such a relationship practically and systematically, the author proposes to create a hierarchy and rearrange the elements in a relation diagram by applying the Fuzzy Structural Modeling, which is developed on the basis of the fuzzy sets theory.

In order to demonstrate how the proposed method works, a practical example is given in which the structure of the relation diagram has been successfully identified and the usefulness of applying the Fuzzy Structural Modeling to relation diagrams is confirmed.

Keywords: fuzzy sets theory, fuzzy structural modeling quality control(QC), relation diagram method

1. INTRODUCTION

Fuzzy sets theory is relatively little known to the sphere of quality control(QC)/quality management(QM). In the clerical, sales and research and development departments, however, objective numerical data are sometimes hard to gather, and the need to handle subjective data arises quite often.

Since 1987 the author has proposed the following applications of fuzzy sets theory in QC/QM [1]:

1) evaluation of quality by fuzzy integrals
2) sensory evaluation by fuzzy structural modeling
3) statistical modeling by fuzzy decision making
4) Matrix Diagram Method by fuzzy composition
5) Relation Diagram Method by fuzzy structural modeling

The Seven Management Tools for QC, i.e., Affinity Diagram, Relation Diagram, Tree Diagram, Matrix Diagram, Matrix-Data Analysis, Process Decision Program Chart (PDPC) and Arrow Diagram, were proposed in 1977 by the Committee for Developing QC Tools (the Union of Japanese Scientists and Engineers, Chairman: Dr. Nayatani) by selecting various conventional techniques for creativity and management. These Seven Management Tools for QC are currently being used in Japan to implement total quality management (TQM) or company wide quality control (CWQC), not only by managers and staff personnel but also by some QC circles which have attained expertise in problem solving.

The objective of the Seven Management Tools for QC is to promote TQM more efficiently by organizing and analyzing language data, collected from all internal and external fields of the company, and applying a method based on language data to complete a plan or lead them to an idea [2].

The Relation Diagram Method, one of the Seven Management Tools for QC, is employed to solve a complex problem, such as out-of-control caused by human factors in production process, by means of arrangement of cause and effect relationships of a problem [3]. In searching for the cause of such a problem by applying the Relation Diagram Method, it is often desirable and sometimes essential to create hierarchies. But the process of rearranging elements in a hierarchy is not involved in the original Relation Diagram Method and is usually dealt with intuitively.

In order to deal systematically with such a problem described in a relation diagram, the author et al.[4] proposed to create a hierarchy and to rearrange elements in a relation diagram by the Interpretive Structural Modeling (ISM) method [5], which is developed on the basis of graph theory. It is considered that ISM is especially effective to get a structure of a relation diagram. In ISM, a matrix representing a subordination relationship among the elements of a problem is called a subordination matrix and the elements are given in a binary relation. However, it is not always reasonable to use the binary relation among many elements of causes and/or results in the problem to be solved in a relation diagram, because such a relation is often weighted in practice and connected by a thick arrow in a relation diagram.

On the other hand, the fuzzy sets concept gives an important mathematical clue for an approach to studies of such relationships having vague boundaries. In order to deal with these practically and systematically, the author proposes to create a hierarchy and rearrange elements in a relation diagram by the Fuzzy Structural Modeling (FSM) method [6][7], which is developed on the basis of the fuzzy sets theory. The entries in the fuzzy reachability matrix are given values in the interval [0,1] by virtue of a fuzzy binary relation. According to the matrix, a fuzzy digraph describes a contextual relationship among the elements of the relation diagram and can be transformed into an interpretive structural model with respect to the relationship.

In order to demonstrate how the proposed method works, a practical example is given.

2. RELATION DIAGRAM METHOD[3]

The Relation Diagram Method, by using the relation diagram and considering a complex problem through an organization process, has the following functions:

1) Grasping the problem as a poor result and thereby searching for the cause of the problem.
2) Elucidating the relationship between the phenomenon and the factor and thereby clarifying the structure of the problem.
3) Developing the means to achieve the purpose.

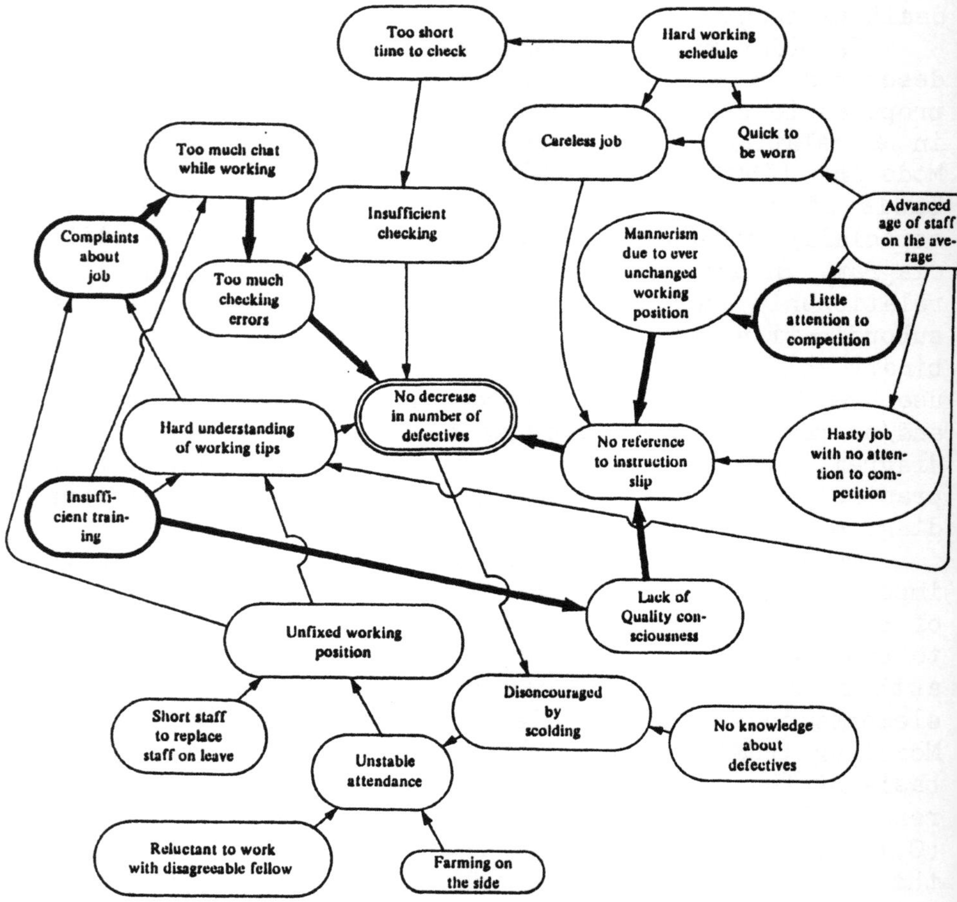

Fig.1 Relation Diagram Made for Searching for the Cause for Defectives in Assembly Line [3]

Fig.1 is a relation diagram made for searching for the cause for "never decreased number of defectives" in the assembly line of a certain household electric appliances manufacture.

The problem of "never decrease in number of defectives" means deterioration of consequence, and one of possible causes is, for example, "too much chat while working". When a relation diagram was made for the first time, topics of chatting were thought to be daily gossips. However, the result of study revealed that "complaints about job" accounted for the majority of topics of their chatting. Then, causes of "complaints about job" were searched by a relation diagram further, and each of the causes was removed. Consequently, it became possible to decrease defectives.

This example tells us how to find out major causes to decrease defectives from a production process to make a relation diagram or finised one. These major causes should be enclosed by thick lined and the process and the result should be connected by a thick arrow.

3. FUZZY STRUCTURAL MODELING [6][7]

On the basis of the fuzzy sets theory, Tazaki and Amagasa [6][7] proposed a method for structuring a hierarchy for problems in social science, and called it the Fuzzy Structural Modeling (FSM) method. An important requirement for the structural modeling of complex relation is that the necessary data is acquired and organized into a form which a structural model can be developed. A fuzzy reachability matrix is one such form. The main purpose of the Fuzzy Structural Modeling method is to describe and illustrate a formal procedure for constructing a graphical representation of the hierarchical arrangement, given the necessary information concerning the relationship between each element. The procedure permits automatic development of a graphical structure portraying the hierarchy. The entries in the matrix are given values in the interval [0,1], by virtue of a fuzzy binary relation. According to the matrix, a fuzzy digraph describes a contextual relationship among the elements of the system and can be transformed into an interpretive structural model of the system with respect to relation.

A significant step in the development of fuzzy structural modeling consists of the relaxation in the rules of the construction of a subordination matrix with respect to transitivity, irreflexivity and asymmetry. This relaxation extends the flexibility and applicability for the structural modeling of systems.

3.1 Preliminaries of Fuzzy Structural Modeling

Several properties with respect to fuzzy subsets and fuzzy relations are shown as mathematical preliminaries for fuzzy structural modeling.

A fuzzy binary relation and its complement in a direct product space $S \times S$ are denoted by f_R and $f_{\bar{R}}$, respectively, and are defined as follows:

$$f_R,\ f_{\bar{R}} : S \times S \to [0,1] \tag{1}$$

where the relation between f_R and $f_{\bar{R}}$ is given by

$$f_{\bar{R}}(s_i,s_j) = (1 - f_R(s_i,s_j))/(1 + \lambda f_R(s_i,s_j)) \tag{2}$$

where λ is a real number in the range in $-1 < \lambda < \infty$.

Let p be a real number given in the half-open interval $(0,1]$. According to such a value of p, some definitions with respect to the fuzzy binary relations are given in the followings.

Definition 1.(fuzzy irreflexive law) When $f_R(s_i,s_i) < p$ for $\forall (s_i,s_i) \in S \times S$, the relation is called fuzzy irreflexive.

Definition 2.(fuzzy asymmetric law) When either $f_R(s_i,s_j)$ or $f_R(s_j,s_i)$ for $\forall s_i,s_j \in S$ $(i \neq j)$ is than p, the relation is called fuzzy asymmetric.

Definition 3.(fuzzy semi-transitive law) Let $m_{ik} = \bigvee_{j=1}^{n} (f_R(s_i,s_j) \wedge f_R(s_j,s_k)) \geq p$ for $\forall (s_i,s_j),(s_j,s_k) \in S \times S$ $(i \neq j \neq k)$. When $f_R(s_i,s_k) \geq m_{ik}$ for any (s_i,s_k), the relation is called fuzzy semi-transitive.

3.2 Rule for Structural Modeling and Algorithm

Let a system object be $S = \{s_1, s_2, \ldots, s_n\}$. A fuzzy subordination matrix A is constructed which represents a fuzzy subordination relation among the elements of S on the basis of a certain contextual relationship

$$A = [a_{ij}],\ i,j = 1,2,\ldots,n \tag{3}$$

where A is a square $n \times n$ matrix and the element a_{ij} of A is given by the fuzzy binary relation f_R as follows:

$$a_{ij} = f_R(s_i,s_j),\ 0 \leq a_{ij} \leq 1,\ i,j = 1,2,\ldots,n \tag{4}$$

This shows the grade of which s_i is subordinate to s_j. In order to show that the grade of subordination is greater than a certain given grade, a parameter p as a threshold is introduced. p must be given on the semi-open interval $(0,1]$.

Definition 4. When the fuzzy semi-transitive law is satisfied in a fuzzy subordination matrix, A is called a fuzzy semi-reachability matrix.

Definition 5. A row (column) is said to be regular if it contains only a single a_{ij} which satisfies $a_{ij} \geq p$ in $A^{(k)}$, where a superscript k shows kth hiearachy.

The algorithm to obtain the fuzzy semi-reachability matrix is given by Tazaki and Amagasa [6].

The procedure of the structural modeling algorithm is given by the following steps.

Step 1. Give a fuzzy subordination matrix $A = [a_{ij}]$ and construct the fuzzy semi-reachability matrix A' satisfying the fuzzy semi-transitive law from A.
Step 2. Identfy the top level set $L_t(s)$, the bottom level set $L_b(s)$ and the isolation level set $L_{is}(s)$ on the basis of A'. Furthermore, determine the subordination relation sets $B(s_i)$ between $L_t(s)$ and $L_b(s)$ $(s_i \in L_b(s))$ and the block sets $\{Q_j\}$.
Step 3. Eliminate all of the rows including elements belonging to $L_t(s)$, the columns including elements belonging to $L_b(s)$ and the rows and columns including elements belonging to $L_{is}(s)$. The fuzzy subordination matrix, consisting of the remaining rows and columns, is reconstructed as A'.
Step 4. From A' obtained in Step 3, construct the single hierarchy matrix $A^{(j)}$ corresponding to each block set Q_j.
Step 5. Set up a fuzzy structure parameter λ and identfy the graphical structure concerned with each single hierarchy matrix $A^{(j)}$ according to the flowchart given by Tazaki and Amagasa [6].

At this point, assume that the regular rows corresponding to s_j are s_{ik}, $k = 1,2,\ldots,m$ $(m < n)$. For such regular rows, the following operation is carried out:

$$[a_{\cdot j}^*] = [a_{\cdot j}] \wedge [\overline{a_{\cdot i1}}] \wedge \ldots \wedge [\overline{a_{\cdot im}}] \tag{5}$$

This enables all of the rows s_{ik}, $k = 1,2,\ldots,m$ to be eliminated by replacing $a_{\cdot j}$ with $a_{\cdot j}^*$. The same operation can be also applied to the columns.

4. ILLUSTRATIVE EXAMPLE

The following example is given in order to demonstrate how the proposed method works.

Let the relation diagram in Fig.2 represent a certain problem to be solved. The relation among many of its elements of causes and/or results is practically weighted and connected by a thick arrow. The thicker the arrow, the stronger a subordination relation. A matrix representing a subordination relation among the elements of the problem is described for convenience as shown in (6), which corresponds to the subordination matrix in ISM. The weighted entries between × and ◉ in the matrix are to be given values in the interval [0,1] by virtue of a fuzzy binary relation.

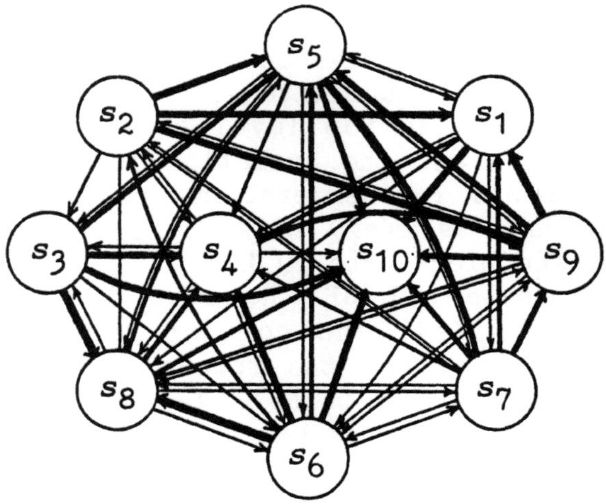

Fig.2 Example of Relation Diagram of a Specific Problem

$A =$

	s_1	s_2	s_3	s_4	s_5	s_6	s_7	s_8	s_9	s_{10}
s_1	×	×	×	○	△	△	△	△	×	◎
s_2	◎	×	△	△	◎	×	△	△	◎	◎
s_3	×	×	×	◎	◎	△	×	◎	×	◎
s_4	△	△	△	×	×	△	×	△	△	△
s_5	△	×	△	○	×	△	△	△	△	**○**
s_6	×	○	×	◎	**○**	×	△	◎	△	◎
s_7	**○**	△	×	○	◎	△	×	△	**○**	**○**
s_8	×	×	△	◎	**○**	△	△	×	△	**○**
s_9	◎	△	×	○	◎	△	×	○	×	◎
s_{10}	×	×	×	×	×	×	×	×	×	×

(6)

By applying the algorithm described in the previous section, we construct the structural model for A.

<u>Step 1.</u> Let $A = [a_{ij}]_{10\times10}$ be the fuzzy subordination matrix for the given problem to be solved as shown in (7). A is equivalent to the fuzzy semi-reachability matrix A' satisfying the fuzzy semi-transitive law.

$A' =$

	s_1	s_2	s_3	s_4	s_5	s_6	s_7	s_8	s_9	s_{10}
s_1	.0	.0	.0	.3	.1	.1	.1	.1	.0	.7
s_2	.7	.0	.1	.1	.7	.0	.1	.1	.7	.7
s_3	.0	.0	.0	.7	.7	.1	.0	.7	.0	.7
s_4	.1	.1	.1	.0	.0	.1	.0	.1	.1	.1
s_5	.1	.0	.1	.3	.0	.1	.1	.1	.1	.5
s_6	.0	.3	.0	.7	.5	.0	.1	.7	.1	.7
s_7	.5	.1	.0	.3	.7	.1	.0	.1	.5	.5
s_8	.0	.0	.1	.7	.5	.1	.1	.0	.1	.5
s_9	.7	.1	.0	.3	.7	.1	.0	.3	.0	.7
s_{10}	.0	.0	.0	.0	.0	.0	.0	.0	.0	.0

(7)

Table 1 Subordination Relation between $L_b(s)$ and $L_t(s)$

$s_i \in L_b(s)$	$B(s_i) \subset L_t(s)$
s_2	s_{10}
s_3	s_4 , s_{10}
s_6	s_4 , s_{10}
s_7	s_{10}

Step 2. The top level set $L_t(s)$ is identified as $\{s_4, s_{10}\}$, the bottom level set $L_b(s)$ as $\{s_2, s_3, s_6, s_7\}$ and the isolation level set $L_{is}(s)$ as null on the basis of A'. Furthermore, the four subordination relation sets $B(s_i)$ between $L_t(s)$ and $L_b(s)$ $(s_i \in L_b(s))$ are determined as shown in Table 1 and unique block set Q_1 as $\bigcup_{i=1}^{4} B(s_i) = \{s_4, s_{10}\}$.

Step 3. All of the rows including elements belonging to $L_t(s) = \{s_4, s_{10}\}$ and the columns including elements belonging to $L_b(s) = \{s_2, s_3, s_6, s_7\}$ are eliminated. The fuzzy subordination matrix consisting of the remaining rows and columns is reconstructed as A' as shown in (8).

$$A' = \begin{array}{c} \\ s_1 \\ s_2 \\ s_3 \\ s_5 \\ s_6 \\ s_7 \\ s_8 \\ s_9 \end{array} \begin{array}{c} \begin{array}{cccccc} s_1 & s_4 & s_5 & s_8 & s_9 & s_{10} \end{array} \\ \left[\begin{array}{cccccc} .0 & .3 & .1 & .1 & .0 & .7 \\ .7 & .1 & .7 & .1 & .7 & .7 \\ .0 & .7 & .7 & .7 & .0 & .7 \\ .1 & .3 & .0 & .1 & .1 & .5 \\ .0 & .7 & .5 & .7 & .1 & .7 \\ .5 & .3 & .7 & .1 & .5 & .5 \\ .0 & .7 & .5 & .0 & .1 & .5 \\ .7 & .3 & .7 & .3 & .0 & .7 \end{array}\right] \end{array} \tag{8}$$

Step 4. The single hierarchy matrix $A^{(1)}$, corresponding to the unique block set Q_1, is equal to A' obtained in Step 3.

Step 5. We identify the graphic structure concerned with the unique single hierarchy matrix $A^{(1)}$ according to the flowchart given by Tazaki and Amagasa [6].

The regular rows corresponding to s_{10} are s_1 and s_5. For these regular rows, the following operation is carried out according to formula (5).

$$[a_{\cdot 10}{}^*] = [a_{\cdot 10}] \wedge [\overline{a_{\cdot 1}}] \wedge [\overline{a_{\cdot 5}}] \tag{9}$$

Next, all of the rows $s_1.$ and $s_5.$ are eliminated by replacing $a_{\cdot 10}$ with $a_{\cdot 10}{}^*$. As a result of this operation, the following matrix is obtained.

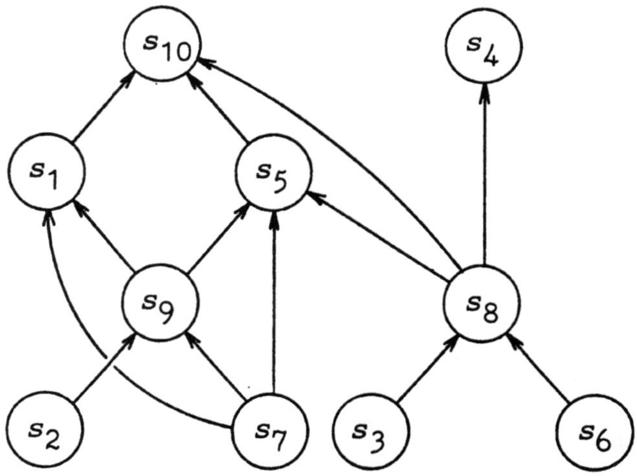

Fig.3 Graphical Structure of the Example (p = 0.5, λ =-0.3 (FSM))

$$A_1^{(1)} = \begin{array}{c} \\ s_2 \\ s_3 \\ s_6 \\ s_7 \\ s_8 \\ s_9 \end{array} \begin{array}{c} \begin{array}{cccccc} s_1 & s_4 & s_5 & s_8 & s_9 & s_{10} \end{array} \\ \left[\begin{array}{cccccc} .7 & .1 & .7 & .1 & .7 & .38 \\ .0 & .7 & .7 & .7 & .0 & .38 \\ .0 & .7 & .5 & .7 & .1 & .59 \\ .5 & .3 & .7 & .1 & .5 & .38 \\ .0 & .7 & .5 & .0 & .1 & .5 \\ .7 & .3 & .7 & .3 & .0 & .38 \end{array} \right] \end{array} \qquad (10)$$

Then, identfy the graphical structure concerned with the unique single hierarchy matrix $A^{(1)}$ in the same way until all of the rows and columns of $A^{(1)}$ are eliminated.

The regular rows (columns) denote the unique subordination relation between elements of causes and/or results in a relation diagram. The subordination relation and its direction correspond to node and branch in graph theory, respectively. Therefore, the graphical structure of this case is as shown in Fig.3.

The constructed graphical representation of the hierarchical rearrangement shown in Fig.3 by applying the Fuzzy Structural Modeling method given the necessary information concerning the relationship between each element in the original relation diagram shown in Fig.2.

A practical example has been given and the structure of the given relation diagram has been successfully identified. Thus, the proposed method works well [8][9].

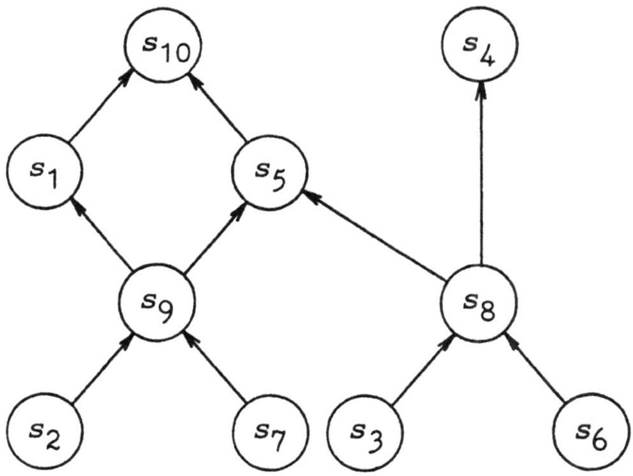

Fig.4 Graphical Structure of the Example
(p = 0.5, λ = 0.5 (FSM) and p = 0.5 (ISM))

5. DISCUSSIONS

In order to investigate the effect of changing the fuzzy structure parameter λ, a further case of $p = 0.5$ and $\lambda = 0.5$ is illustrated using the same matrix A. The graphical structure of this case is shown in Fig.4. The change of the fuzzy structure parameter λ produces no effect on the basic structure of the relation diagram but has an effect on the subordination of each element to each other. On the other hand, the change of the threshold p produces an effect on the basic structure of the relation diagram. Consequently, the structure of a relation diagram can be uniquely determined depending on a threshold p and a structure parameter λ.

Furthermore, in order to compare the result of the Fuzzy Structural Modeling (FSM) method with that of the Interpretive Structural Modeling (ISM) method, we put $p = 0.5$ ($\lambda = 0$) in Fig.4. The obtained structure coincides with that of the FSM case ($p = 0.5$, $\lambda = 0.5$). As a result, the same structure given by ISM can be also obtained from special values of the threshold p and the structure parameter λ [10].

The method proposed in this paper for structuring a hierarchy practically and systematically for a complex relation diagram by using the FSM method is applicable even if a cyclic path appears in the structure [11].

Furthermore, many man-hours are required to generate the subordination matrix. For the purpose of saving time, Tazaki and Amagasa [6] proposed an automatic matrix generating method.

6. CONCLUSION

Hierarchy is fundamental to many fields of science, engineering and human situations. It is one of several basic concepts needed in working with complex systems. In this paper a method is proposed for structuring such a hierarchy practically and systematically for a complex relation diagram by applying the Fuzzy Structural Modeling (FSM) method, developed on the basis of the fuzzy sets theory. Fuzzy structural modeling is a method for developing hierarchy from a fuzzy semi-reachability matrix, which is the fuzzy subordination matrix satisfying the fuzzy semi-transitive law.

A significant step in the development of fuzzy structural modeling consists of the relaxation in the rules of the construction of a subordination matrix with respect to transitivity, irreflexivity and asymmetry. This relaxation extends the flexibility and applicability for the structural modeling of a relation diagram. Furthermore, it has been shown that the structure of a relation diagram can be uniquely determined, depending on a threshold p and a structure parameter λ, and in particular the same structure given by Interpretive Structural Modeling (ISM) can be be also obtained from special values of p and λ.

In order to demonstrate how the proposed method works, a practical example has been given and the structure of the system has been successfully identified and the usefulness of applying the Fuzzy Structural Modeling to relation diagrams is confirmed.

ACKNOWLEDGEMENTS

The author would like to show his deep appreciation to Dr. E.Tazaki and Dr. M Amagasa for their suggestions to him relating to the Fuzzy Structural Modeling. Also, many thanks to the editors for their useful advice.

This study is financially supported by the Grant-in-Aid for Scientific Research on Priority Areas, the Japanese Ministry of Education, Science and Culture.

REFERENCES

[1] Nagasawa,S.: "Fuzzy Sets - Theory and Application -" (in Japanese), *Hinshitsu Kanri (Statistical Quality Control)*, Vol.43, No.8, pp.83-89, 1992.
[2] Nayatani,Y.:"Seven Management Tools for QC", *Reports of Statistical Application Reseach, JUSE* (The Union of Japanese Scientists & Engineers), Vol.33, No.2, pp.1-6, 1986.
[3] Futami,R.: "The Outline of Seven Management Tools for QC", *ibid.*, Vol.33, No.2, pp.7-26, 1986.
[4] Ootaki,A., Nagasawa,S., Uehara,M. and Ichikawa,T.: "Hierarchical Structuring of Relation Diagram by Interpretive Structural Modeling" (in Japanese), *Proceedings of JSQC 26th Research Congress* (Japanese Sciety for Quality Control), pp.31-35, 1984.
[5] Warfield,J.N. et al.: *A Unified Systems Engineering Concept*, Battelle M. Innst, 1972.
[6] Tazaki,E. and Amagasa,M.: "Structural Modeling in a Class of Systems Using Fuzzy Sets Theory", *Fuzzy Sets and Systems*, Vol.2, No.1, pp.1-17, 1978.
[7] Tazaki,E. and Amagasa,M.: "Structural Modeling with Consensus for Societal Systems Using Fuzzy Sets Theory", *Proceedings of IFAC/81 Congress*, pp.V-35-40, 1981.
[8] Nagasawa,S.: "Application of Fuzzy Theory to Quality Control (4) - Hierarchical Structuring of Relation Diagram by the Fuzzy Structural Modeling -" (in Japanese), *Proceedings of JSQC 18th Annual Congress*, pp.33-36, 1988.
[9] Nagasawa,S.: "An application in Quality Control for Performance Improvement", *Proceedings of 1st Asian Fuzzy Systems Symposium*, pp.1-6, 1993.
[10]Nagasawa,S.:"Application of Fuzzy Theory to Relation Diagram", *Transactions of 45th Annual Quality Congress* (American Society for Quality Contol), pp. 573-578, 1990.
[11]Nagasawa,S.: "Relation Among Evaluation Methods by Paired Comparisons - On AHP, Fuzzy Structural Modeling and Scheffé's Method -" (in Japanese), *Proceedings of 9th Fuzzy Systems Symposium*, pp.89-92, 1993.

6

RELIABILITY OF MAN–MACHINE SYSTEMS

HUMAN RELIABILITY ANALYSIS WITH FUZZY INTEGRAL

Takashi Washio* and Masaharu Kitamura**

*Mitsubishi Research Institute, Inc.
1-8-1, Shimomeguro, Meguro-ku, Tokyo 153, Japan
**Tohoku University
Aramaki-aza-Aoba, Aoba-ku, Sendai 980, Japan

Abstract

A method of human reliability analysis (HRA) in the field of large scale plant operations is presented in this work. The fuzzy integral methodology was applied to provide a clear mathematical basis of the expert judgement required in human reliability analysis (HRA). Also, a method named the analytic hierarchy process (AHP) was introduced for supporting overall decision-making associated with the improvement of human reliability. The effectiveness of the proposed method is demonstrated through an application to decision-making on operability improvement of a process cooling system.

Key words: fuzzy integral, analytic hierarchy process, human factors, human reliability Analysis, man-machine interface

1. INTRODUCTION

The reliability of hardware utilized in large scale systems requiring high safety, e.g., aerospace and nuclear power, has been highly improved in recent years. As a consequence, the relative influence of human errors on the operation reliability of those systems is becoming higher. Hence, the human reliability analysis (HRA) has an important role in the decision-making associated with the system management including the operation and the maintenance.

However, the following difficulties prevent the direct utilization of the HRA in practical fields.

(1) The large scale systems generally involve many facilities based on new high technology. However, the empirical reliability data related with the new facilities have not been accumulated sufficiently.

(2) As the operators of the large scale systems are highly educated and trained, and have quite high moral, the human reliability data from conventional engineering domain should not be applied directly.

Accordingly, the data modification and the reliability estimation on the basis of subjective evaluation by human experts are widely conducted in human reliability analysis (HRA) of large scale systems. The representative methods for the subjective evaluation are THERP developed by Swain et al. (1990), SLIM by Embrey et al. (1984), STAHR by Phillips et al. (1983), and a method using Fuzzy Integral by Onisawa (1986). All of these methods evaluate the success probability (Ps) of each human task based on dominant factors, "performance shaping factors (PSFs)", influencing the human reliability in the task. THERP utilizes many empirical and subjectively modified tables relating Ps with PSFs. As many tasks in practical fields, especially tasks associated with new facilities as aforementioned, are not covered by the tables sufficiently, the applicability is limited to some conventional systems. The methods of SLIM and STAHR use human reliability models (HRMs) of linear functions to evaluate Ps from PSFs which coefficients are determined by subjective evaluation of human for each task. An advantage of these approaches is the relaxation of the limit on the applicable task domain, because the HRMs can be evaluated in wide area as far as the effective human expertise are available. However, the linear HRMs reduces its general appropriateness due to the mutual strong correlations among PSFs' contributions to Ps in practical fields. Onisawa's method provides an effective measure to resolve the issue of this nonlinearlity by introducing a fuzzy integral for HRMs which defines a weight of each contribution from multiple PSFs to Ps. However, as his method requires many empirical data, again the limited area of the applicable task becomes a significant problem.

Furthermore, the decision-making associated with the human reliability improvement requires the consideration of the multiple criteria, e.g., required cost and efficiency, besides the evaluated human error probability. This type of the synthetic decision-making has the following difficulty.

(3) Some of the decision criteria are qualitative, and have different scales not to be mutually compared directly.

Because of this difficulty, the decision-making are also often carried out subjectively by human experts. As such a subjective procedure lacks of any clear mathematical basis, logical foundations, assumptions, and justifications behind the analysis, the decision remains unclear to the people outside the decision procedure. In other words, the transparency of the decision-making is spoiled.

This work summarizes our attempts to alleviate these difficulties. A new approach of the fuzzy-integral-based HRA using human expertise was proposed for the application to the nonlinear human performance in wide task domain. Also, a systematic method of decision-support under multiple criteria, named as "analytic hierarchy process (AHP)" proposed by Satty (1980), was introduced.

2. DEMONSTRATION EXAMPLE

For the preparation to explain the proposed method comprehensively, the demonstration example used in this paper is described at first. The example is a decision problem to select the most appropriate option to improve the operation reliability of a task under multiple decision criteria. The task is to control coolant flow rate supplied to a water cooling tank as depicted in fig.1. The detail of the task is as follows.

Objective:

Settle the coolant flow rate at a certain appropriate level Fs.

State Indication:

The inlet flow rate Fi, the inlet temperature Ti, and the outlet temperature To of coolant are indicated on a control panel.

Restriction:

The difference between the inlet and the outlet temperature, Ti and To must be kept within a certain limit ΔT.

Manipulation Facility:

The valve opening is controlled by the electric motor rotation corresponding to the rotation angle of a rotary switch on a control panel.

Operator:

Currently, only one operator performs this task.

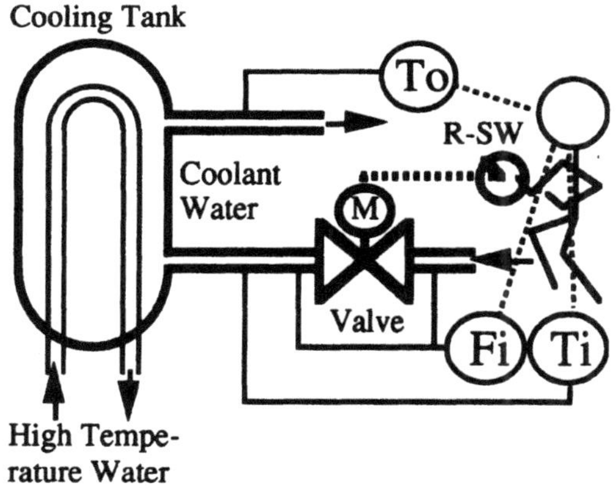

Fig.1 A task to control coolant supply to a water cooling tank.

This task enforces an operator to check multiple process signals and to manipulate a facility concurrently. Because of the complexity of this task, the frequency of operation error occurrence is known to be significant. Consequently, the administrative managers of this plant decided to improve the operation reliability of this task by selecting the most appropriate countermeasure from the following three options.

Option1 (O1): Introduce a reliable automated control system that the operator should just set a target flow rate once and monitor accidental machine faults.

Option2 (O2): Introduce a plant simulator to train the operator periodically.

Option3 (O3): Employ an assistant operator to monitor the process state.

3. OUTLINE OF METHOD

The example aforementioned is a typical decision problem associated with the human reliability of plant operation. This type of decision-making requires the consideration of multiple criteria, e.g., cost and operator's acceptance, besides the operation reliability. Accordingly, the HRA and some method to evaluate each option based on various criteria are needed for the practical human reliability improvement in the system operation.

Figure 2 shows the entire framework of the method proposed in this work. The advantage value of each option associated with each criterion is evaluated based on each option's features. Especially, the HRA requires complex procedures and expert judgements under various conditions. For the systematic and reasonable evaluation of human reliability of a task, a human reliability model (HRM) using

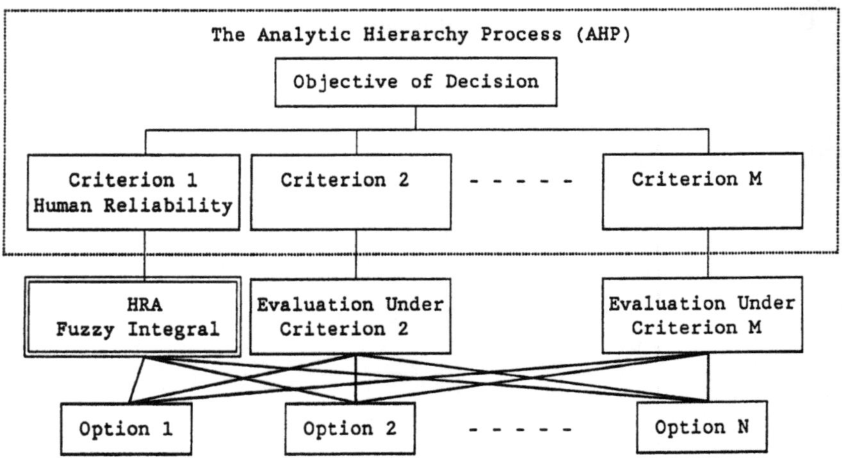

Fig.2 The entire framework of the proposed decision support method.

the framework of fuzzy integral is introduced. The idea of the fuzzy integral was originated by Ishii et al. (1985), and its application to HRMs has been researched by Onisawa (1986) and Washio et al. (1991). In addition, the analytic hierarchy process (AHP) originated by Satty (1980) evaluates the total value of each option in terms of the different types of criteria based on the expert judgements within simple and transparent procedures. The fuzzy integral model may by also applicable to this decision-support stage. However, the simpler AHP is adopted, because the transparency of the decision process is our important concern. The application of AHP to human factor management has been assessed by Kitamura et al. (1991) and Washio et al. (1991).

4. HRA BASED ON FUZZY INTEGRAL

The outline of the proposed HRA is depicted in fig.3. The HRA consists of the two stage. In the first stage, the HRM of an objective task is developed in the framework of fuzzy integral. In the next stage, the success likelihood index (SLI) representing the subjective success grade in the analyst's judgement and the success probability (Ps) of the task are evaluated.

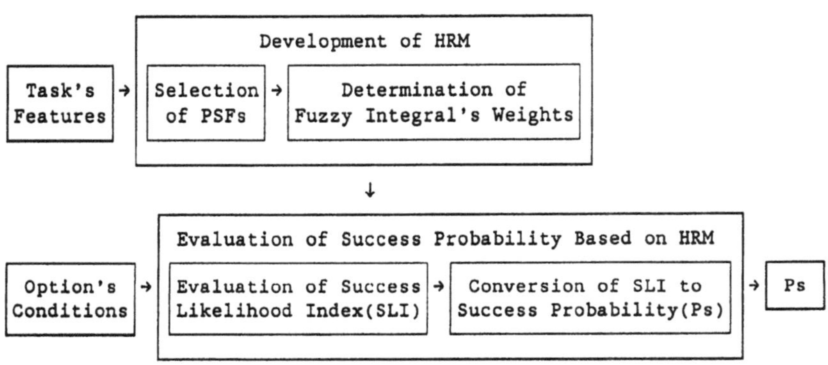

Fig.3 The outline of HRA procedure using HRM based on fuzzy integral.

(1) Development of HRM

This stage involves two more detailed steps. Initially, the analysts have to identify some dominant factors, "performance shaping factors (PSFs)", influencing the human reliability in the task. An efficient way to identify appropriate PSFs is to consider each factor influencing one of the stages of the situation cognition K1, the action judgement K2, and the manipulation K3 in the task as proposed by Washio et al. (1991). The PSFs considered in HRA is usually limited to only dominant ones, because the subjective evaluations associated with the contributions of minor PSFs are quite difficult and erronious for human experts. In the demonstration example, the following five PSFs are identified.

K1 = {PSFs influencing the situation cognition stage}
  = {PSF1:readability of the temperature indicators,
   PSF2:readability of the flow rate indicator},
K2 = {PSFs influencing the action judgement stage}
  = {PSF3:appropriateness of the operator(s)'s stress,
   PSF4:experience of the operator(s)},
K3 = {PSFs influencing the manipulation stage}
  = {PSF5:controllability of the valve opening}.   (1)

In the next step, the following relation between the success likelihood index (SLI) and PSFs represented by a fuzzy integral is introduced as an HRM.

$$SLI = \int_K h(PSF)g(*) = \max_{A\in 2^K} [\ \min_{PSFi\in A} \{h(PSFi)\}\wedge g(A)\ ], \tag{2}$$

where

h(PSFi) : a grade value of PSFi,

g(A) : a weight of PSFs' subset A,

$$K = K1 \cup K2 \cup K3, \tag{3}$$

$$0 \le h(PSFi) \le 1,\ 0 \le SLI \le 1,\ 0 = g(\phi) \le g(A) \le g(K) = 1, \tag{4}$$

$$A \subseteq B \subseteq K \Rightarrow g(A) \le g(B). \tag{5}$$

In the most of tasks, the influences of PSFs to SLI and Ps are mutually correlated. As the HRM of the fuzzy integral defines weights g(A) on the combination of multiple PSFs in A, the proper relation between SLI and PSFs can be modeled even under the existence of strong correlations among PSFs.

To obtain the HRM of a task, all weights g(A) ($\forall A \in 2^K$) must be determined. A systematic procedure based on expert judgements is used to determine the value of each g(A). First, for every Ai ($\in 2^{Ki}$, Ki=K1,K2 and K3), the following situation of PSFs' grades is assumed.

$$\left.\begin{aligned} &h(PSFj) = 1 \quad (\forall PSFj \in Ai), \\ &h(PSFj) = 0 \quad (\forall PSFj \in K\text{-}Ai). \end{aligned}\right] \tag{6}$$

This is the situation that all PSFs belonging to Ai are the best, and the rest are the worst. Then, the experts obtain their subjective evaluation of SLI value under this situation. The value of the SLI represents the expert's subjective evaluation of the possibility of success in terms of relative comparison. Because the value of SLI can be considered to represent the inherent influence of Ai to SLI, the same value is assigned to g(Ai). In this process, the g(Ai)s' satisfaction of the expressions (4) and (5) must be examined, and the iterative corrections must be conducted, if needed. When the number of PSFs is increasing, the number of fuzzy measures of Ais becomes large due to their combinatorial explosion, and hence the proposed method does not seem to be able to deal with the determination of fuzzy measures. However, this difficulty is not significant for the application of HRA, because the number of PSFs to be considered is quite limited as mentioned before. After obtaining the values of all g(Ai) for every i (i=1,2,3), the weight g(A) of every A which is the superset of Ais is determined under the assumption of the independency among Kis' influences on SLI as follows.

$$g(A) = \sum_{i=1}^{3} g(Ai), \tag{7}$$

where

$$A = \sum_{i=1}^{3} Ai \qquad (\forall A \in 2^{K},\ \exists Ai \in 2^{Ki}). \tag{8}$$

Strictly speaking, this assumption does not hold in the subjective point of experts' view. However, as the HRA usually takes only dominant contributions of PSFs to SLI into account, each PSF group in the expression (1) can be approximately considered to affects the different portion of the task independently.

The entire procedure described here has derived the weights indicated in Table 1 for the demonstration example. For the step K1, PSF1:readability of the temperature indicators and PSF2:readability of the flow rate indicator are considered to have almost independent contributions to the human reliability. Hence, the linear combination of each weight was assigned to the weight of their merged contribution. Contrary, for the step K2, the weight of the merged effect of PSF3:appropriateness of the operator(s)'s stress and PSF4:experience of the operator(s)'s stress was set to be lower than the linear sum of each weight, because the large amount of the experience of the operator(s) may reduce the inappropriateness of the operator(s)'s stress even under emergency conditions.

(2) Evaluation of Success Probability Based on HRM

This stage also involves two detailed steps. In the first step, the experts evaluate the grade values of PSFis (i=1,2,3,4,5) for each countermeasure's option. The measure of the PSFs' values are subjective and relative as well as the SLI. The PSFs' values evaluated for the demonstration example are summarized in Table 2. As the option O1 improves the man-machine interface of the control system, it

Table 1 The weights g(A) in the example.

A	g(A)	A	g(A)
φ	0.00	{PSF1,PSF2,PSF3}	0.60
{PSF1}	0.15*	{PSF1,PSF2,PSF4}	0.70
{PSF2}	0.15*	{PSF1,PSF2,PSF5}	0.50
{PSF3}	0.30*	{PSF1,PSF3,PSF4}	0.65
{PSF4}	0.40*	{PSF1,PSF3,PSF5}	0.65
{PSF5}	0.20*	{PSF1,PSF4,PSF5}	0.75
{PSF1,PSF2}	0.30*	{PSF2,PSF3,PSF4}	0.65
{PSF1,PSF3}	0.45	{PSF2,PSF3,PSF5}	0.65
{PSF1,PSF4}	0.55	{PSF2,PSF4,PSF5}	0.75
{PSF1,PSF5}	0.35	{PSF3,PSF4,PSF5}	0.70
{PSF2,PSF3}	0.45	{PSF1,PSF2,PSF3,PSF4}	0.80
{PSF2,PSF4}	0.55	{PSF1,PSF2,PSF3,PSF5}	0.80
{PSF2,PSF5}	0.35	{PSF1,PSF2,PSF4,PSF5}	0.90
{PSF3,PSF4}	0.50*	{PSF1,PSF3,PSF4,PSF5}	0.85
{PSF3,PSF5}	0.50	{PSF2,PSF3,PSF4,PSF5}	0.85
{PSF4,PSF5}	0.60	{PSF1,PSF2,PSF3,PSF4,PSF5}	1.00

The values labeled by "*"s have been determined by expert judgements. The others are derived by the expressions (7) and (8).

Table 2 The PSF's values evaluated by experts for each option.

Option	PSF1	PSF2	PSF3	PSF4	PSF5
O1:Introduce an automated system	0.90	0.90	0.70	0.50	0.90
O2:Introduce a training simulator	0.20	0.20	0.10	0.90	0.20
O3:Employ an assistant operator	0.70	0.70	0.50	0.50	0.70

enhances the PSFs' values except PSF4 (the operator's experience). On the contrary, the O2 keeps the PSFs' values low except the operator's experience. The evaluation of the O3 takes the intermediate position between the O1 and O2. These PSFs' values are then substituted to the expression (2) characterized by the weights in Table 1, and the SLI of each option is calculated. The result is shown on the second column of Table 3. The value in each parentheses is calculated by the following linear sum which is standard in the aforementioned SLIM.

$$SLI = \sum_{i=1}^{3} g(\{PSFi\})h(PSFi). \tag{9}$$

Each linear sum is overestimated, because the nonlinear suppression between PSF3 and PSF4 is not considered.

Table 3 The result of the value evaluation for the decision criteria's indices.

Option	SLI	Ps	Cost	Accept.
O1:Introduce an automated system	0.70	0.998	$1500k	low
O2:Introduce a training simulator	0.40	0.978	$ 400k	medium
O3:Employ an assistant operator	0.50	0.990	$ 800k*	high

"*":For employment until the plant life is terminated (20 years).

In the second step, the values of SLI are converted into the success probability Ps. As described by "Fechner's law", a subjective sense of humans is generally proportional to the logarithm of a physical quantity. Hence, the human sense associated with the success likelihood is considered to be proportional to the logarithms of the success and fault probability. Accordingly, the following conversion rule proposed by Washio et al. (1991) is employed in this study.

$$\log(Ps)-\log(1-Ps) = aSLI+b. \qquad (10)$$

The constants a and b are dependent to the objective task and the analyst. They are determined by the least squares fitting of the expression (10) to small amount of available reference data. The reference data can be obtained from the past operation records. In this demonstration, the data in the literature of Ref.1 which includes typical human error probability values in many representative tasks are utilized, and the following values of a and b are obtained.

$$\left.\begin{array}{l} a = 8.58, \\ b = 0.30. \end{array}\right] \qquad (11)$$

Based on the expressions (10) and (11), the values of Ps indicated in the third column on Table 3 are derived.

As exemplified by the sample problem, the HRA using the fuzzy integral model provides an efficient measure to enable systematic evaluation of proper success probability based on human experts' knowledge and experience.

5. THE ANALYTIC HIERARCHY PROCESS

The overall decision-making requires the consideration of multiple criteria. The analytic hierarchy process (AHP) provides an efficient remedy to derive reasonable and consistent solutions of a decision problem under conflicting requirements. The basic principle of the AHP is to decompose the problem into a structured layer (hierarchy) of criteria, each representing one of the multiple issues to be considered in the decision making. In fig.2, a total problem is decomposed into several criteria within one layer. Each criterion can be further decomposed, whenever needed. In our example, the problem has been decomposed into three criteria in a layer.

Criterion 1 (C1): success probability of the task in each option.

Criterion 2 (C2): cost required by each option.

Criterion 3 (C3): acceptance of operator(s) for each option.

The performance of each option for the C1 has been evaluated in the previous section. The fourth and the fifth columns of Table 3 represent the each performance for the C2 and C3. The former values are the summation of all required cost in each option. The latter qualitative expressions have been obtained by the interview with the operator.

The key issue of this methodology is in the weighted structural identification of influence relationships among the multiple criteria. For example, the experts must quantitatively define the relative importance of the above criteria C1, C2, and C3. The paired comparison is carried out through step-by-step determination of the components of a matrix denoted A. Suppose we have n criteria in a level. For all possible pairs of ith and jth criteria, the one with higher importance (priority, preference, etc., depending on the problem at hand) is needed to be decided. For the (i,j) element of the A matrix, a(i,j), an integer k ($1 \leq k \leq 9$) is assigned, when ith criterion is judged to be more important than jth. If jth criterion is more important than ith, then 1/k is assigned to a(i,j): The relative importance of all n criteria is given by the n components of normalized eigenvector X of the A matrix corresponding to the largest eigenvalue as described bellow.

$$X = (x(1),\ldots, x(n))^t$$
$$= (\text{importance of 1st criterion},\ldots,\text{importance of nth criterion})^t, \qquad (12)$$

In our demonstration example, the experts resulted the paired comparison matrix shown in Table 4 (a) indicating that the success probability (C1) is regarded as the most important criteria among three. The relative importance of the options in the lowest layer of fig.2 is also evaluated as indicated in Table 4 (b)(c)(d). The relative advantage of each option in the sense of higher success probability can be determined by either of a conversion function from Ps to a(i,j) as proposed by Washio et al. (1991) or a subjective conversion of human experts. The latter method is used in this study.

Finally, the total value of each option is evaluated by taking the average of each advantage weighted by the criterion's importance for a option. In case of the demonstration example, the following expression gives the total advantage.

Table 4 The paired comparison matrices resulted for task improvement.

(a) For C1, C2, and C3.

	C1	C2	C3	X_C
C1	1	5	5	0.72
C2	1/5	1	1	0.14
C3	1/5	1	1	0.14

(b) For O1, O2, and O3 under C1.

C1	O1	O2	O3	X_{C1}
O1	1	7	3	0.65
O2	1/7	1	1/5	0.07
O3	1/3	5	1	0.28

(c) For O1, O2, and O3 under C2.

C2	O1	O2	O3	X_{C2}
O1	1	1/7	1/5	0.07
O2	7	1	3	0.65
O3	5	1/3	1	0.28

(d) For O1, O2, and O3 under C3.

C3	O1	O2	O3	X_{C3}
O1	1	1/3	1/7	0.09
O2	3	1	1/3	0.24
O3	7	3	1	0.67

$$X_T(Oi) = \sum_{j=1}^{3} X_C(Cj) X_{Cj}(Oi), \tag{13}$$

where

xT(Oi) : total advantage of option Oi,

xC(Cj) : relative importance of criterion Cj,

xCj(Oi) : relative advantage of option Oi under criterion Cj.

The total value of the three options has been resulted as follows.

$$\left.\begin{array}{ll} \text{O1 : Introduce an automated system} & ;\ xT(O1) = 0.49, \\ \text{O2 : Introduce a training simulator} & ;\ xT(O2) = 0.18, \\ \text{O3 : Employ an assistant operator} & ;\ xT(O3) = 0.33. \end{array}\right] \tag{14}$$

Because of the high importance of the success probability (C1), the option to introduce an automated control system (O1) has the largest preference value even under the conditions of the high required cost and the low operator's acceptance.

6. CONCLUSION

A method for supporting complex decision-making incorporating the fuzzy-integral-based HRA and the AHP has been proposed. The applicability of this method has been demonstrated through an example of a decision problem to improve operation reliability of a plant task. The main features of the proposed method are summarized as follows.

(1) The appropriate values of nonlinear human reliability under various task conditions are evaluated systematically by complementing a few empirical data with human experts' knowledge and experience.
(2) A reasonable and consistent solution for a decision problem under various mutually conflicting criteria is systematically obtained.
(3) The foundations behind the analysis and decision are visualized to allow re-examinations and adjustments for better consensus formulation.

REFERENCES

Swain A.D. and Guttmann H.E. (1990) "Handbook of Human Reliability Analysis with Emphasis on Nuclear Power Plant Applications", Final Report, NUREG/CR-1278.

Embrey D.E., et al. (1984) "SLIM-MAUD (Success Likelihood Index Methodology Multi-Attribute Utility Decomposition); An Approach to Assessing Human Error Probabilities Using Structured Expert Judgement", NUREG/CR-3518, Vol.2.

Phillips L.D., et al.(1983) "A Socio-Technical Approach to Assessing Human Reliability".

Ishii K. and Sugeno M. (1985) "A Model of Human Evaluation Process Using Fuzzy Measure", Int. J. Man-Machine Studies, Vol.32, pp.19-38.

Onisawa T. (1986) "Performance Shaping Factors Modeling Using Human Error Possibility and Fuzzy Integrals", Human Engineering, Vol.22, No.2, pp.82-89 (in Japanese).

Washio T., et al. (1991) "Human Reliability Analysis of Performing Tasks in Plants Based on Fuzzy Integral", J. At. Energy Soc. Japan, Vol.33, No.10, pp.983-993 (in Japanese).

Satty T.L. (1980) "The Analytic Hierarchy Process", McGrow-Hill.

Kitamura M., et al. (1991) "Incorporation of Analytic Hierarchy Process Methodology with Probabilistic Safety Assessment for Consistent Decision Making", Proc. Int. Conf. on Probabilistic Safety Assessment and Management, Vol.2, pp.1445-1450.

Washio T., et al. (1991) "Decision Support for Operability Improvement and Maintenance Planning by Analytic Hierarchy Process Methodology", Proc. Int. Conf. on Probabilistic Safety Assessment and Management, Vol.2, pp.1451-1456.

Fuzzy Reliabilty Analysis of Labour (Man-Machine) Systems

A. Rotshtein
Department of Computer-Based Information and Management System Vinnitsa Polytechnical Institute, Khmelnitsky Shosse,95, Vinnitsa,286021,Ukraine

Abstract. This article discloses an approach to model construction for assessment of reliability and quality of labour (man-machine) systems functioning using probabilistic-time figures.

The suggested approach is based upon composition of two types of models: (a) probabilistic-algorithmic models of labour process with parameters of labour operations in fuzzy number form; (b) logical- linguistic models of labour operations in fuzzy logical proposition.
Keywords: labour (man-machine) system, reliability and quality functioning, extension principle, fuzzy probabilistic-algorithmic models, fuzzy reasoning.

1. INTRODUCTION

By labour (man-machine) systems we mean an object of the form (Rotshtein, 1991): <X,H,S,F,Y>,

where $X=\{x_1,x_2,...\}$ - a set of labour objects;

$H=\{h_1,h_2,...\}$ - a set of labour subjects;

$S=\{s_1,s_2,...\}$ - a set of labour tools;

Y- a labour product gained as a result of labour system functioning;

F- labour process in the course of which the conversion of X into Y is realized using H and S.

Typical examples of labour system would be:

a) a "work-machine" complex: X- a part blank; H- a worker; S- a machine; Y- a part; F- a technological operation;

b)an "operator-display" complex: X- source data; H- a man-operator; S- a display; Y- information loaded into the computer from a display; F- information input process.

Analogic denotation of an ergonomic system is introduced by Karwowski (1991).

When designing labour system it is expedient to have some quantitative assessment of the following interconnected figures:

P_Y-probability of defects (or fault) absence in a labour product;

T_Y-spendings (of time or other resources) on gaining a labour product.

Interconnection between these figures is stipulated by the various checking operations in the labour process, which serve for probability P_Y increase. It leads to figures T_Y increasing because of additional spendings (of time or other resources) for execution of checking operations.

Depending upon specifics of the labour system, figures P_Y can be interpreted as zero-defect probability, zero-error probability or reliability functioning of the system.

Figure T_Y can be used to predict labour consumption of the technological process or to predict the probability of system task execution within required time limits.

Contemporary theory of man-machine system reliability and design (Gubinsky, 1982; Rotshtein, 1992) allows to assess figures P_Y and T_Y on the basis of probabilistic-algorithmic and regression models. The probabilistic-algorithmic models bind figures P_Y and T_Y with the labour process structure (F) and the labour operations implementation taking into account the dependencies of the labour operations implementation quality upon influencing factors: man qualification, ambient conditions, operating stress and other factors.

The main drawback of the existing theory are introduced by the fact that:

(1) as rule, notcorrect source data is stuffed into strict probabilistic models;

(2) gaining of correct source data about operations implementation quality and source data dependencies upon influencing factors demands carrying out labour consuming and expensive experiments.

The approach suggested in this article allows to eliminate the abovementioned drawback using fuzzy sets theory (Zadeh,1973). The suggested approach is based upon the following principles:

(a) logic-algorithmic description of the labour system functioning process;

(b) transfer from logic-algorithmic description to probability models;

(c) generalization of probabilistic-algorithmic models for a fuzzy case;

(d) representation of labour operations characteristics (source data for simulation) in a fuzzy number form;

(e) taking into consideration factors influencing labour operations implementation quality by using fuzzy logical propositions.

2. LABOUR SYSTEMS SIMULATION LANGUAGE

Gluskov's language of the algorithmic algebras (Gluskov et al.,1978) is a useful tool for formal specification of the labour system functioning process, and this language completely corresponds to the conception of structured programming (Linger et. al,1979).

The operators in this language are designated by capital or small Latin letters (A,B,C,...,a,b,c...) and logic conditions- by small Greek letters (α, β, γ...). According to the Gluskov's theorem of regularization a random algorithm can be represented in the form of the next operator structures superposition:

a) *linear structure*- $B=A_1A_2$, consisting of consecutive operators A_1 and A_2 execution;

b) *branching structure (if-then-else)*- $C=(A_1 \vee A_2)_{\omega}$ consisting of operator A_1 execution if condition ω is true (ω=1), or of operator A_2 execution if condition ω is false (ω=0);

c) *iterative structure do-while*- $D=\{A\}_{\omega}$ consisting of repeat processing of operation A till condition ω is true.

Application of this language allows us to represent labour process structure (content and logical interconnection of the main and check operations) in the form of concise algebraic formulae.These formulae, on one side, are quite vivid, and, on the other side, they are convenient for computer input to solve analysis and synthesis problems (Rotshtein, 1992). In this case the algorithm operators are the main labour operations done by a man and/or a machine and the logic conditions are check procedures used for reliability inprovement.

Example 1. Let's consider the "user-abonent-computer" information labour system used for booking plane tickets. Graph-scheme of this system process functioning is shown in Fig.1.

Algebraic analog of this scheme has this form:

$$Y = \{A_1\}_{\omega_1} \{\{A_2(E \vee A_3)_{\omega_2} A_4\}_{\omega_3} A_5 A_6\}_{\omega_4} A_7. \qquad (1)$$

where: A_1 pass of the request from the client to the systems operator;

ω_1- checking the validity of the request from the client;

A_2- keying of the request by the systems operator;

ω_2- visual control of the keying validity on the screen;

E- registration of the control results;

A_3- error correction;

A_4- request transfer to the computer;

ω_3- programmed control of the request validity;

A_5- computer request processing;

A_6- transfer of the reply from the computer to the screen;

ω_4- reception and control of the reply by the abonent.

A_7- reply documenting and transfer of the ticket to the client.

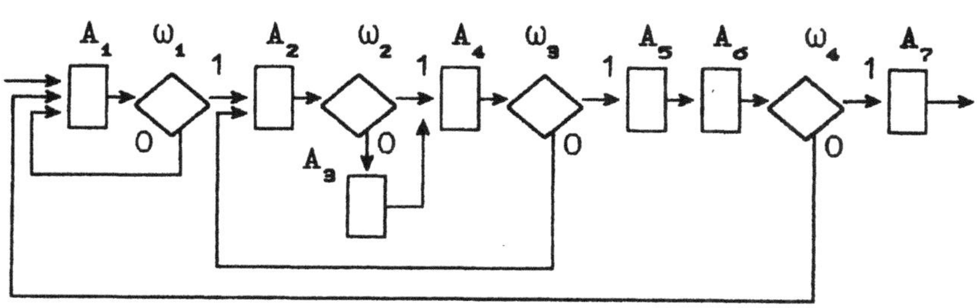

Figure 1. Graph-scheme of algorithm (1)

Each operator A_i in the formula (1) can be represented by an algorithm on a lower level of detail.

Algorithmic models of production systems are discussed in Rotshtein, 1992.

3. PROBABILISTIC-ALGORITHMIC MODELS

Transfer from algorithmic description of a labour

process to the assessment of the figures of its implementation reliability and quality is carried out according to the followin formal rules:

$$
\left.
\begin{array}{l}
B=A_1A_2 \quad \Rightarrow \quad P_B=P_{A_1}P_{A_2}; \quad t_B=t_{A_1}+t_{A_2}; \\[2ex]
C=(A_1 \underset{\omega}{\vee} A_2) \Rightarrow \left\{
\begin{array}{l}
P_C=P_\omega K_\omega^{11}P_{A_1}+\ (1-P_\omega)K_\omega^{00}P_{A_2} \\
t_C=t_\omega+[P_\omega K_\omega^{11}+(1-P_\omega)(1-K_\omega^{00})]t_{A_1}+ \\
\quad +[P_\omega(1-K_\omega^{11})+(1-P_\omega)K_\omega^{00}]t_{A_2};
\end{array}
\right. \\[6ex]
D=\{A\}_\omega \quad \Rightarrow \left\{
\begin{array}{l}
P_D=\dfrac{P_A K_\omega^{11}}{1-[P_A(1-K_\omega^{11})+(1-P_A)K_\omega^{00}]} \\[3ex]
t_D=\dfrac{t_A+t_\omega}{1-[P_A(1-K_\omega^{11})+(1-P_A)K_\omega^{00}]}
\end{array}
\right.
\end{array}
\right\} \quad (2)
$$

where:

P_A - probability of correct operation A execution;

K^{11} and K^{00} - probabilisties of the first and second type of errors absence when checking condition ω, respectively;

P_ω - probability of condition ω truthfulness;

t_A and t_ω are time spending (cost) of operator A and condition ω implemention, respectively.

These rules are proved in Appendix.

4. CONVERSION OF THE UNDEFINED SOURCE DATA INTO FUZZY NUMBERS

Let q be a labour operation characteristic described by an operator or by a condition (probability, time or cost) in (2). Suppose that for this q there are known:

a) interval of changing $(\underline{q},\bar{q})$, where $\underline{q}$ is lower bound and $\bar{q}$ is upper bound;

b) linguistic assessment in the interval $(\underline{q},\bar{q})$ set at one of five levels: 0- low(L); 1- below average (bA); 2- average (A); 3- upper average (uA); 4- high (H).

Let's also suppose that linguistic assessments have membership functions $\mu(q)$ shown in Fig.2 for all characteristics.

To present a fuzzy number $\tilde{q}$ let's get use of representation in the form of α-level sets system:

$$\tilde{q} = \bigcup_{\alpha\in[0,1]} (\underline{q}_\alpha, \bar{q}_\alpha)_\alpha,$$

where $(\underline{q}_\alpha, \bar{q}_\alpha)_\alpha$- interval of characteristic q values of the α-level membership.

Taking into account Fig.2 information about undefined characteristic q can be transformed into the fuzzy number $\tilde{q}$ using the next rules:

$$(\underline{q},\bar{q},\text{low})=(\underline{q},\bar{q})_0 \cup (\underline{q},\underline{q}+\Delta)_{0.5} \cup (\underline{q},\underline{q})_1;$$

$$(\underline{q},\bar{q},\text{below average})=(\underline{q},\bar{q})_0 \cup (\underline{q},\underline{q}+2\Delta)_{0.5} \cup (\underline{q}+\Delta,\underline{q}+\Delta)_1;$$

$$(\underline{q},\bar{q},\text{average})=(\underline{q},\bar{q})_0 \cup (\underline{q}+\Delta,\bar{q}-\Delta)_{0.5} \cup (\underline{q}+2\Delta,\bar{q}-2\Delta)_1;$$

$$(\underline{q},\bar{q},\text{upper average})=(\underline{q},\bar{q})_0 \cup (\bar{q}-2\Delta,\bar{q})_{0.5} \cup (\bar{q}-\Delta,\bar{q}-\Delta)_1;$$

$$(\underline{q},\bar{q},\text{high})=(\underline{q},\bar{q})_0 \cup (\bar{q}-\Delta,\bar{q})_{0.5} \cup (\bar{q},\bar{q})_1,$$

where $\Delta = \dfrac{\bar{q} - \underline{q}}{4}$.

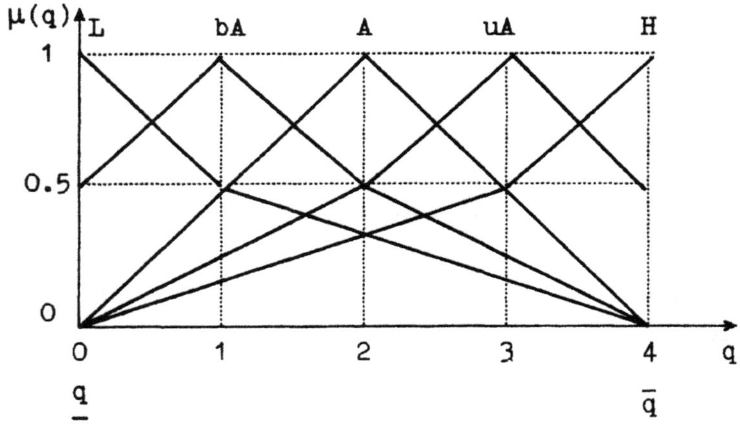

Figure 2. Membership functions of q

Note that the difference between the suggested above method of undefined source data representation and the analogic method applied in fuzzy-fault tree analysis (Tanaka et al., 1983; Lai et al, 1988) consists in the possibility of taking into consideration the additional information about characteristic q value in the interval $(\underline{q},\bar{q})$. This additional information is given in the form of linguistical values: L, bA, A, uA, and H (see Fig.2).

Example 2. Let's consider operation A_2 in process (1). Suppose that the next information about this operation is know:

P_{A_2} = (0.95÷0.99,high)- probability of correct operation A_2 execution;

t_{A_2} = (6÷14 sec,upper average)- time of operation A_2 execution.

Using the suggested above method we can obtain the next fuzzy numbers:

$$\tilde{P}_{A_2} = (0.95,0.99)_0 \cup (0.98,0.99)_{0.5} \cup (0.99,0.99)_1;$$

$\tilde{t}_{A_2} = (6,14)_0 \cup (10,14)_{0.5} \cup (12,12)_1.$

5. FUZZY PROBABILISTIC-ALGORITHMIC MODELS

To apply probabilistic-algorithmistic models with fuzzy source data let's use the next principle.

5.1. Extension principle

Let $y=f(q_1,q_2,\ldots,q_n)$- function of n independent variables. If $\tilde{q}_l$- fuzzy number represented in the form of α-level sets

$$\tilde{q}_l = \bigcup_{\alpha\in[0,1]} (\underline{q}_\alpha, \bar{q}_\alpha)_\alpha, \quad l=\overline{1,n},$$

then fuzzy number $\tilde{y}=f(\tilde{q}_1,\tilde{q}_2,\ldots,\tilde{q}_n)$ is defined this way:

$$\tilde{y}_l = \bigcup_{\alpha\in[0,1]} (\underline{y}_\alpha, \bar{y}_\alpha)_\alpha,$$

where $\underline{y} = \inf \{f(q_1,q_2,\ldots,q_n)\}$,

$\bar{y} = \sup \{f(q_1,q_2,\ldots,q_n)\}$

for every α-level.

This principle is given here as a definition.

Using extension principle and three-level representation of fuzzy numbers we can obtain the fuzzy analogs of the probabilistic-algorithmic models, described in the section 3.

To find the upper $(\bar{y})$ and lower $(\underline{y})$ bounds on function $y=f(q_1,q_2,\ldots,q_n)$ it is necessary to use the next rule:

$$\text{If } \frac{\partial y}{\partial q_l} \geqslant 0, \text{ then } = \begin{cases} \underline{y}, & \text{when } q_l = \underline{q}_l, \\ \bar{y}, & \text{when } q_l = \bar{q}_l \end{cases}$$

$$\text{If } \frac{\partial y}{\partial q_l} < 0, \text{ then } = \begin{cases} \underline{y}, & \text{when } q_l = \bar{q}_l \\ \bar{y}, & \text{when } q_l = \underline{q}_l . \end{cases}$$

5.2. Linear structure $B=A_1A_2$

Fuzzy probability of operator B correct execution is calculated by formulas:

$$\tilde{P}_B = (\underline{P}_B, \bar{P}_B)_0 \cup (\underline{P}_B, \bar{P}_B)_{0.5} \cup (\underline{P}_B, \bar{P}_B)_1 ,$$

where for all α-levels:

$$\underline{P}_B = \underline{P}_{A1}\underline{P}_{A2}; \ \bar{P}_B = \bar{P}_{A1}\bar{P}_{A2} .$$

Fuzzy time taken for operator B execution is calculated by formula:

$$\tilde{t}_B = (\underline{t}_B, \bar{t}_B)_0 \cup (\underline{t}_B, \bar{t}_B)_{0.5} \cup (\underline{t}_B, \bar{t}_B)_1 ,$$

where for all α-levels:

$$\underline{t}_B = \underline{t}_{A1} + \underline{t}_{A2}; \ \bar{t}_B = \bar{t}_{A1} + \bar{t}_{A2} .$$

5.3. Branching structure $C=(A_1 \underset{\omega}{v} A_2)$

Fuzzy probability of correct operator C execution is calculated by formulas:

$$\tilde{P}_C = (\underline{P}_C, \bar{P}_C)_0 \cup (\underline{P}_C, \bar{P}_C)_{0.5} \cup (\underline{P}_C, \bar{P}_C)_1 ,$$

where for all α-levels:

$$\underline{P}_C = P^* \underline{K}_\omega^{11} \underline{P}_{A_1} + (1-P^*) \underline{K}_\omega^{00} \underline{P}_{A_2} ,$$

$$\text{in which } P^* = \begin{cases} \underline{P}_\omega, & \text{when } \underline{K}_\omega^{11}\underline{P}_{A_1} - \underline{K}_\omega^{00}\underline{P}_{A_2} \geq 0, \\ \bar{P}_\omega, & \text{otherwise} \end{cases}$$

$$\bar{P}_C = P^{**} \bar{K}_\omega^{11} \bar{P}_{A_1} + (1-P^{**}) \bar{K}_\omega^{00} \bar{P}_{A_2} ,$$

$$\text{in which } P^{**}=\begin{cases}\bar{P}_{\omega}, & \text{when } \bar{K}^{11}_{\omega}\bar{P}_{A_1}-\bar{K}^{00}_{\omega}\bar{P}_{A_2}\geqslant 0,\\ \underline{P}_{\omega}, & \text{otherwise;}\end{cases}$$

Fuzzy time taken for operator C execution is calculated by formula:

$$\tilde{t}_c=(\underline{t}_c,\bar{t}_c)_0\cup(\underline{t}_c,\bar{t}_c)_{0.5}\cup(\underline{t}_c,\bar{t}_c)_1,$$

where for all α-levels:

$$\underline{t}_c=\underline{t}_{\omega}+\underline{t}_{A_1}\left[P^{*}_{\omega}K^{11*}_{\omega}+(1-P^{*}_{\omega})(1-K^{00*}_{\omega})\right]+$$
$$+\underline{t}_{A_2}\left[P^{*}_{\omega}(1-K^{11*}_{\omega})+(1-P^{*}_{\omega})K^{00*}_{\omega}\right],$$

$$\text{in which } K^{11*}_{\omega}=\begin{cases}\bar{K}^{11}_{\omega}, & \text{when } \underline{t}_{A_2}-\underline{t}_{A_1}\geqslant 0,\\ \underline{K}^{11}_{\omega}, & \text{otherwise,}\end{cases}$$

$$K^{00*}_{\omega}=\begin{cases}\underline{K}^{00}_{\omega}, & \text{when } \underline{t}_{A_2}-\underline{t}_{A_1}\geqslant 0,\\ \bar{K}^{00}_{\omega}, & \text{otherwise,}\end{cases}$$

$$P^{*}_{\omega}=\begin{cases}\underline{P}_{\omega}, & \text{when } (\underline{t}_{A_2}-\underline{t}_{A_1})(1-K^{11*}_{\omega}-K^{00*}_{\omega})\geqslant 0,\\ \bar{P}_{\omega}, & \text{otherwise;}\end{cases}$$

$$\bar{t}_c=\bar{t}_{\omega}+\bar{t}_{A_1}\left[P^{**}_{\omega}K^{11**}_{\omega}+(1-P^{**}_{\omega})(1-K^{00**}_{\omega})\right]+$$
$$+\bar{t}_{A_2}\left[P^{**}_{\omega}(1-K^{11**}_{\omega})+(1-P^{**}_{\omega})K^{00**}_{\omega}\right],$$

$$\text{in which } K^{11**}_{\omega}=\begin{cases}\underline{K}^{11}_{\omega}, & \text{when } \bar{t}_{A_2}-\bar{t}_{A_1}\geqslant 0,\\ \bar{K}^{11}_{\omega}, & \text{otherwise,}\end{cases}$$

$$K_\omega^{oo*} = \begin{cases} \overline{K}_\omega^{oo}, \text{when } \overline{t}_{A_2} - \overline{t}_{A_1} \geqslant 0, \\ \underline{K}_\omega^{oo}, \text{otherwise}, \end{cases}$$

$$P_\omega^{*} = \begin{cases} \overline{P}_\omega, \text{when } (\overline{t}_{A_2} - \overline{t}_{A_1})(1 - K_\omega^{11**} - K_\omega^{oo**}) \geqslant 0, \\ \underline{P}_\omega, \text{otherwise}. \end{cases}$$

5.4. Iterative structure $D = \{A\}_\omega$

Fuzzy probability of correct operator D execution is calculated by formulas:

$$\tilde{P}_D = (\underline{P}_D, \overline{P}_D)_0 \cup (\underline{P}_D, \overline{P}_D)_{0.5} \cup (\underline{P}_D, \overline{P}_D)_1,$$

where for all α-levels:

$$\underline{P}_D = \frac{\underline{P}_A \underline{K}_\omega^{11}}{1 - [\underline{P}_A(1 - \underline{K}_\omega^{11}) + (1 - \underline{P}_A)\underline{K}_\omega^{oo}]}$$

$$\overline{P}_D = \frac{\overline{P}_A \overline{K}_\omega^{11}}{1 - [\overline{P}_A(1 - \overline{K}_\omega^{11}) + (1 - \overline{P}_A)\overline{K}_\omega^{oo}]} .$$

Fuzzy time taken for operator D execution is calculated by formula:

$$\tilde{t}_D = (\underline{t}_D, \overline{t}_D)_0 \cup (\underline{t}_D, \overline{t}_D)_{0.5} \cup (\underline{t}_D, \overline{t}_D)_1,$$

where for all α-levels:

$$\underline{t}_D = \frac{\underline{t}_A + \underline{t}_\omega}{1 - [P_A^*(1 - \overline{K}_\omega^{11}) + (1 - P_A^*)\underline{K}_\omega^{oo}]},$$

$$\text{in which } P^{*}_{A} = \begin{cases} \underline{P}_{A}, & \text{when } (1-\bar{K}^{11}_{\omega} - \underline{K}^{oo}_{\omega}) \geqslant 0, \\ \bar{P}_{A}, & \text{otherwise;} \end{cases}$$

$$t_{D} = \frac{\bar{t}_{A}+\bar{t}_{\omega}}{1-[P^{**}_{A}(1-\underline{K}^{11}_{\omega})+(1-P^{**}_{A})\bar{K}^{oo}_{\omega}]},$$

$$\text{in which } P^{**}_{A} = \begin{cases} \bar{P}_{A}, & \text{when } (1-\underline{K}^{11}_{\omega} - \bar{K}^{oo}_{\omega}) \geqslant 0, \\ \underline{P}_{A}, & \text{otherwise.} \end{cases}$$

Example 3. Using the suggested above models we can obtain the probability of algorithm (1) execution and time for algorithm (1) execution with source data from Table 1.

The sequence of calculation steps is shown on Fig.3 in the form of graph-tree. At each step we fulfil the reduction of any algorithmic structure in the corresponding rectangle.

Step 1. Inital algorithm: formula (1).

Reducing substitution: $(E \vee_{\omega_2} A) = C_1$.

Source data:

$\tilde{K}^{11}_{\omega_2} = (0.93,\ 0.97)_0 \cup (0.94,\ 0.96)_{0.5} \cup (0.95,\ 0.95)_1$,

$\tilde{K}^{oo}_{\omega_2} = (0.6,\ 0.7)_0 \cup (0.625,\ 0.675)_{0.5} \cup (0.65,\ 0.65)_1$,

$\tilde{T}_{\omega_2} = (0,\ 2)_0 \cup (0,\ 1)_{0.5} \cup (0.5,\ 0.5)_1$,

$\tilde{P}_{E} = 1$; $\tilde{T}_{E} = 0$.

Result of reduction:

$\tilde{P}_{C_1} = (0.93,\ 0.97)_0 \cup (0.94,\ 0.96)_{0.5} \cup (0.95,\ 0.95)_1$,

$\tilde{T}_{C_1} = (0.2,\ 0.28)_0 \cup (0,\ 1.12)_{0.5} \cup (0.55,\ 0.55)_1$.

Table 1
Propabilistic- time characteristics of process (1) elements

Element	Propability of correct execution		Time of execution	
A_1	$P_{A_1}=0.90\div0.95$	low	$t_{A_1}=0.8\div2$	low
ω_1	$K^{11}_{\omega_1}=0.93\div0.97$ $K^{oo}_{\omega_1}=0.6\div0.7$	average average	$t_{\omega_1}=0\div2$	below average
A_2	$P_{A_2}=0.95\div0.99$	high	$t_{A_2}=6\div14$	upper average
ω_2	$K^{11}_{\omega_2}=0.93\div0.97$ $K^{oo}_{\omega_2}=0.6\div0.7$	average average	$t_{\omega_2}=0\div2$	below average
A_3	$P_{A_3}=0.95\div0.99$	upper average	$t_{A_3}=0\div4$	below average
A_4	$P_{A_4}=0.99990\div0.99998$	upper average	$t_{A_4}=0.5$	-
ω_3	$K^{11}_{\omega_3}=0.99990\div0.99998$ $K^{oo}_{\omega_3}=0.4\div0.6$	upper average below average	$t_{\omega_3}=0\div2$	below average
A_5	$P_{A_5}=0.99990\div0.99998$	low	$t_{A_5}=10\div30$	low
A_6	$P_{A_6}=0.99990\div0.99998$	average	$t_{A_6}=1$	-
ω_4	$K^{11}_{\omega_4}=0.93\div0.97$ $K^{oo}_{\omega_4}=0.6\div0.7$	upper average average	$t_{\omega_4}=0\div4$	below average
A_7	$P_{A_7}=0.9999$	-	$t_{A_7}=4\div8$	average

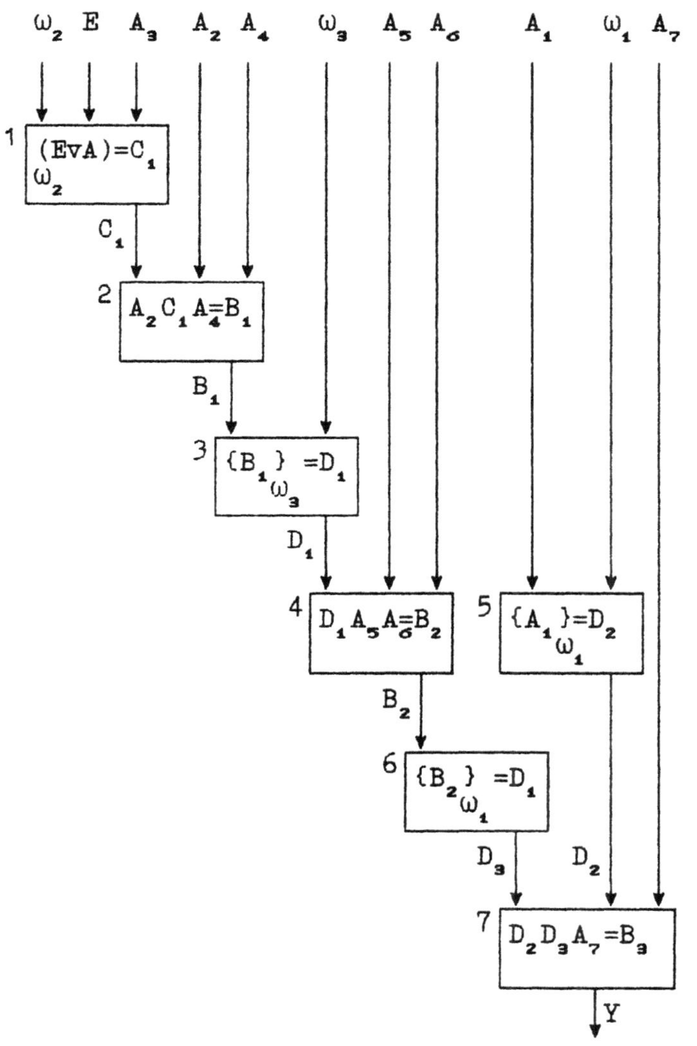

Figure 3. The sequence of calculation steps in the example 3

Step 2. Inital algorithm:

$$Y=\{A_1\}_{\omega_1}\{\{A_2C_1A_4\}_{\omega_3}A_5A_6\}_{\omega_4}A_7.$$

Reducing substitution: $A_2C_1A_4=B$.

Source data:

$\tilde{P}_{A_2}=(0.95, 0.99)_0 \cup (0.97, 0.99)_{0.5} \cup (0.99, 0.99)_1$,

$\tilde{T}_{A_2}=(6,14)_0 \cup (10, 14)_{0.5} \cup (12, 12)_1$,

$\tilde{P}_{A_4}=(0.9999, 0.99998)_0 \cup (0.99994, 0.99998)_{0.5} \cup$
$\cup (0.99996, 0.99996)_1$,

$\tilde{T}_{A_4}=0.5$.

Result of reduction:

$\tilde{P}_{B_1}= (0.8834, 0.96037)_0 \cup (0.9211, 0.9504)_{0.5} \cup$
$\cup (0.9405,0.9405)_1$,

$\tilde{T}_{B_1}=(6.5, 16.78)_0 \cup (10.5, 15.62)_{0.5} \cup (13.05, 13.05)_1$.

Step 3. Inital algorithm:

$Y=\{A_1\}_{\omega_1} \{\{B_1\}_{\omega_3} A_5A_6\}_{\omega_4} A_7$.

Reducing substitution: $\{B_1\}_{\omega_3}=D_1$.

Source data:

$\tilde{K}^{11}_{\omega_3}= (0.9999,0.99998)_0 \cup (0.9999,0.99994)_{0.5} \cup$
$\cup (0.99992, 0.99992)_1$,

$\tilde{K}^{00}_{\omega_3}=(0.4, 0.6)_0 \cup (0.4, 0.5)_{0.5} \cup (0.45, 0.45)_1$,

$\tilde{T}_{\omega_3}=(1, 3)_0 \cup (1.5, 2.5)_{0.5} \cup (2,2)_1$.

Result of reduction:

$\tilde{P}_{D_1}= (0.9266, 0.9837)_0 \cup (0.9589, 0.9795)_{0.5} \cup$
$\cup (0.9723,0.9723)_1$,

$\tilde{T}_{D_1}=(7.62, 21.27)_0 \cup (12.3, 19.02)_{0.5} \cup (15.56, 15.56)_1$.

Step 4. Inital algorithm:

$Y=\{A_1\}_{\omega_1}\{D_1A_5A_6\}_{\omega_4}A_7.$

Reducing substitution: $D_1A_5A_6=B_2.$

Source data:

$\tilde{P}_{A_5}=(0.9999,\ 0.99998)_0 \cup (0.9999,\ 0.99992)_{0.5} \cup$

$\cup\ (0.9999,\ 0.9999)_1,$

$\tilde{T}_{A_5}=(10,30)_0 \cup (10,\ 15)_{0.5} \cup (10,\ 10)_1,$

$\tilde{P}_{A_6}=(0.9999,\ 0.99998)_0 \cup (0.99992,\ 0.99996)_{0.5} \cup$

$\cup\ (0.99994,\ 0.99994)_1,$

$\tilde{T}_{A_6}=1.$

Result of reduction:

$\tilde{P}_{B_2}=(0.9264,\ 0.9836)_0 \cup (0.9587,\ 0.9794)_{0.5} \cup$

$\cup\ (0.9721,0.9721)_1,$

$\tilde{T}_{B_2}=(18.62,\ 52.26)_0 \cup (23,\ 35)_{0.5} \cup (26.56,\ 26.56)_1.$

Step 5. Inital algorithm:

$Y=\{A_1\}_{\omega_1}\{B_2\}_{\omega_4}A_7.$

Reducing substitution: $\{A_1\}_{\omega_1}=D_2.$

Source data:

$\tilde{K}^{11}_{\omega_1}=(0.93,0.97)_0 \cup (0.94,0.96)_{0.5} \cup (0.95,\ 0.95)_1,$

$\tilde{K}^{00}_{\omega_1}=(0.6,\ 0.7)_0 \cup (0.625,\ 0.675)_{0.5} \cup (0.65,\ 0.65)_1,$

$\tilde{T}_{\omega_1}=(0,\ 2)_0 \cup (0,\ 1)_{0.5} \cup (0.5,\ 0.5)_1.$

$\tilde{P}_{A_1}=(0.9,\ 0.94)_0 \cup (0.9,\ 0.91)_{0.5} \cup (0.9,\ 0.9)_1,$

$\tilde{T}_{A_1}=(0.8,2)_0 \cup (0.8, 1.1)_{0.5} \cup (0.8, 0.8)_1$,

Result of reduction:

$\tilde{P}_{D_2}= (0.9544, 0.9806)_0 \cup (0.9575, 0.9678)_{0.5} \cup$

$\cup (0.9606,0.9606)_1$,

$\tilde{T}_{D_2}=(0.85, 4.61)_0 \cup (0.88, 2.39)_{0.5} \cup (1.46, 1.46)_1$.

Step 6. Inital algorithm:

$Y=D_2\{B_2\}_{\omega_4} A_7$.

Reducing substitution: $\{B_2\}_{\omega_4}=D_3$.

Source data:

$\tilde{K}^{11}_{\omega_4}= (0.93,0.97)_0 \cup (0.95,0.97)_{0.5} \cup (0.96, 0.96)_1$,

$\tilde{K}^{00}_{\omega_4}=(0.6, 0.7)_0 \cup (0.625, 0.675)_{0.5} \cup (0.65, 0.65)_1$,

$\tilde{T}_{\omega_4}=(0, 4)_0 \cup (0, 2)_{0.5} \cup (1, 1)_1$.

Result of reduction:

$\tilde{P}_{D_3}= (0.9679, 0.9948)_0 \cup (0.9829, 0.9928)_{0.5} \cup$

$\cup (0.9894,0.9894)_1$,

$\tilde{T}_{D_3}=(19.38, 63.67)_0 \cup (24.84,40.91)_{0.5} \cup (29.84,29.84)_1$.

Step 7. Inital algorithm:

$Y=D_2D_3A_7$.

Reducing substitution: $D_2D_3A_7=B_3$.

Source data:

$\tilde{P}_{A_7}= 0.999$,

$\tilde{T}_{A_7}=(4, 8)_0 \cup (5, 7)_{0.5} \cup (6, 6)_1$.

Result of reduction:

$\tilde{P}_{B_3} = \tilde{P}_Y = (0.9228,\ 0.9755)_0 \cup (0.9411,\ 0.9608)_{0.5} \cup$

$\cup\ (0.9504, 0.9504)_1,$

$\tilde{T}_{B_3} = \tilde{T}_Y = (24.24, 76.29)_0 \cup (30.72, 50.3)_{0.5} \cup (37.3, 37.3)_1.$

Thus, at the seventh step we obtain fuzzy estimates of algorithmical process (1) reliability and quality functioning ($\tilde{P}_Y$ and $\tilde{T}_Y$) which are shown in Fig.4. Using the membership functions in Fig.2 these fuzzy estimates can be interpreted in the next way:

$\tilde{P}_Y = (0.9228 \div 0.9755$, average);

$\tilde{T}_Y = (24.24 \div 76.29$, below average).

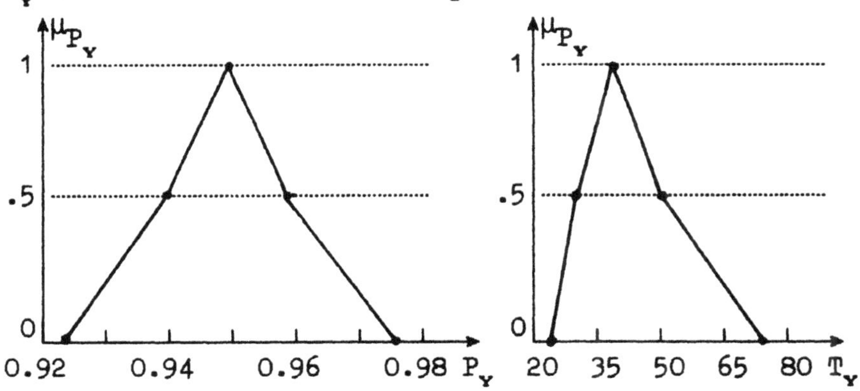

Figure 4. Fuzzy estimates of process (1) reliability and quality functioning

6. TAKING INTO CONSIDERATION SAME FACTORS INFLUENCING LABOUR OPERATION IMPLEMENTATION QUALITY

Let $x_1, x_2, \ldots, x_n$ are factors influencing parameter q value of a labour operation. Expert information about this influence is represented in the form of a system with m fuzzy logical expressions:

$$\left.\begin{array}{l}
\text{IF } x_1 = \tilde{x}_{11} \text{ and } x_2 = \tilde{x}_{21} \text{ and } \ldots x_1 = \tilde{x}_{n1} \text{ THEN } q = \tilde{q}_1 \\
\text{IF } x_1 = \tilde{x}_{12} \text{ and } x_2 = \tilde{x}_{22} \text{ and } \ldots x_1 = \tilde{x}_{n2} \text{ THEN } q = \tilde{q}_2 \\
\cdot \quad \cdot \quad \cdot \quad \cdot \quad \cdot \quad \cdot \quad \cdot \quad \cdot \quad \cdot \quad \cdot \\
\text{IF } x_1 = \tilde{x}_{1m} \text{ and } x_2 = \tilde{x}_{2m} \text{ and } \ldots x_1 = \tilde{x}_{nm} \text{ THEN } q = \tilde{q}_m
\end{array}\right\} \quad (3)$$

where $\tilde{x}_{ij}$ and $\tilde{q}_i$ ($i=\overline{1,n}$; $j=\overline{1,m}$) fuzzy estimates with membership functionins from Table 2, which corresponds to Fig.1.

The system of fuzzy logical expressions (3) and compositional rule of inference (Baldin and Pilsworth, 1979) allows to predict fuzzy of parameter q by this formula:

$$\tilde{q} = \bigcap_{j=1}^{n} \left\{ \tilde{x}_j \circ \left[\bigcup_{i=1}^{m} (\tilde{x}_{ji} \times \tilde{q}_i) \right] \right\},$$

where:

o is an operation of max-min matrices product;

$\cup$ - corresponds to operation of max;

$\cap$ - corresponds to operation of min;

$\times$ - Cartesian product.

Table 2

Membership functionins

Linguistic assessment	0	1	2	3	4
Low (L)	1.0	0.5	0.34	0.17	0.0
Below average (bA)	0.5	1.0	0.5	0.25	0.0
Average (A)	0.0	0.5	1.0	0.5	0.0
Upper average (uA)	0.0	0.25	0.5	1.0	0.5
High	0.0	0.17	0.34	0.5	1.0

Formula (4) is free of operations with multidimentional matrices.

Thus, using expert knowledge base (3) and fuzzy logical inference we can obtain linguistic values of parameter q inside the interval $(\underline{q}, \bar{q})$ taking into consideration some influencing factors.

Example 4. Let us consider operation A_2 from process (1). We study zero-error request keyning probability dependence (P_{A_2}) upon the next factors: operator qualification (x_1) and operating stress (x_2).

The system of expert propositions has the form of:

$$\left.\begin{aligned} &\text{IF } x_1 = H \text{ and } x_2 = L, \text{ THEN } P_{A_2} = H \\ &\text{IF } x_1 = A \text{ and } x_2 = A, \text{ THEN } P_{A_2} = A \\ &\text{IF } x_1 = L \text{ and } x_2 = L, \text{ THEN } P_{A_2} = L \end{aligned}\right\} \qquad (5)$$

Formula (4) taking into consideration the system (5) will have this form:

$$P_{A_2} = \tilde{x}_1 \circ \Big[(H{\times}H) \cup (A{\times}A) \cup (L{\times}L)\Big] \cap \tilde{x}_2 \Big[(L{\times}H) \cup (A{\times}A) \cup (L{\times}L)\Big] \qquad (6)$$

Using membership function from Table 2, we can represent the formula (6) in this form:

$$\tilde{P}_{A_2} = \tilde{x}_1 \circ \begin{bmatrix} 1 & .5 & .34 & .17 & 0 \\ .5 & .5 & .5 & .5 & .17 \\ .34 & .5 & 1 & .5 & .34 \\ .17 & .5 & .5 & .5 & .5 \\ 0 & .17 & .34 & .5 & 1 \end{bmatrix} \cap \tilde{x}_2 \circ \begin{bmatrix} 1 & .5 & .34 & .5 & 1 \\ .5 & .5 & .5 & .5 & 1 \\ 0 & .5 & 1 & .5 & .34 \\ .17 & .5 & .5 & .5 & .17 \\ 0 & .5 & 0 & 0 & 0 \end{bmatrix}.$$

Application of this obtained formula allows to predict linguistical probability estimate P_{A_2} inside interval of $(\underline{P}_{A_2}, \bar{P}_{A_2})$, for example:

a) if x_1=L and x_2=H, then $\tilde{P}_{A_2}$ =(0.17 0.5 0.5 0.34

0.34), which could be approximated by "average";

b)if x_1=uA and x_2=bA, then $\tilde{P}_{A_2}$=(0.34 0.5 0.5 0.5 0.5), which could be approximated by "not low" or "more or less high".

Gained assessments are quite consistent with expert opinion. The assessment quality encreases with the correctness of propositions and their quantity.

7. CONCLUSIONS

The above described method and designed mathematical models are implemented in a dialog expert-simulation system for man-machine technological processes and man-operator activities algorithm design by probability-time criteria.

Application of fuzzy models was found to be quite acceptable at early stages of labour systems design when there is no strict source data about new labour operations.

REFERENCES

Baldin T.P. and Pilsworth B.W. (1979) A model of fuzzy reasoning through multi-valued logic and set theoty. Internat. J.Man-Machine Stud.11, 351-380

Gluskov,V.M., Tseitlin ,G.E. and Justcenko ,E.L. (1978) Algebra. Languages. Programming. Kiev. Naukova Dumka Publishing House. (In Russian).

Gubinsky A.I. 1982 Reliability and quality of ergatic system. Leningrad. Nauka Publishing House. (In Russian).

Karwowski W., Complexity, fuzziness, and ergonomic incompatibility issues. Ergonomics ,Vol. 34, 671-686.

Lai F.S., Shenoi S. and Fan L.T. (1988) Fuzzy fault tree anakysis: Theoty and applications. In A.Kandel and E.Avni (Eds): Engineering Risk and Hazard Assessment, Vol.1, CRC Press, Boca Raton, FA, 117-137.

Linger R.C., Mills H.D. and Witt B.I. (1979) Structured programming: theory and practice. Addison-Wesly Publishing Company.

Rotshtein A.P. (1992) Design of non-defective man-machine technological process. Kiev, Technica Publishing House. (In Russian).

Rotshtein A.P. (1991) Formalized design of non-defective labour process. In Proceeding of the 11-th Congress of the IEA, Taylor and Francis, 1239-1241.

Tanaka H.,Fan L.T., Lai F.S., and Taguchi K. (1983) Fault tree analysis by fuzzy probability, IEEE Trans. Reliability, R-32(5), 453-457.

Zadeh L. (1973) Outline of a new approach to the analysis of complex systems and decision processes, IEEE Trans. of System, Man and Cybernet. 3, 28-44.

APPENDIX

Proof of rules (2)

To describe the conditions of correct and noncorrect algorithm structures execution logical operations "Λ"-AND, "V"-OR are used.

The following events can take place when operator A_i is performed:

$A_i^1(A_i^0)$- correct (noncorrect) operator A execution.

Propabilities of these events are:

$Pr(A_i^1)=P_{A_i}$, $Pr(A_i^0)=1-P_{A_i}$.

The following events can take place when logical condition ω is checked:

$\omega^1(\omega^0)$- condition ω is true (false),

$\omega^{11}(\omega^{10})$- true condition ω is recognized as true (as false),

$\omega^{00}(\omega^{01})$- false condition ω is recognized as false (as true).

The probabilities of these events:

$Pr(\omega^1)=P_\omega$, $Pr(\omega^0)=1-P_\omega$,

$Pr(\omega^{11})=K_\omega^{11}$, $Pr(\omega^{10})=1-K_\omega^{11}$,

$Pr(\omega^{00})=K_\omega^{00}$, $Pr(\omega^{01})=1-K_\omega^{00}$.

A.1. Linear structure $B=A_1A_2$

The condition for correct operator B execution: $B^1=A_1^1 \wedge A_2^1$.

Therefore, probability of operator B correct execution is calculated using this formula:

$Pr(B^1)=Pr(A_1^1)Pr(A_2^1)$ or $P_B=P_{A_1}P_{A_2}$.

The formula for spending t_B is evident.

A.2. Branching structure $C=(A_1 \underset{\omega}{\vee} A_2)$

The condition for correct execution of operator C is defined by:

$C^1=(\omega^1 \wedge \omega^{11} \wedge A_1^1) \vee (\omega^0 \wedge \omega^{00} \wedge A_2^1)$.

Therefore,

$Pr(C^1)= Pr(\omega^1)Pr(\omega^{11})Pr(A_1^1)+Pr(\omega^0)Pr(\omega^{00})Pr(A_2^1)$ or

$P_C=P_\omega K_\omega^{11}P_{A_1}+ (1-P_\omega)K_\omega^{00}P_{A_2}$.

Spending for operator $C=(A_1 \underset{\omega}{V} A_2)$ execution is calculated by this formula:

$t_C=P_1(t_\omega+t_{A_1})+P_2(t_\omega+t_{A_2})= (P_1+P_2)t_\omega+P_1t_{A_1}+P_2t_{A_2}=$

$= t_\omega+P_1t_{A_1}+P_2t_{A_2}$,

where $P_1(P_2)$- probability of transfer to operator A_1 (operator A_2) and $P_1+P_2=1$.

The condition for transfer to operator A_1 is:

$(\omega^1 \wedge \omega^{11}) \vee (\omega^0 \wedge \omega^{01})$.

The condition for transfer to operator A_2 is:

$(\omega^1 \wedge \omega^{10}) \vee (\omega^0 \wedge \omega^{00})$.

Therefore,

$P_1=P_\omega K_\omega^{11}+(1-P_\omega)(1-K_\omega^{00})$

$P_2=P_\omega(1-K_\omega^{11})+(1-P_\omega)K_\omega^{00}$.

A.3. Iterative structure $D=\{A\}_{\omega}$

Logic condition ω in this structure is interpreted as control of operator A correct execution. The following complete group of events takes place in each cycle of operator A execution:

$S_1=A^1\Lambda\omega^{11}$- correct completion of iterative structure execution;

$S_2=A^0\Lambda\omega^{01}$- noncorrect completion of iterative structure execution;

$S_3=(A^1\Lambda\omega^{10})V(A^0V\omega^{00})$-transfer to operator A repeat.

The number of operator A repeats is assumed as unlimited.

Probabilities of these events are:

$Pr(S_1)=a=P_A K_{\omega}^{11}$,

$Pr(S_2)=c=(1-P_A)(1-K_{\omega}^{00})$,

$Pr(S_3)=b=P_A(1-K_{\omega}^{11})+(1-P_A)K_{\omega}^{00}$,

$(a+b+c)=1$.

The probability of correct iterative structure $D=\{A\}_{\omega}$ execution after l cycles is:

$$P_D(l)= a+ba+b^2a+...+b^{l-1}a= a\,\frac{1-b^l}{1-b}.$$

When $l \rightarrow \infty$ we obtain:

$$P_D= \frac{a}{1-b}\ .$$

The number of operator A repeat cycles in structure $D=\{A\}_{\omega}$ is a random value with average significance of $(1-b)^{-1}$. It is stipulated by geometrical law of cycles number distribution. Therefore,

$$t_D= \frac{t_A+t_{\omega}}{1-b}\ .$$

7

SAFETY AND RISK ANALYSES

RISK INDEX AND APPLICATION OF FUZZY LOGIC

Christian Preyssl* and Yasushi Nishiwaki**

* European Space Agency ESA
2200 AG Noordwijk
The Netherlands

** University of Vienna
1010 Vienna
Austria

Abstract

A "risk index method" is proposed. This method provides the framework for different risk considerations and supports decision making under uncertainty. The simple concept is based on probabilistic risk analysis and uses some elements of fuzzy logic. Risk results are displayed on an index scale which is both qualitative and quantitative.

Keywords

fuzzy logic, risk analysis, uncertainty, risk index

1. INTRODUCTION

In light of the increasing number of complex technical systems the awareness of risks has been increasing. Different types of risks are associated with these complex technical systems. These risks need to be identified and managed in a systematic and professional way. The identification process is supported by probabilistic risk analysis. The risk analysis results are used as input to risk management. Risk management comprises risk reduction, tracking and risk acceptance. The overall process involves the communication of risk results to specialists, to managers, to political executives and also to the public. Very often "risk communication" is confronted with severe difficulties because of the lack of an accepted and standardized "risk language".

Risk is not an empirical reality. It is in itself not a measurable quantity but it has quantifiable components like likelihood, consequence severity and uncertainties. There are other aspects of risk which will influence risk considerations such as the utility of taking the risk.

In order to support decision making associated with risk management and in particular risk communication a simple "risk index" method is proposed. The "risk index" constitutes a measuring stick for the magnitude of risk.

2. REQUIREMENTS FOR "RISK INDEX METHOD"

A set of requirements for the "risk index method" is defined.

The "risk index method" shall:

* be simple, robust and transparent
* be compatible with conventional probabilistic risk analysis
* able to model and represent uncertainties
* integrate objective and subjective data
* be easy to interpret and use within decision making associated with risk management

3. RISK INDEX METHOD - THEORY

3.1 Overview

An overview of the "risk index method", that satisfies the requirements above is provided in figure 1.

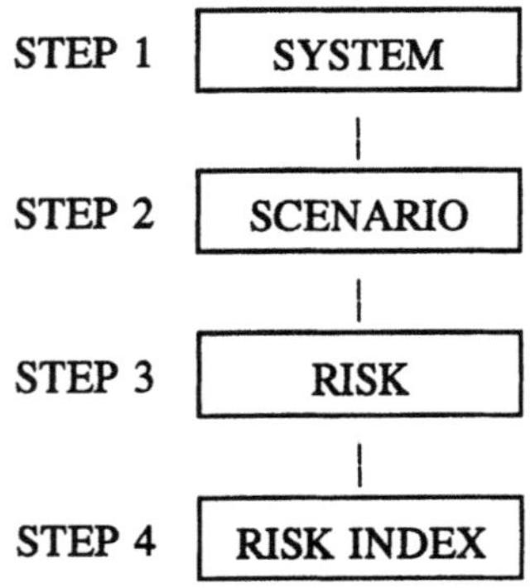

Figure 1: "Risk Index Method - Overview"

The individual theoretical elements and steps of the risk index method are defined below.

3.2 Step 1 - "System Definition"

A "system" is characterized by the complex and dynamic interaction of the system constituents. Hardware, software, human and environment are linked via functional and physical dependencies.

3.3 Step 2 - "Identification of Scenarios"

"Risk analysis" when applied to a system answers the following questions:

- what can go wrong
- how likely is it
- how good do we know about it

The identification of undesirable scenarios corresponds to answering the question "what can go wrong".

A "scenario" consists of the following elements:

- undesirable events
- consequence

Undesirable events are defined as the initiators and propagation phenomena induced by the effects of hazardous system characteristics and functional failures in the system. A consequence is defined as the last event of an event chain that makes a scenario. With each consequence a consequence severity is associated. The consequence severity can be measured in qualitative or quantitative terms.

The identification of the likelihood of the undesirable scenario corresponds to answering the question "how likely is it". The likelihood is expressed by a qualitative or quantitative statement. The degree of subjectivity of the statement is determined by the degree of pure expert judgement involved in the assessment of the likelihood.

The identification of the uncertainties associated with the likelihoods corresponds to answering the question "how good do we know about it". The uncertainty is represented as a distribution function over the likelihood.

3.4 Step 3 - "Identification of Risk"

The "risk" R implied by a scenario S is defined as a function r

$$R = r[S(C),P(C),U(P)]$$

where S denotes the consequence severity of the consequence C, P the likelihood of occurrence and U the associated uncertainty. Risk is an increasing function of S, P and U.

Three sources of uncertainty associated with likelihoods can be defined. The first source is the randomness of occurrence of events leading to randomness of occurrence of the consequence. Randomness is introduced by stochastic physical phenomena. The second source is the fuzziness in event definitions and propagation phenomena. Fuzziness is introduced by imprecision in description of event characteristics and of physical modeling of system response to undesired events. The third source is lack of knowledge. Uncertainty measures the state of knowledge on the scenario under consideration and it becomes possible to treat "no information" as "important information".

3.5 Step 4 - "Determination of Risk Index"

The risk R is converted into a "risk index" I(R) in two steps.

In the first step the risk R associated with a scenario is mapped into the "risk grid" as defined in figure 2. This requires the transformation of S into the severity index I(S) and P together with U into the likelihood index I(P')

$$S \rightarrow I(S)$$
$$P \text{ with } U(P) \rightarrow I(P')$$

The uncertainty U associated with P - represented as a distribution over P - is "concentrated" into one representative point value P' of the distribution which is then mapped into I(P'). A typical concentration operation is the mean operation.

Both indices are based on fuzzy logic because they are defined as

$$0.0 < I(S) < 1.0$$
$$0.0 < I(P') < 1.0$$

The value 1.0 of the index corresponds to the maximum of the severity or likelihood and 0.0 to the minimum. Quantitative and qualitative scales with quantitative and qualitative expressions can be associated with the indices.

To each risk R corresponds a point "o" in the grid.

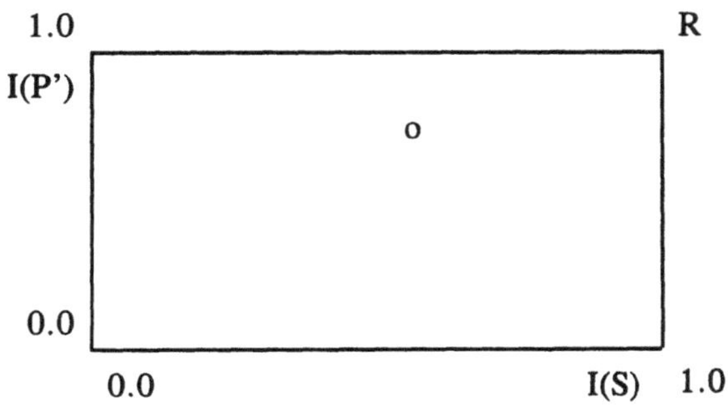

Fig. 2: Risk Grid

The transformation law for the consequence severity S into I(S) has to be defined on a case by case by basis due to the different possible qualitative and quantitative scales for the consequence severity.

The transformation of the likelihood P into I(P') has to be defined on a case by case basis due to the different possible qualitative and quantitative scales for the likelihood.

In the second step the risk R represented in the grid is converted into the "risk index" I(R) using the general formula

$$I(R) = I(S) * I(P')$$

$$0.0 < I(R) < 1.0$$

The risk index is based on fuzzy logic because the value 1.0 of the index corresponds to the "maximum risk" and 0.0 to "no risk" or "minimum risk". The various degrees of risk range between 1.0 and 0.0.

This index summarizes all information including uncertainty about the risk.

The operator "*" has to be defined on a case by case by basis. In some cases "*" will be the product, in other cases "*" will be of a qualitative form.

The risk index concept also allows the cumulation "&" of risks Ri, Rj into the total risk Rt using the general formula

$$I(Rt) = I(Ri \,\&\, Rj) = I(Ri) \# I(Rj)$$

The operator "#" has to be defined on a case by case by basis. A general constraint for "#" is

$$Rt > \max\{Ri, Rj\} \rightarrow I(Rt) > \max\{I(Ri), I(Rj)\}$$

Finally the risk R is displayed on the "risk index scale" as defined in figure 3.

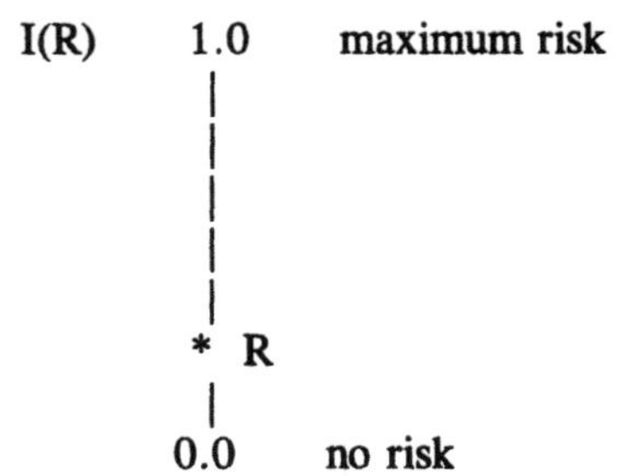

Fig. 3: Risk Index Scale

The upper end of the scale corresponds to the cumulation of all highest possible risks.

4. RISK INDEX METHOD - CASE STUDY APPLICATION

4.1 Overview of Case Study

A simple case study is used to illustrate the general theoretical concept by a specific application.

During the development of the space craft a preliminary design review is held and top management is requested to authorize the go ahead after risk management decisions are made. To support the decision process and communicate the risk results the risk index method is used.

4.2 Step 1 - "System Definition"

The system is a manned space craft for a mission to Mars.

Major design problems exist with respect to the propulsion subsystem during the launch phase.

4.3 Step 2 - "Identification of Scenarios"

A risk analysis is performed and many credible accident scenarios are found. Most of the dominating scenarios involve the propulsion subsystem during on orbit operation.

Three associated scenarios are:

a engine explosion leading to explosion of space craft and loss of crew

b leak in propellant line with loss of fuel leading to loss of mission

c burst of propellant tank leading to severe damage of space craft

Five qualitative consequence severity categories are introduced and are used to classify the scenarios:

scenario	severity		consequence
a	V	catastrophic	loss of total crew and/without spacecraft
	IV	very serious	loss of crewmember and/without damage to spacecraft
b, c	III	serious	major damage
	II	critical	damage
	I	marginal	small damage to spacecraft

The consequences and severities are defined such that one level always includes the associated lower levels.

The likelihoods of the scenarios are determined using mainly expert judgement and use is made of a combined qualitative - quantitative scale:

scenario	likelihood		
	5	extremely high	0.0 - 0.1
a	4	high	0.1 - 0.01
a, b	3	medium	0.01 - 0.001
c	2	low	0.001 - 0.0001
	1	extremely low	0.0001 - 0.0

4.4 Step 3 - "Identification of Risk"

The risks associated with the three scenarios are :

scenario	severity	likelihood with uncertainty	
a	V	4,3	0.1 - 0.001
b	III	3	0.01 - 0.001
c	III	2	0.001 - 0.0001

4.5 Step 4 - "Determination of Risk Index"

Using the definition of the consequence severity and likelihood categories the risk grid is introduced. As both scales are divided in five categories the risk grid consists of 25 boxes with the corresponding pairs of severity/likelihood categories:

1.0

I(P)					
	I 5	II 5	III 5	IV 5	V 5
	I 4	II 4	III 4	IV 4	V 4
	I 3	II 3	III 3	IV 3	V 3
	I 2	II 2	III 2	IV 2	V 2
0.0	I 1	II 1	III 1	IV 1	V 1

0.0 I(S) 1.0

The mean operation is used to concentrate the likelihood with uncertainty of scenario "a" into the point value P'(a). The result is in category 4.

Now the risks of the individual scenarios can be plotted in the risk grid:

5					
4					a
3			b		
2			c		
1					
	I	II	III	IV	V

The risk index scale for the scenarios is defined. The upper end of the scale corresponds to the highest possible risk of a scenario and the lower end corresponds to the lowest possible risk of a scenario:

risk category		index scale
		1.0
A	maximum	
B	very high	
C	high	
D	medium	
E	low	
F	very low	
G	minimum	
		0.0

In addition to the numeric risk index scale a qualitative risk categorization was introduced.

Using the risk categorization the 25 boxes are ranked in the risk grid:

5	F	D	C	B	A
4	F	E	D	C	B
3	G	F	E	D	C
2	G	F	F	E	D
1	G	G	F	F	F
	I	II	III	IV	V

The ranking corresponds to a large extent to a risk index that is defined by the product of the severity index and the likelihood index. Asymmetry reflects lower ranking of very low probability scenarios.

Now the risk indices of the individual scenarios can be identified and used for comparison and ranking:

risk category		index scale	
		1.0	
A	maximum	|	
B	very high	a	scenario
C	high	|	
D	medium	|	
E	low	b	scenario
F	very low	c	scenario
G	minimum	|	
		0.0	

Scenario a is characterized by a much higher risk than scenarios b and c. No cumulative risks evaluations are performed.

Based on this result top management requests risk reduction with respect to scenario a before the authorization to go ahead is given.

5. CONCLUSION

The concept of risk index is not new. For example concepts similar to the risk grid have been in use by the aerospace industry.

What has been missing is a standard method that constitutes an upgrade and integration of some of the already available concepts. The proposed risk index method provides a general framework for such a standard concept and aims at supporting decision making and risk communication. In order to keep the method simple and mathematically sound some meaningful basic elements of fuzzy logic and established probability concepts have been applied. The risk index method involves mapping risk parameters into a non physically measurable scale with the aim to reduce misinterpretation and misuse of mainly probabilistic information.

REFERENCES

C.Preyssl, et al. "Safety Risk Assessment for ESA Space Programs" ESA Symposium 'Space Product Assurance for Europe in the 1990s', ESA SP-316, Noordwijk - The Netherlands, 1991

C.Preyssl "Fuzzy Risk Analysis for Space Flight Systems" Int. Workshop on Fuzzy System Applications, Kyushu Institute of Technology, Iizuka - Japan, 1988

RISK-BASED RANKING IN INSPECTION OF INDUSTRIAL FACILITIES

Ahmed A. Ibrahim[1] and Bilal M. Ayyub[2]
[1]Department of Construction Technology and Environmental Engineering, Higher Technological Institute, 10th City of Ramadan, P.O.Box: 228, Egypt
[2]Department of Civil Engineering, University of Maryland, College Park, MD 20742, USA

ABSTRACT

Inspection can play a significant role in reducing the likelihood of unexpected structural failures. However, for many critical components and systems that are required to maintain pressure boundary integrity or that are subjected to severe service conditions, inspection requirements for these vital components are either established based upon prior experience and engineering judgment or are non-existent. Most inspection requirements or guidelines, if they exist, are usually established with only an implicit consideration of risk.

Recent catastrophic structural failures over the past decade highlight the societal need to relate more explicitly risk and uncertainty with inspection programs. In this study, risk-based methods with uncertainty evaluation and propagation were developed for the purpose of developing inspection strategies. Interval analysis was used for this purpose which is a form of fuzzy arithmetic. The methods result in establishing priority ranking lists for failure modes and components where actions need to be taken accordingly. The ranking priority lists for inspection purposes were based on the assessments of the probabilities of failure, resulting consequences, expected human and economic risks and the uncertainties associated with these assessments.

KEYWORDS

consequences; fuzzy; inspection; interval; probability; ranking; risk; and uncertainty

1. INTRODUCTION

A major goal of reliability and safety analyses is to control the probability of failure of a complex system, and any potential human, economic and environmental losses. Losses occur when one or more failure events create a system hazard. There are three basic failure events which lead to a system hazard. They are: (1) events related to human beings, e.g., operator error, design error and maintenance error; (2) events related to hardware, e.g., leakage of toxic fluid from a valve, loss of lubrication in a motor and incorrect measurement by a sensor; and (3) events related to the environment, e.g., earthquakes or ground subsidence, storm, flood, tornado and ignition caused by sparks or lightning. System hazards are frequently caused by a combination of failure events, e.g., hardware failures plus human error and/or environmental fault events. Some typical policies used to minimize and control hazards and risk include: (1) redundancy; (2) inspection and maintenance; (3) protective systems; and (4) alarm displays. These policies can result in changing the failure probability or the losses. For example, the inspection operation detects component degradation, e.g., crack growth, such that repair or replacement actions can be instituted prior to component failure; resulting in controlling and improving the component reliability and system reliability as well as reducing losses.

In this study, risk-based ranking methods were developed for inspection purposes. The methods were based on the assessments of failure modes, failure probabilities, consequences and the uncertainties associated with them. In these methods, the consequences can include property damages, injuries and fatalities. Single and multiple component failure modes were considered, as for some components several failure modes may exist. The uncertainties associated with the failure of components and the consequences were propagated to the risk measures which include human and economic risks. The human and economic risks were modeled as the products of the probabilities of failure and the corresponding magnitudes of human fatality and property damage, respectively.

The purpose of the uncertainty analysis and propagation is to combine the uncertainties of the various input quantities in an analysis, e.g., probabilities of failure of components and magnitudes of damages or fatalities resulting from failures, to yield a probability distribution for the desired output quantity, e.g., human and economic risks. The uncertainties for the input quantities are typically obtained as a result of a data analysis. There are four basic methods of uncertainty propagation as modeled by Ang and Tang (1975, 1984) and Bier (1987). They are the method of moments, Monte Carlo simulation, method of discrete probability distribution and integration method. In addition, interval risk estimates were considered as a suggested method for risk estimation and propagation of uncertainties. The analysis according to this method was based on the mathematics of intervals proposed by Moore (1966).

The objective of the methods is to construct an inspection priority list where action can be taken accordingly. The priority list is determined based on human and economic risks for all the failure modes and components of the system under investigation. The ranking is based on the severity potential of human and/or economic risks with uncertainty considerations, and it is performed from highest to lowest. In the case of tied rankings, the statistically appropriate approach for determining the ranking of the tied components or failure modes is to average their ranks. However, in this study the ranking of ties is taken as the least rank among them. This rank is assigned to all of them. The next available rank is then taken as the least rank plus the number of tied components. This practice is commonly used in the developments of ranks. Several examples were used to illustrate the suggested methods and comparatively discuss their results.

2. UNCERTAINTY PROPAGATION

As mentioned in the previous section, there are four basic techniques for uncertainty propagation available in the literature. The method of moments propagates the means and variances of the input quantities through the risk model to yield the means and variances of the risk output quantities. Monte Carlo simulation is based on random sampling from the probability distributions for the input quantities, whereas the method of discrete probability distributions is based on systematic sampling. The integration method deals with obtaining the distribution of the output from the distributions of the input quantities. In the interval analysis, uncertainty is propagated by modeling input quantities as intervals and using the mathematics of intervals, resulting in interval output quantities. In this section some of these methods are briefly outlined along with a discussion about their advantages and disadvantages.

2.1. Method of Moments

The method of moments is quite simple in the case where the input quantities are all mutually uncorrelated and expressed in the form of first and second moments as described in Ang and Tang (1975). For example, if X represents a probability of failure and Y represents a magnitude of consequence and they are mutually uncorrelated random variables, then the risk of failure Z = X Y is the desired output quantity, therefore the mean value and variance of the output quantity are given by:

$$\mu_Z = \mu_X \, \mu_Y \tag{1}$$

$$\mathrm{Var}(Z) = \mathrm{Var}(X)\,\mathrm{Var}(Y) + \mathrm{Var}(X)\,[\mu_Y]^2 + \mathrm{Var}(Y)\,[\mu_X]^2 \tag{2}$$

where μ = mean value and Var = variance of a random variable.

In the case where the input quantities are mutually correlated, the expressions for the mean and variance of the output quantity are more complicated. However, for many expressions and many types of input distributions, the mean and variance of the desired output quantity can still be evaluated analytically as described in Ang and Tang (1975). The case of correlated input quantities is not considered as a part of this study.

The advantage of this method is that in many cases it yields closed-form analytical results. Therefore, it is possible to see exactly how the results depend on the means and variances of the input quantities. However, the method of moments yields only the mean and variance of the desired output quantity, not the underlying distribution. There are exceptions to this general rule, for example, for X and Y that are lognormally distributed, the distribution for Z is Lognormal. For cases with unknown distribution, a distribution type can be assumed for the output quantity, introducing a possible source of inaccuracy.

2.2. Integration Method

The integration method described in Ang and Tang (1975) and Mann, et al (1974), deals with obtaining the distribution of the output quantity based on their relationship and the distributions of the input quantities. For example, let X and Y be known continuous input random variables with probability density functions $f_X(x)$ and $f_Y(y)$, respectively, and $Z = g(x, y) = X\,Y$ be the continuous output random variable; the cumulative probability distribution function of Z is:

$$F_Z(z) = \int\int f_{X,Y}(x,y)\, dx\, dy \tag{3}$$

where $f_{X,Y}(x,y)$ = joint probability density function of X and Y. The probability density function of Z can be expressed by:

$$f_Z(z) = \int f_{X,Y}(g^{-1},y) \left| \frac{\partial g^{-1}}{\partial z} \right| dy \tag{4}$$

where $g^{-1} = g^{-1}(z,y)$ = inverse of g(z,y) in terms of z and y. For independent X and Y, $f_Z(z)$ can be expressed as:

$$f_Z(z) = \int f_X(g^{-1})\, f_Y(y) \left| \frac{\partial g^{-1}}{\partial z} \right| dy \tag{5}$$

In the case of $Z = X\,Y$, $g^{-1} = \frac{z}{y}$ and $\left| \frac{\partial g^{-1}}{\partial z} \right| = \frac{1}{y}$.

The advantage of this method is that it yields the complete probability distribution function of the output quantity, and therefore the mean and variance can be calculated based on this distribution. The drawback of this method is in performing the integration specially if the input variables are not independent. In such cases, a numerical integration might be required.

2.3. Interval Analysis

An interval value is a range of numbers with lower and upper limits that is assigned to some or all of the input quantities in a risk model to account for uncertainties within these parameters. For example, I = [a,b] is an interval value ranging from a lower limit (a) to an upper limit (b). The operations of interval analysis as proposed by Ibrahim And Ayyub (1990), can be used to determine uncertainty propagation in the analytical process of risk estimate. The algebraic operations of interval values are extensions of operations on real numbers as discussed by Moore (1966) and Dong et al (1987). For example, if $I_1 = [a,b]$ and $I_2 = [c,d]$ are two interval values where, d > c and b > a, then the following operations are defined:

$$[a,b] \times 0 = 0 \tag{6}$$

$$[a,b] + [c,d] = [a+c, b+d] \tag{7}$$

$$[a,b] - [c,d] = [a-d, b-c] \quad (8)$$

$$[a,b] \times [c,d] = [ac, bd] \quad (9)$$

$$[a,b] / [c,d] = [a,b] \times [1/d, 1/c] \quad \text{if } 0 \notin [c,d] \quad (10)$$

These operations are considered a special case of fuzzy arithmetic as shown by Kaufman and Gupta (1985). In order to rank two quantities where their magnitudes are given by interval values, for example $I_1 = [a,b]$ and $I_2 = [c,d]$, the following logic operations are suggested:

$$\text{IF } d > b \text{ THEN } I_2 > I_1. \quad (11)$$

$$\text{IF } d = b \text{ AND IF } c > a \text{ THEN } I_2 > I_1. \quad (12)$$

$$\text{IF } d = b \text{ AND } c = a \text{ THEN } I_2 = I_1 \quad (13)$$

$$\text{Otherwise } I_2 < I_1 \quad (14)$$

where $I_2 <$ (or $>$) I_1 means that the quantity I_1 is ranked at a higher (or lower) severity in magnitude level than the quantity I_2, and $I_2 = I_1$ means that the quantities I_1 and I_2 are ranked at the same level.

3. EXAMPLE

In this section, an example of risk estimation is discussed. The economic and the human risks were calculated for the system components, the component failure modes and the system as a whole. The system example is based on components with single failure modes and components with multiple failure modes. The components were assumed to be in series and the probability of more than one element failure is assumed to be negligible. The example was evaluated both by the method of integration and by the method of interval analysis. In addition, a simple risk evaluation was performed based on a single point estimate where the input values were represented by single values. In the latter case there was no uncertainty analysis or propagation in evaluating the risk. In all cases, components and failure modes were ranked and a comparison was performed. The human and economic risks were modeled as the products of the probabilities of failure and the corresponding magnitudes of human fatality and property damage, respectively.

3.1. Single Point Estimate

Table I shows a system composed of 6 components in column 1. Some of the components exhibit single failure modes, others exhibit multiple failure modes as shown in column 2. Each mode of failure had a single point estimate for the probability of failure, as shown in column 3. The probabilities of failure were selected to rang from $1.00\text{x}10^{-13}$ to $1.00\text{x}10^{-4}$. The magnitudes of casualty were selected to range form 0 to 100000 in some units as shown in column 4. The magnitudes of damage were selected to range from 45 to 10000 in some units as shown in column 5. The human risk, column 6, due to a component failure mode was determined as the product of its probability of failure, column 3, and the corresponding magnitude of casualty, column 4. The economic risk, column 7, due to a component failure mode was determined as the product of its probability of component failure, column 3, and the corresponding magnitude of damage, column 5.

The failure probability of the system is bounded by upper and lower limits as shown at the end of column 3. The lower and upper limits correspond to statistically correlated (positive) and statistically independent failure events of the failure modes of the components, respectively. The limits were based on the assumption that the failure modes of the components are in series. Therefore, the limits for the probability of system failure were determined as follows:

$$\max_{i=1}^{i=10} P_{fci} \leq P_{fs} \leq 1 - \prod_{i=1}^{i=10} (1 - P_{fci}) \tag{15}$$

where P_{fci} = the probability of component failure mode i and P_{fs} = the probability of system failure. The estimated probability of system failure can be assumed as the geometric average of the two system failure limits as shown at the end of column 3. The geometric average of (n) values, e.g., y_1, y_2, ..., and y_n, is defined as:

$$\text{geometric average} = (y_1 \, y_2 \ldots y_n)^{1/n} \tag{16}$$

In this case the single point estimate for the system probability of failure was determined as the square root of the product of the two failure limits for the system.

The risk estimates for the system were determined as the summation of all the risks of its components failure modes as shown in Table I. The human risk values are shown at the end of column 6. The economic risk values are shown at the end of column 7. The components and their failure modes can be ranked according to human and economic risks as shown in columns 8 and 9, respectively.

3.2. Interval Analysis

In order to perform an interval analysis, Table II was constructed similar to Table I to facilitate a comparison between the two solutions. The probabilities of component failure modes, and the magnitudes of casualty and damage were expressed in the form of intervals as shown in Table II. The single point estimates for probability of failure modes of Table I are shown in Table II as the best estimates in column 4, and interval values were assumed for each best estimate as a range bounded by a lower and upper values as shown in columns 3 and 5, respectively. Similarly, lower and upper values for the magnitudes of casualty were assumed as shown in columns 6 and 8, respectively, and lower and upper values for the magnitudes of damage were assumed as shown in columns 9 and 11, respectively. The resulting estimates of human and economic risks were expressed in the form of intervals. The mathematical operation that was used for this purpose is given by Equation 9. Columns 12 and 13 show the lower and upper estimate values for the human risk. Columns 14 and 15 show the lower and upper estimate values for the economic risk.

Table I. Point Estimate Method

Component	Failure Mode	Probability of Faliure	Magnitude of				Ranking according to	
			Casualties	Damage	Human Risk	Economic Risk	Human Risk	Economic Risk
(1)	(2)	(3)	(4)	(5)	(6)	(7)	(8)	(9)
1	Mode 1	1.00E-13	0.00E+00	2.00E+02	0.00E+00	2.00E-11	8	10
	Mode 2	1.00E-04	0.00E+00	1.00E+02	0.00E+00	1.00E-02	8	1
2	Mode 1	1.00E-07	1.00E+05	1.00E+04	1.00E-02	1.00E-03	1	2
3	Mode 1	1.00E-10	1.00E+03	3.50E+03	1.00E-07	3.50E-07	7	7
	Mode 2	1.00E-08	3.00E+02	2.00E+02	3.00E-06	2.00E-06	3	5
4	Mode 1	1.00E-08	3.40E+01	7.00E+02	3.40E-07	7.00E-06	6	4
	Mode 2	1.00E-12	0.00E+00	5.00E+02	0.00E+00	5.00E-10	8	9
	Mode 3	1.00E-08	5.00E+01	9.00E+01	5.00E-07	9.00E-07	4	6
5	Mode 1	1.00E-05	7.50E+02	4.50E+01	7.50E-03	4.50E-04	2	3
6	Mode 1	1.00E-10	4.00E+03	4.30E+02	4.00E-07	4.30E-08	5	8
System Upper Limit		1.10E-04						
System Lower Limit		1.00E-04						
System Average		1.05E-04			1.75E-02	1.15E-02		

Table II. Interval Analysis

Component	Failure Mode	Probability of Failure			Magnitude of Casualties			Magnitude of Damage			Human Risk		Economic Risk		Ranking according to	
		Lower	Best Estimate	Upper	Lower	Best Estimate	Upper	Lower	Best Estimate	Upper	Lower	Upper	Lower	Upper	Human Risk	Economic Risk
(1)	(2)	(3)	(4)	(5)	(6)	(7)	(8)	(9)	(10)	(11)	(12)	(13)	(14)	(15)	(16)	(17)
1	Mode 1	1.00E-16	1.00E-13	1.00E-08	0.00E+00	0.00E+00	3.00E+01	7.00E+01	2.00E+02	3.00E+03	0.00E+00	3.00E-07	7.00E-15	3.00E-05	8	9
	Mode 2	1.00E-06	1.00E-04	1.00E-03	0.00E+00	0.00E+00	0.00E+00	9.00E+01	1.00E+02	1.50E+02	0.00E+00	0.00E+00	9.00E-05	1.50E-01	10	2
2	Mode 1	1.00E-09	1.00E-07	1.00E-05	3.00E+03	1.00E+05	5.00E+05	2.00E+03	1.00E+04	1.00E+05	3.00E-06	5.00E+00	2.00E-06	1.00E+00	2	1
3	Mode 1	1.00E-12	1.00E-10	1.00E-07	3.00E+01	1.00E+03	5.00E+04	7.00E+02	3.50E+03	4.00E+03	3.00E-11	5.00E-03	7.00E-10	4.00E-04	4	7
	Mode 2	1.00E-10	1.00E-08	1.00E-06	2.00E+02	3.00E+02	3.50E+02	2.00E+01	2.00E+02	5.00E+02	2.00E-08	3.50E-04	2.00E-09	5.00E-04	5	6
4	Mode 1	1.00E-09	1.00E-08	1.00E-05	2.00E+01	3.40E+01	5.00E+02	5.00E+02	7.00E+02	7.00E+03	2.00E-08	5.00E-03	5.00E-07	7.00E-02	3	3
	Mode 2	1.00E-12	1.00E-12	1.00E-09	0.00E+00	0.00E+00	1.20E+01	4.00E+02	5.00E+02	5.50E+02	0.00E+00	1.20E-08	4.00E-10	5.50E-07	9	10
	Mode 3	1.00E-12	1.00E-08	1.00E-06	4.00E+01	5.00E+01	6.00E+01	1.00E+01	9.00E+01	1.00E+02	4.00E-11	6.00E-05	1.00E-11	1.00E-04	7	8
5	Mode 1	1.00E-10	1.00E-05	1.00E-04	5.00E+02	7.50E+02	7.00E+04	3.00E+01	4.50E+01	3.00E+02	5.00E-08	7.00E+00	3.00E-09	3.00E-02	1	4
6	Mode 1	1.00E-13	1.00E-10	1.00E-09	2.00E+03	4.00E+03	8.00E+04	4.00E+02	4.30E+02	1.00E+06	2.00E-10	8.00E-05	4.00E-11	1.00E-03	6	5
System Upper Limit		1.00E-06		1.12E-03												
System Lower Limit		1.00E-06		1.00E-03												
System Best Estimate		1.00E-06		1.12E-03							3.09E-06	1.20E+01	9.25E-05	1.25E+00		

The probability of the system failure is bounded by upper and lower limits as given by Equation 15. Equations 6 to 10 were used to perform the operations of Equation 15. the result is an interval for each limit. For example, the upper limit of the probability of failure of the system is the right hand side of Equation 15, and since P_{fci} of each failure mode was expressed as an interval value the result is an interval with lower and upper limits. This interval is shown in the cells identified as (system upper limit-column 3) and (system upper limit-column 5) for lower and upper interval values, respectively. On the other hand, the lower limit of the system probability of failure is the left hand side of Equation 15, and since the maximum component probability of failure is an interval, the lower limit of the system is an interval with lower and upper values as shown in the cells identified as (system lower limit-column 3) and (system lower limit-column 5), respectively.

In order to cover all the uncertainties and correlation levels of the components, a best estimate for the lower limit of the system probability of failure was selected as the smaller value of the lower limit interval as shown in the cell (system best estimate-column 3), and the best estimate for the upper system limit was taken as the larger value of the upper limit interval as shown in the cell (system best estimate-column 5).

The resultant risk estimates for the system is in the form of an interval. The system risk was determined as the summation of all the risks of its components as shown in Table II. The lower and upper human risk values of the system are shown in the cells identified as (system best estimate-column 12) and (system best estimate-column 13), respectively. The lower and upper economic risk values for the system are shown in the cells identified as (system best estimate-column 14) and (system best estimate-column 15), respectively. The rankings of components and/or failure modes were determined using the logic provided by Equations 11 to 14, as shown in columns 16 and 17 for human and economic risks, respectively.

3.3. Integration Method

In this case, density function were assigned to the probabilities of failure and the magnitudes of fatality or damage for the components. Each density function $f_X(x)$ for a random variable X was assumed to be uniform with a finite range defined by the lower and upper limits for each component as used in the interval analysis of the previous section. As an example, the magnitude of the density function of component 2 of the system is $\frac{1}{1.00\mathrm{x}10^{-5} - 1.00\mathrm{x}10^{-9}}$ as shown in Figure 1.

The mean value of a uniform continuous probability distribution function can be expressed as:

$$\mu_X = \int_a^b X\, f_X(x)\, dx = \frac{b+a}{2} \quad (17)$$

where a and b are the lower and upper limits of the uniform distribution, respectively. The variance can be expressed as:

$$Var(X) = E(X^2) - \mu_X{}^2 = \frac{(b-a)^2}{12} \tag{18}$$

where $E(X^2)$ is

$$E(X^2) = \int_a^b X^2 f_X(x)\, dx \tag{19}$$

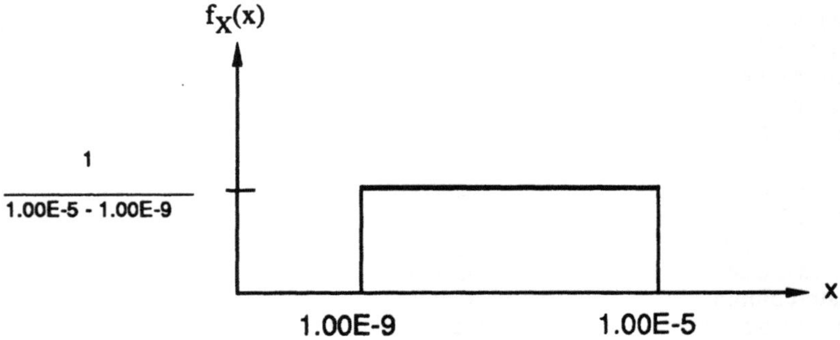

Fig. 1. Density function of component 2 of the system

In Table III, columns 3 and 6 were considered as the boundaries of the assigned uniform distribution to the probability of failure for each component failure mode. Columns 7 and 10 were taken as the boundaries of the assigned uniform distribution to the corresponding magnitudes of casualty. Columns 11 and 14 were considered as the boundaries of the assigned uniform distribution to the corresponding magnitudes of damage. The means and variances of the probability of failure, magnitudes of casualty and damage are shown in columns 4, 5, 8, 9, 12, and 13, respectively. These values were calculated using Equations 17 and 18.

The human and economic risks for the component failure modes were defined as the product of the respective probability of failure and the magnitudes of casualty and damage, respectively. Equation 5 was used to calculate the distribution functions of the risks, in which $f_X(x) = 1/(b-a)$ and $f_Y(y) = 1/(d-c)$ are continuous uniform probability density functions representing the probability of failure and magnitude of consequence, respectively. Then the risk $Z = X\,Y$ has a probability density function $f_Z(z)$ which can be computed using Equation 5 as follows:

$$f_Z(z) = \int_{y_{lower}}^{y_{upper}} \left[\frac{1}{(b-a)}\right] \left[\frac{1}{(d-c)}\right] \left(\frac{1}{Y}\right) dy \tag{20}$$

Table III. Distribution Analysis

Component	Failure Mode	Probability of Failure				Magnitude of Casualties				Magnitude of Damage			
		Lower	Mean	Variance	Upper	Lower	Mean	Variance	Upper	Lower	Mean	Variance	Upper
(1)	(2)	(3)	(4)	(5)	(6)	(7)	(8)	(9)	(10)	(11)	(12)	(13)	(14)
1	Mode 1	1.00E-16	5.00E-09	8.33E-18	1.00E-08	0.00E+00	0.00E+00	0.00E+00	3.00E+01	7.00E+01	0.00E+00	0.00E+00	3.00E+03
	Mode 2	1.00E-06	5.01E-04	8.32E-08	1.00E-03	0.00E+00	0.00E+00	0.00E+00	0.00E+00	9.00E+01	1.20E+02	3.00E+02	1.50E+02
2	Mode 1	1.00E-09	5.00E-06	8.33E-12	1.00E-05	3.00E+03	2.52E+05	2.06E+10	5.00E+05	2.00E+03	5.10E+04	8.00E+08	1.00E+05
3	Mode 1	1.00E-12	5.00E-08	8.33E-16	1.00E-07	3.00E+01	2.50E+04	2.08E+08	5.00E+04	7.00E+02	2.35E+03	9.08E+05	4.00E+03
	Mode 2	1.00E-10	5.00E-07	8.33E-14	1.00E-06	2.00E+02	2.75E+02	1.88E+03	3.50E+02	2.00E+01	2.60E+02	1.92E+04	5.00E+02
4	Mode 1	1.00E-09	5.00E-06	8.33E-12	1.00E-05	2.00E+01	2.60E+02	1.92E+04	5.00E+02	5.00E+02	3.75E+03	3.52E+06	7.00E+03
	Mode 2	1.00E-12	5.01E-10	8.32E-20	1.00E-09	0.00E+00	0.00E+00	0.00E+00	1.20E+01	4.00E+02	0.00E+00	0.00E+00	5.50E+02
	Mode 3	1.00E-12	5.00E-07	8.33E-14	1.00E-06	4.00E+01	5.00E+01	3.33E+01	6.00E+01	1.00E+01	5.50E+01	6.75E+02	1.00E+02
5	Mode 1	1.00E-10	5.00E-05	8.33E-10	1.00E-04	5.00E+02	3.53E+04	4.03E+08	7.00E+04	3.00E+01	1.65E+02	6.08E+03	3.00E+02
6	Mode 1	1.00E-13	5.00E-10	8.33E-20	1.00E-09	2.00E+03	4.10E+04	5.07E+08	8.00E+04	4.00E+02	5.00E+05	8.33E+10	1.00E+06

Human Risk			Economic Risk			Ranking according to			
						Mean		90% Percentile	
Mean	Variance	90% Percentile	Mean	Variance	90% Percentile	Human Risk	Economic Risk	Human Risk	Economic Risk
(15)	(16)	(17)	(18)	(19)	(20)	(21)	(22)	(23)	(24)
0.00E+00	0.00E+00	0.00E+00	0.00E+00	0.00E+00	0.00E+00	8	9	8	9
0.00E+00	0.00E+00	0.00E+00	6.01E-02	1.30E-03	1.10E-01	8	2	8	2
1.26E+00	1.21E+00	3.05E+00	2.55E-01	4.84E-02	6.10E-01	2	1	2	1
1.25E-03	1.22E-06	3.10E-03	1.18E-04	7.63E-09	2.52E-04	4	7	3	7
1.38E-04	6.93E-09	2.54E-04	1.30E-04	1.20E-08	3.05E-04	5	6	5	6
1.30E-03	1.20E-06	3.05E-03	1.88E-02	2.35E-04	4.31E-02	3	3	4	3
0.00E+00	0.00E+00	0.00E+00	0.00E+00	0.00E+00	0.00E+00	8	9	8	9
2.50E-05	2.19E-10	4.53E-05	2.75E-05	4.77E-10	6.20E-05	6	8	7	8
1.76E+00	2.38E+00	4.27E+00	8.25E-03	4.29E-05	1.85E-02	1	4	1	4
2.05E-05	3.09E-10	4.88E-05	2.50E-04	4.86E-08	6.20E-04	7	5	6	5

resulting into

$$f_Z(z) = \left[\frac{1}{(b-a)}\right]\left[\frac{1}{(d-c)}\right]\left[\mathrm{Ln}\left(\frac{Y_{upper}}{Y_{lower}}\right)\right] \tag{21}$$

where Y_{upper} = smaller of $\left\{d\,,\,\frac{z}{a}\right\}$, and Y_{lower} = larger of $\left\{c\,,\,\frac{z}{b}\right\}$

The resulting distributions of the risks are not necessarily uniform. For example, the distribution of the economic risk for component 2 of the system has the shape shown in Figure 2. The mean and variance of the human and economic risks are shown in columns 15, 16, 18 and 19 of Table III, respectively. They were calculated based on the resulting distribution according to Equations 17 and 18. These values can be determined alternatively using Equations 1 and 2.

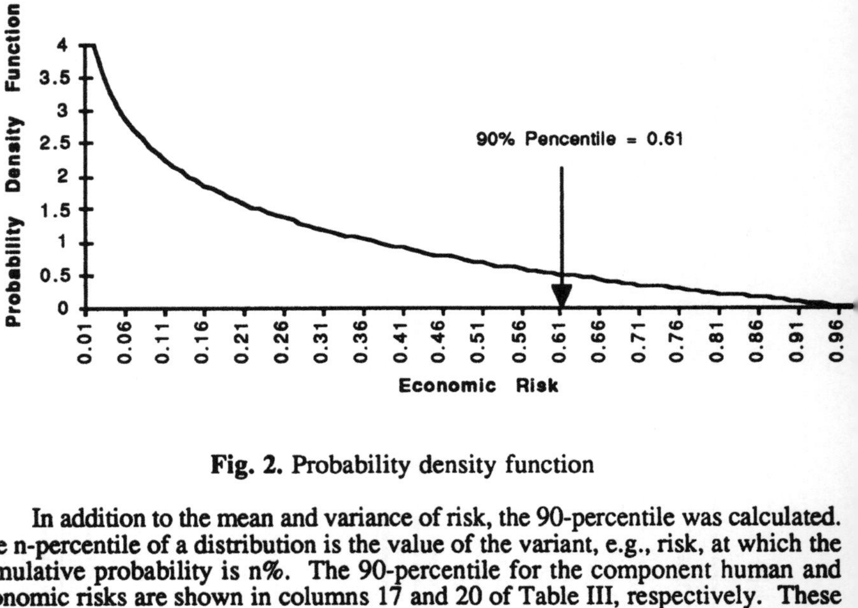

Fig. 2. Probability density function

In addition to the mean and variance of risk, the 90-percentile was calculated. The n-percentile of a distribution is the value of the variant, e.g., risk, at which the cumulative probability is n%. The 90-percentile for the component human and economic risks are shown in columns 17 and 20 of Table III, respectively. These values include uncertainty information since they depend on the variances. Ranking the components failure modes according to the mean value and the 90-percentile for the human and economic risks are shown in columns 21, 22, 23 and 24, respectively.

3.4. Comparative Discussion

Figures 3 and 4 illustrate the ranking comparison for the system components and their failure modes according to human and economic risks, respectively. In these Figures, the components failure modes were ranked based on the single point estimate method, interval estimate method and integration method. The results as shown in these Figures demonstrate the effect of the method of analysis on the ranking results. The inclusion of uncertainty in the evaluation by performing the

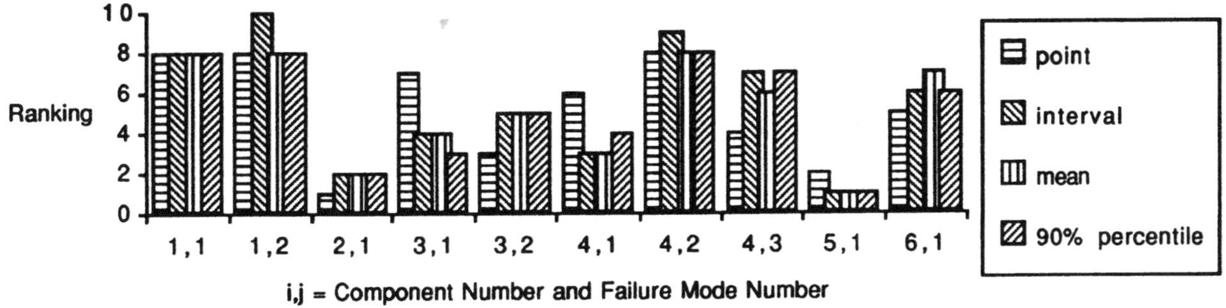

Fig. 3. Ranking of component failure modes based on human risk

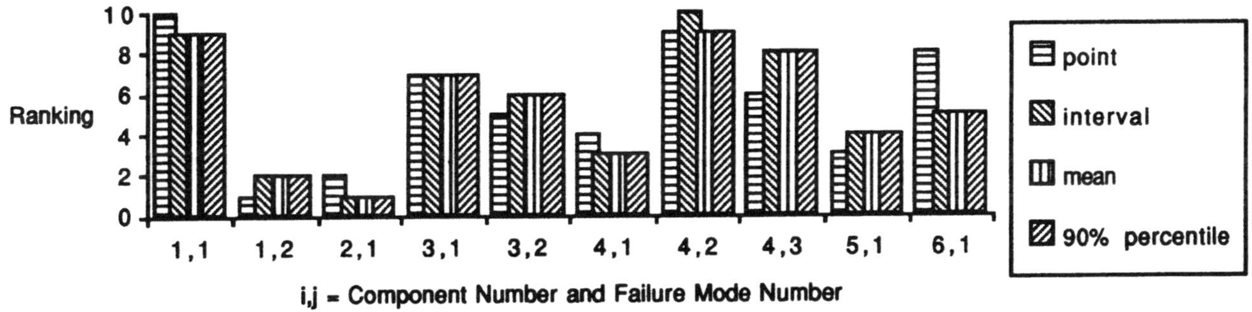

Fig. 4. Ranking of component failure modes based on economic risk

interval analysis can result in a different ranking compared to the single point estimate method. In the case where some components exhibit different failure modes, component ranking can be based on the complete ranking of all failure modes of the components. In such cases, failure modes of all the components are compiled in one evaluation process, and they are ranked based on either the single point estimate, interval analysis or integration method. After ranking the failure modes of the components, one can establish an inspection priority list for the components. Establishing this component list can be based on ranking the components according to the inspection priority of their failure modes. Therefore, a component with a higher ranked failure mode would have a higher ranking than other components.

4. GROUP RANKING

In many cases, as described in Henley and Kumamoto (1981) and by Farmer (1981), analysts are more interested in ranking components, sub-systems or systems in groups rather than ranking them individually, where the failure modes, components, sub-systems or systems in each group have relatively equal estimated attributes, e.g., risks. The group ranking can be based on the probability of failure, consequences, economic and/or human risks. The group ranking results in establishing a priority list for inspection.

As an example of group ranking, the component failure modes of the introduced system example shown in table I for single point estimates were ranked in groups according to the probability of failure, magnitude of damage and economic risk as shown in Figures 5, 6 and 7, respectively. For example, in Figure 7 the top inclined arrow line separated and grouped failure modes 1 and 2 of components 1 and 4, respectively in group C which has an economic risk less than 5.0×10^{-9}. It should be noted that the inclined arrow lines represent lines of equal (or uniform) economic risk. These lines can be considered as risk contours. Figure 8 is a resultant summary of group ranking based on the probability of failure, magnitude of damage and economic risk.

5. SUMMARY AND CONCLUSIONS

Risk-based ranking methods were developed for establishing priority lists for inspection purposes. The methods were based on the assessments of failure modes, failure probabilities, consequences and propagating the uncertainties associated with them through the risk of failure. The uncertainties associated with the failure probabilities of components and the consequences were propagated to the risk measures which include human and economic risks. Single and multiple component failure modes were considered. The representation of the values of the probability of failure and magnitudes of consequences by interval estimates provided a better risk estimation results in a more reliable ranking judgment for inspection purposes compared to the representations by single point estimates. The results were compared to the uniform probability distribution estimates. The ranking was performed at the failure mode, component, subsystem and system levels.

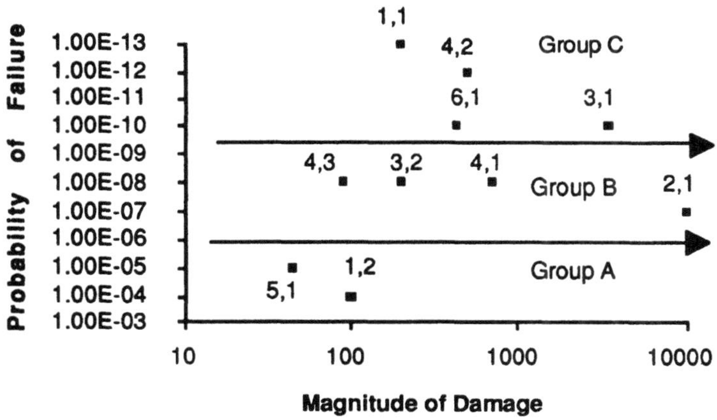

Fig. 5. Group ranking for failure probability

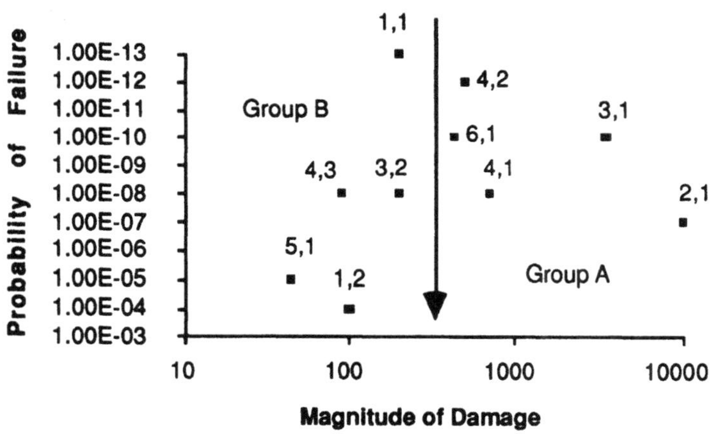

Fig. 6. Group ranking for magnitude of damage

NOTE: i,j = Component Number and Failure Mode Number, respectively

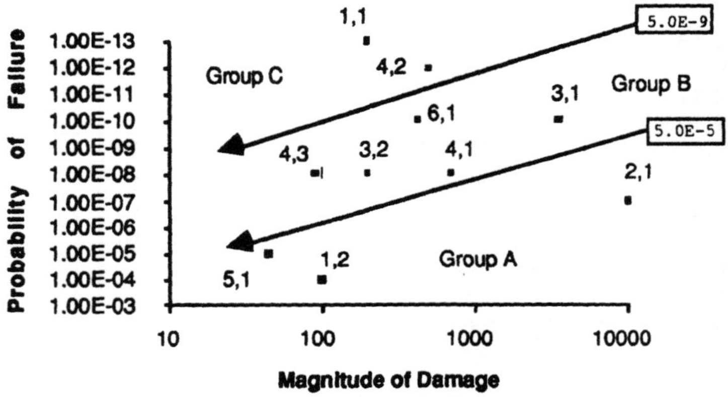

Fig. 7. Group ranking for economic risk

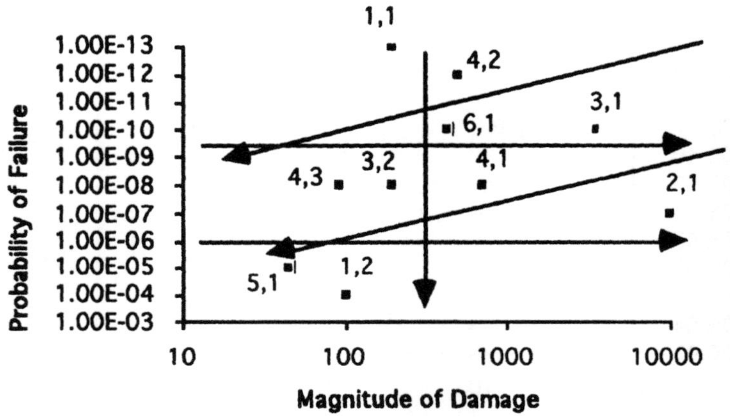

Fig. 8. Group ranking summary

NOTE: i,j = Component Number and Failure Mode Number, respectively

ACKNOWLEDGMENTS

This paper resulted from research projects funded by the American Society of Mechanical Engineers and Nuclear Regulatory Commission.

REFERENCES

Ang A. H-S. and Tang W. H. (1975) Probability concepts in engineering planning and design, Vol. 1-Basic Principles, John Wiley and Sons, Inc., New York, N.Y.

Ang A. H-S. and Tang W. H. (1984) Probability concepts in engineering planning and design, Vol. 2-Decision, Risk, and Reliability, John Wiley and Sons, Inc., New York, N.Y.

Bier V. M. (1987) "Uncertainty analysis as applied to probabilistic risk assessment," Uncertainty in Risk Assessment, *Risk Management, and Decision Making*. Plenum Publishing Corporation, 469-478.

Dong W. M. and Wong F. S. (1987) "From uncertainty to approximate reasoning: part 2:reasoning with algorithmic rules," *Civil Engineering Systems*, **3**(3), 192-202.

Farmer F. K. (1981) "Quantities of hazardous materials," *Proceedings of the NATO Advanced Study Institute on Technological Risk Assessment,* Erice, Sicily, Italy. Series E, Applied sciences, **81**, 85-101.

Henley E. J. and Kumamoto H. (1981) Reliability engineering and risk assessment, Prentice Hall, Englewood Cliffs, New Jersey.

Ibrahim A. and Ayyub B. M. (1990) "Modeling judgment uncertainty in inspection of industrial facilities," *Proceedings of ISUMA'90, The First International Symposium on Uncertainty Modeling and Analysis,* University of Maryland, College Park, Maryland, Published by IEEE-Computer Society, 463-469.

Kaufmann A. and Gupta M. (1985) Introduction to fuzzy arithmetic, Van Nostrand Reinhold.

Mann N. R. ,Schafer R. E. and Singpurwalla N. D. (1974) Methods for statistical analysis of reliability and life data, John Wiley & Sons, Inc, New York.

Moore R. E. (1966) Interval analysis, Printice-Hall, New Jersey.

A PROBABILISTIC-FUZZY MODEL FOR SEISMIC HAZARD

Dan M. Frangopol* and Kappyo Hong**

*Department of Civil Engineering
University of Colorado
Boulder, Colorado 80309-0428, U.S.A.

**Department of Architectural Engineering
Yonsei University
Seoul 120-749, Korea

ABSTRACT. A probabilistic-fuzzy model for seismic hazard analysis is developed. The proposed model is able to reproduce both the randomness and the imprecision in conjunction with earthquake occurrences. Results of this study are (a) membership functions of both peak ground accelerations associated with a given probability of exceedance and probabilities of exceedance associated with a given peak ground acceleration, and (b) characteristic values of membership functions at each location of interest. The proposed probabilistic-fuzzy model for assessment of seismic hazard is successfully applied to the Wasatch Front Range in Utah in order to obtain the seismic maps for different annual probabilities of exceedance, different peak ground accelerations, and different time periods.

Keywords: seismic hazard, fuzzy sets, probability theory.

1. INTRODUCTION

Even though deterministic methods for seismic hazard analysis are still used by many engineers, most of the earthquake occurrence models currently in use are based on the probabilistic Poisson model (i.e., time-independent model), which assumes a memoryless property of seismic occurrences. Cornell's (1968) original work on the time-independent model, also called the point-source model, and his subsequent pioneering efforts with Vanmarcke (1969), can be considered as the beginning of application of probabilistic concepts to time-independent seismic hazard prediction. Point-source models are based on the assumption that the energy released during an earthquake is radiated from the focus and the intensity of the site ground motion is a function of the distance to the source. Although this assumption may be acceptable for certain earthquakes and regions, it would not, however, be valid for large events where the total energy released is distributed along a rupture zone. The "fault-rupture model" developed by Der Kiureghian and Ang (1977) is based on the assumption that the earthquake originates at the focus and propagates symmetrically on each side of the focus along a fault, and the maximum intensity of the ground shaking at a site is determined by the rupture that is closest to the site. Bender (1985) further developed a "finite-fault rupture model" assuming that the entire rupture must be contained within the fault.

Although the Poisson model provides an adequate description for some seismic regions, time-dependent models have recently been developed. These models are based on the observation that the size of and the elapsed time since the last major earthquake are positively correlated. In 1984, Kiremidjian and Anagnos (1984) developed the stochastic slip-predictable model for the estimation of earthquake occurrences based on the premise that strain energy accumulates along a fault and is released during an earthquake event. In this model, the temporal dependence of events is presented through semi-Markovian modeling of the sequence of earthquake occurrences.

In order to improve probabilistic analysis models, several attempts have been made acknowledging the fact that seismic hazard for the same site is quite different depending on the experts and uncertainties they consider. The Yankee Atomic Electric Company (1984) methodology employed a logic tree approach to model the uncertainty. In the Lawrence Livermore National Laboratory (1981) methodology the zonation and estimation of the parameters of the recurrence model were selected by experts, and a multi-method approach was used for determining site specific spectra for Eastern United States based on the opinions of ground motion experts. This methodology encompasses a probabilistic approach for predicting peak acceleration, peak velocity and uniform hazard spectra for different time periods, and an empirical approach which includes the calculation of 50th and 84th percentile spectra from ensembles of real data at different magnitudes, site conditions, and distance ranges. Modeling uncertainties were introduced using alternative zonations and/or models as well as ranged of values for the seismicity parameters. These uncertainties were modeled in terms of probability distributions which were sampled, using Monte Carlo simulation, to describe the resulting uncertainty in the estimation of seismic hazard. In the Electric Power Research Institute (1985) seismic hazard methodology, the logic tree approach was applied to model the uncertainty using input provided by six seismicity teams. In this methodology, no attempt was made to model the uncertainty in the ground motion and only few equally weighted ground motion models were used to develop representative interim results.

Fuzzy mathematics was applied to seismic hazard by Lammarre and Dong (1986), among others. In this methodology, the knowledge necessary to perform this task is extracted from experts' opinions using questionnaires combining various factors involved in a seismic hazard assessment, such as ground motion, soil condition, and ground failure potential. Combination functions are then selected to reproduce the experts' opinions, and fuzzy sets theory (using vertex method) is used to combine the vague information to get a total evaluation of the seismic hazard for a given location.

In this study, probabilistic time-independent models and fuzzy sets theory are combined in an original probabilistic-fuzzy approach to seismic hazard analysis. The proposed approach is able to reproduce both the randomness and the imprecision in conjunction with earthquake occurrences. Results of this research are (a) membership functions of both peak ground accelerations associated with a given probability of exceedance and probabilities of exceedance associated with a given peak ground acceleration, and (b) characteristic values of membership functions at each location of interest. The proposed probabilistic-fuzzy model for assessment of seismic hazard is successfully applied to the Wasatch Front Range in Utah in order to obtain the seismic maps for different annual probabilities of exceedance, different peak ground accelerations, and different time periods.

2 PROBABILISTIC-FUZZY APPROACH

The general seismic hazard analysis (SHA) process for an individual site requires four steps: (a) source modeling (e.g., point, line or area sources), (b) magnitude-frequency relationship, (c) attentuation law, and (d) evaluation of the probability of exceeding a given ground motion level (often peak ground acceleration) within a specified exposure time (often annual). Presently, there are four ways by which the SHA is conducted: (a) using local codes; (b) using a deterministic seismic hazard evaluation; (c) using a probabilistic hazard evaluation; and (d) using a Bayesian seismic hazard evaluation. The advantages and disadvantages of each of these approaches have been summarized by Hong (1987) who pointed out that fuzzy sets theory could be used in order to conduct SHA. There are large uncertainties associated with each step of the SHA process. In general, these uncertainties are distinctive: randomness and fuzziness. Randomness is a variability that is inherent to the unpredictable nature of future earthquakes and is beyond our control, while fuzziness can be attributed to incompleteness of statistical data (i.e., estimation error), model imperfection (i.e., lack of understanding), and differences of opinion among experts. Expert opinions often play a key role in each step of seismic risk analysis, resulting in personal biases and some arbitrariness. Usually the expert opinion is vague, consisting mostly of language descriptions. Hence a method which includes the subjective opinion is needed in order to infer the final results from that information. In this study, fuzzy sets theory is combined with probability theory in order to include the subjective opinion.

Fuzzy sets theory, initiated by Zadeh (1965), deals with the subjective uncertainty factors in a quantitative way. The concept of fuzzy probability is particularly useful in the derivation of a new method of analysis in which the subjective information given by experts is integrated through the membership function instead of correction factors in the calculation of seismic risk. The theory of fuzzy sets deals with a subset A of the possibility space X, also termed the universe of discourse, where the transition between full membership and no membership is gradual rather than abrupt. The fuzzy subset

A has no well-defined boundaries in the possibility space X which covers a definite range of objects. Fuzzy classes of objects are often encountered in the real world. For instance, A may be the set of "enough earthquake data" in a region X. Traditionally, the grade of membership 1 is assigned to those objects that fully and completely belong to A, while 0 is assigned to objects that do not belong to A at all. The more an object x belongs to A, the closer to 1 is its grade of membership $\mu_A(X)$ in fuzzy sets theory.

Consider an example problem of application of probabilistic fuzzy approach to seismic hazard analysis. A line source symmetric with respect to a site is considered. For this source the following assumptions are made: the maximum magnitude m_{max} is 8; the shortest fault-to-site distance r_0 is $20km$; the length of the fault ℓ is $500km$; the Donovan's (1973) attenuation function is used; the slope for the Gutenberg-Richter law of magnitude is prescribed; and the slip-length relationship is given according to Bonilla and Buchanan (1970). In this example problem, the maximum magnitude, M_{max}, the shortest distance to the fault, R_0, and the length of fault, L, are assumed as fuzzy variables with the following membership functions:

$$\begin{aligned} \mu_{M_{max}}(m) &= \\ &= 0.0|7.5 + .36|7.6 + .64|7.7 + .84|7.8 + .96|7.9 + 1.0|8.0 + \\ &+ .96|8.1 + .84|8.2 + .64|8.3 + .36|8.4 + 0.0|8.5 \end{aligned} \tag{1}$$

$$\begin{aligned} \mu_{R_0}(r_0) &= \\ &= 0.0|16 + .5|17 + 1.0|18 + .83|19 + .67|20 + .5|21 + \\ &+ .33|22 + .17|23 + 0.0|24 \end{aligned} \tag{2}$$

$$\begin{aligned} \mu_L(\ell) &= \\ &= 0.0|450 + .33|467 + .67|483 + 1.0|500 + .67|517 + \\ &+ .33|533 + 0.0|550 \end{aligned} \tag{3}$$

The extension principle has been applied to this problem along with the computer program SEISRISK II developed by Bender and Perkins (1982) for obtaining two different types of membership functions: (a) Type 1 - membership function of peak ground acceleration given the annual probability of exceedance, and (b) Type 2 - membership function of annual probability of exceedance given the peak ground acceleration. Results for such membership functions are given in Hong (1987), Frangopol, Ikejima and Hong (1988), Frangopol and Hong (1988), and Hong and Frangopol (1988).

Two computer programs have been developed by Hong (1987) for SHA using a probabilistic-fuzzy approach. The first program is based on the extension principle and the second on the vertex method. Using (a) the subjective input from expert opinions represented by the membership function of fuzzy variables (i.e., maximum magnitude, fault length, distance to the fault), (b) the random uncertainties in the parameters of the Bender-Perkins model, and (c) the program SEISRISK II, the two computer programs developed by Hong use fuzzy integration in order to find the characteristic values of membership functions (e.g., peak ground acceleration given the annual probability of exceedance) at each location of interest.

3. APPLICATION TO UTAH AREA

In this section, the probabilistic-fuzzy approach briefly presented previously is applied to the Wasatch fault zone in Utah in order to obtain the seismic map from $37°N$ latitude, southernmost of Utah, to $42°N$ latitude, northernmost of Utah, and from $114°W$ longitude, westernmost of Utah, to $109°W$ longitude, easternmost of Utah.

The fifteen seismic source zones considered are based on the national map of Algermissen et al. (1982). Strong-motion data recorded with 50 km of the rupture zone were used to develop the attenuation characteristics of horizontal peak ground acceleration for worldwide earthquakes of magnitudes 5.0 to 7.7. The data provided by the U.S. Geological Survey, Golden, Colorado, had been used for the near-source attenuation relationships in terms of earthquake magnitude, source-to-site distance, and several source and site parameters in north central Utah (Campbell 1987). The data base consists of 134 horizontal components of peak ground acceleration recorded from 21 earthquakes. From the total of 134 records, 129 earthquakes are used in this study including only earthquakes in California, acknowledging that the tectonics and recording practices at other sites (Nicaragua, Hawaii, New Zealand, Iran, and Alaska) may be substantially different from those in the Western United States. It is also acknowledged that reducing the number of data may reduce the credibility of statistical data. However, the total number of earthquake data is outweighted by the consideration of uncertainties residing behind the data, such as the accuracy of recordings, different geology conditions, and different tectonics.

For an earthquake occurring randomly with a source, the attenuation relationship

$$Y = c_1 e^{c_2 M}(R + 25)^{-c_3} \tag{4}$$

is used, where Y represents the mean of the peak acceleration scaled from the two horizontal components of each acceleration recording, M is the earthquake magnitude, R is source-to-site distance, and c_1, c_2 and c_3 are regression coefficients which are correlated with each other through the data provided. In order to obtain the membership functions of the coefficients c_1, c_2 and c_3, a membership function is assumed for the coefficient c_2. The membership function of the coefficient c_2 shown in Fig. 1 was suggested in 1987 by researchers from the U.S. Geological Survey, Golden, Colorado. The membership functions of the coefficients c_1 and c_3, also shown in Fig. 1, were obtained by Hong (1987) applying regression analysis to the 129 available earthquake recordings.

Similar to the derivation of the membership functions of the coefficients c_1, c_2 and c_3 in the attenuation relation (4), a membership function is assigned to the coefficient β in the magnitude-frequency relation

$$\ell n \; N(m) = \alpha - \beta M \tag{5}$$

and, using regression analysis, the membership function of the coefficent α is found (Hong 1987). In Eq. 5, $N(m)$ is the number of earthquakes in a given time period (e.g., annual) having magnitudes greater than m. The membership function of the coefficient β, shown in Fig. 1, was also provided in 1987 by reserchers from the U.S. Geological Survey, Golden, Colorado.

An estimation of maximum magnitude M_{max} of earthquakes for each seismic source of interest is needed for probabilistic seismic risk analysis. Typically, maximum magnitudes are estimated in the Western United States based on assessments of various fault characteristics, such as rupture length, total length, maximum displacement event, rupture area, and seismic moment. Assessments of maximum magnitude in the Central and Eastern United States are difficult because of the uncertainties involved in associating earthquakes with faults. The methods most commonly used to estimate maximum magnitudes in Eastern United States are as follows (Coppersmith 1987): (a) addition of an increment to the largest historical magnitude; (b) extrapolation of normalized frequency-magnitude recurrence curves to "rare" events such as 100 year event; (c) statistical treatment of seismicity data; and (d) analogy to other seismic sources or other regions of similar tectonic characteristics. In this study, membership function for the maximum magnitude is assumed, based on addition of an increment (i.e., 1.0) to the largest historical magnitude, as follows

$$\begin{aligned}
\mu_{M_{max}} &= 0 \quad if\ M < M_{h,max} \\
\mu_{M_{max}} &= 1 \quad if\ M_{h,max} \leq M < M_{h,max} + 0.5 \\
\mu_{M_{max}} &= 1 - \ 2(M - (M_{h,max} + 0.5)) \qquad (6) \\
\mu_{M_{max}} & \quad if\ M_{h,max} + 0.5 \leq M < M_{h,max} + 1 \\
\mu_{M_{max}} &= 0 \quad if\ M > M_{h,max} + 1
\end{aligned}$$

where $M_{h,max}$ is the largest historical magnitude. The largest historical magnitude for each of the 15 seismic source zones for Utah and surrounding area considered in this study is given in Hong (1987).

Using as input the membership functions of (a) the coefficients c_1, c_2 and c_3 in Eq. 4, (b) the coefficients α and β in Eq. 5, and (c) the maximum magnitudes of earthquakes associated with the 15 zones considered in this study, a probabilistic-fuzzy computational procedure was developed to calculate the membership functions of both the probability of exceedance and the peak ground acceleration (PGA) for the entire Utah area. The main steps of this procedure, involving α-cut method, vertex method, probabilistic seismic risk analysis (i.e., SEISRISK II program), extension principle, and fuzzy in-

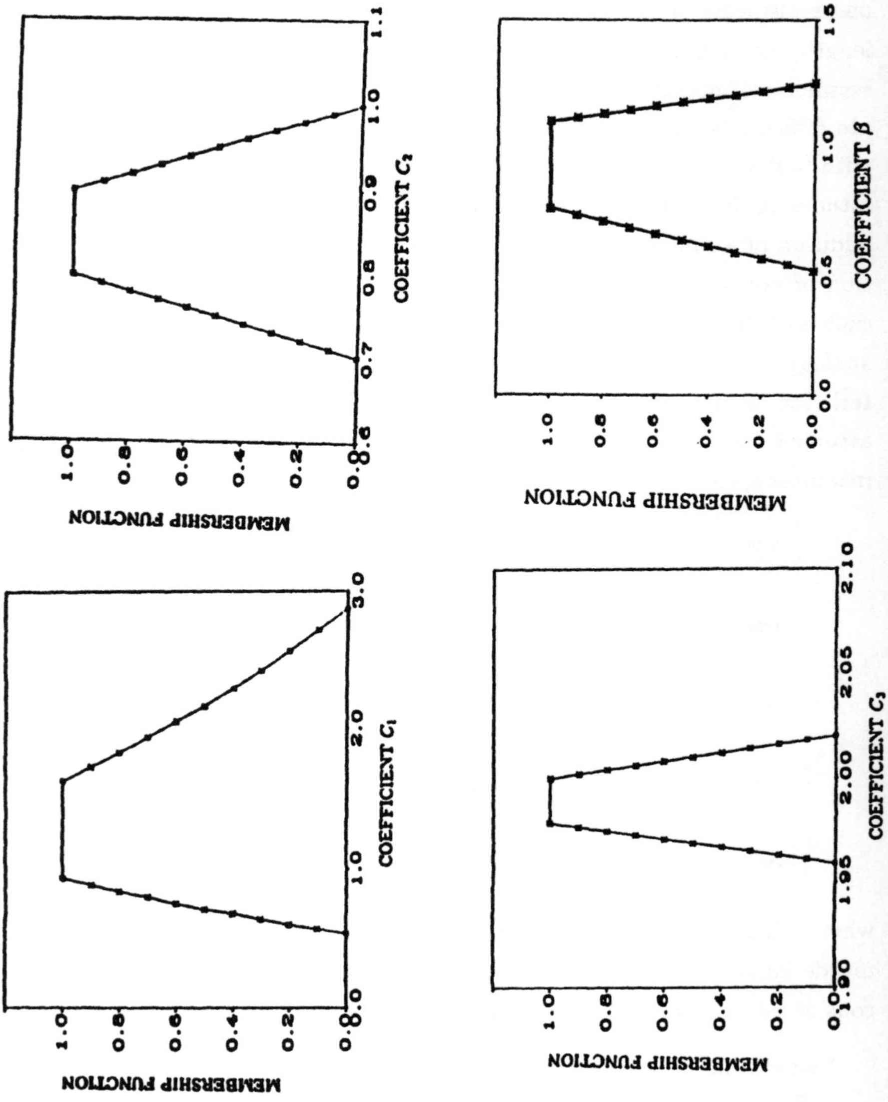

FIGURE 1. Membership functions of coefficients c_1, c_2, c_3 and β.

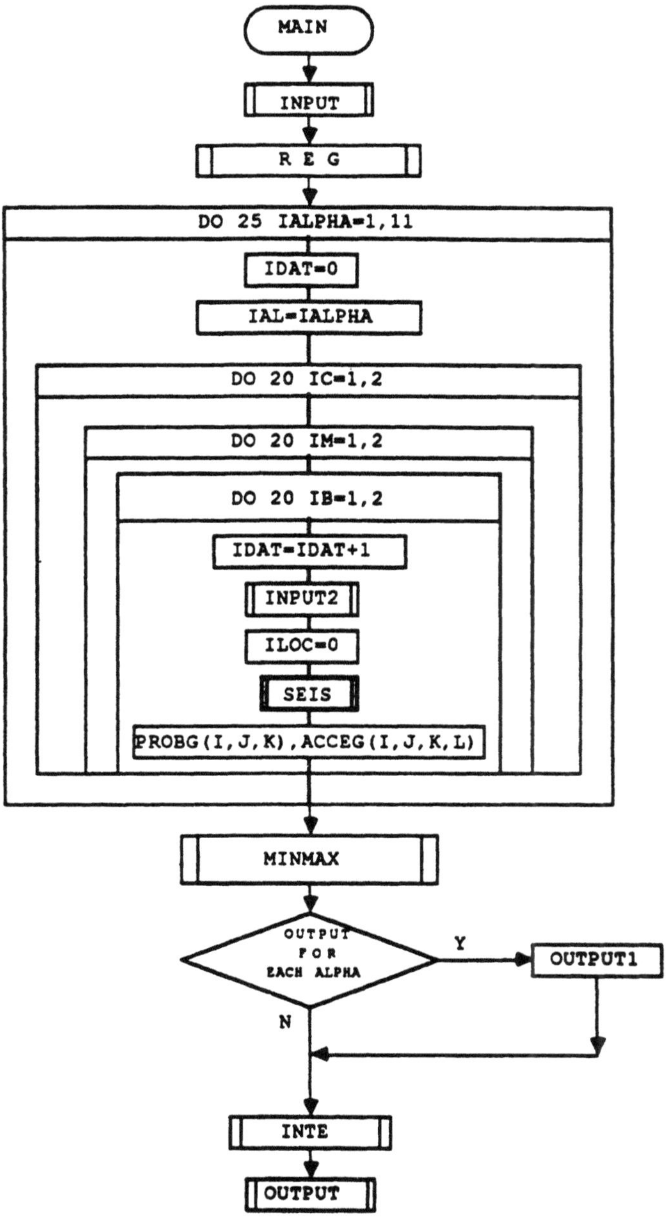

FIGURE 2. Flow-Chart of main program.

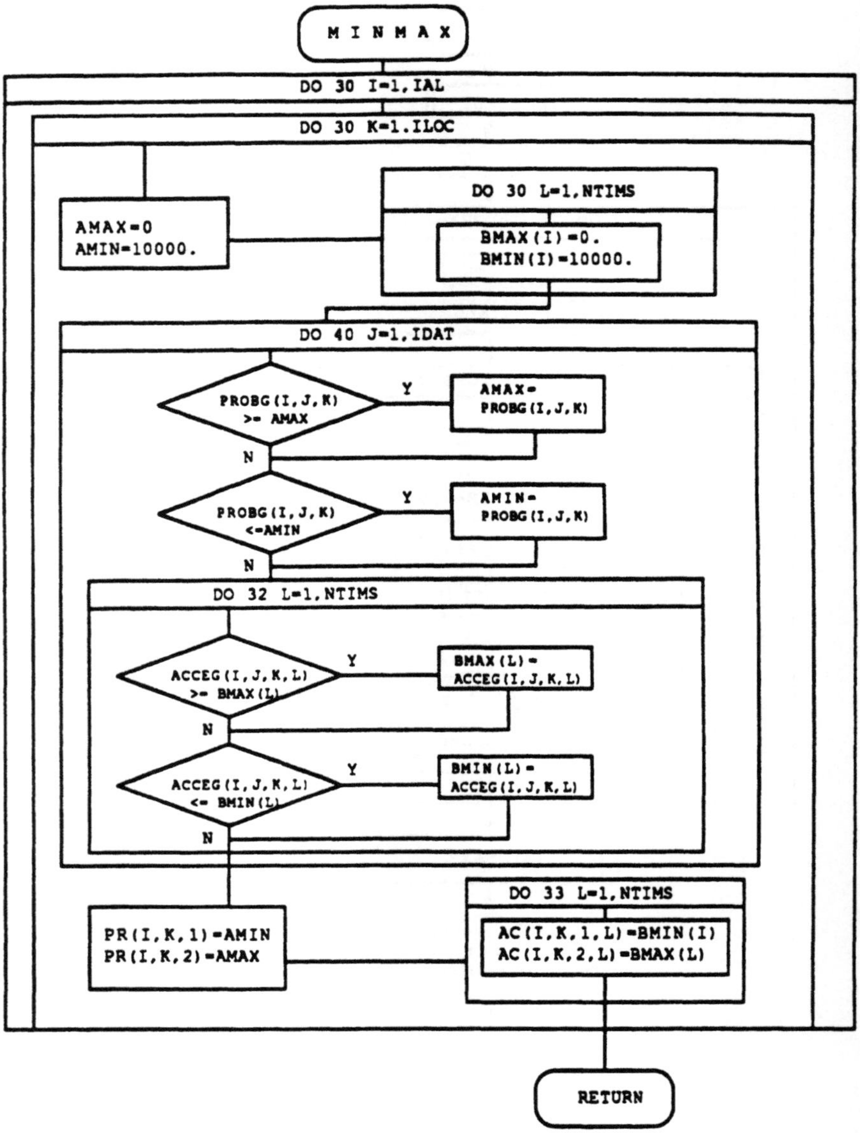

FIGURE 3. Flow-chart of subroutine MINMAX.

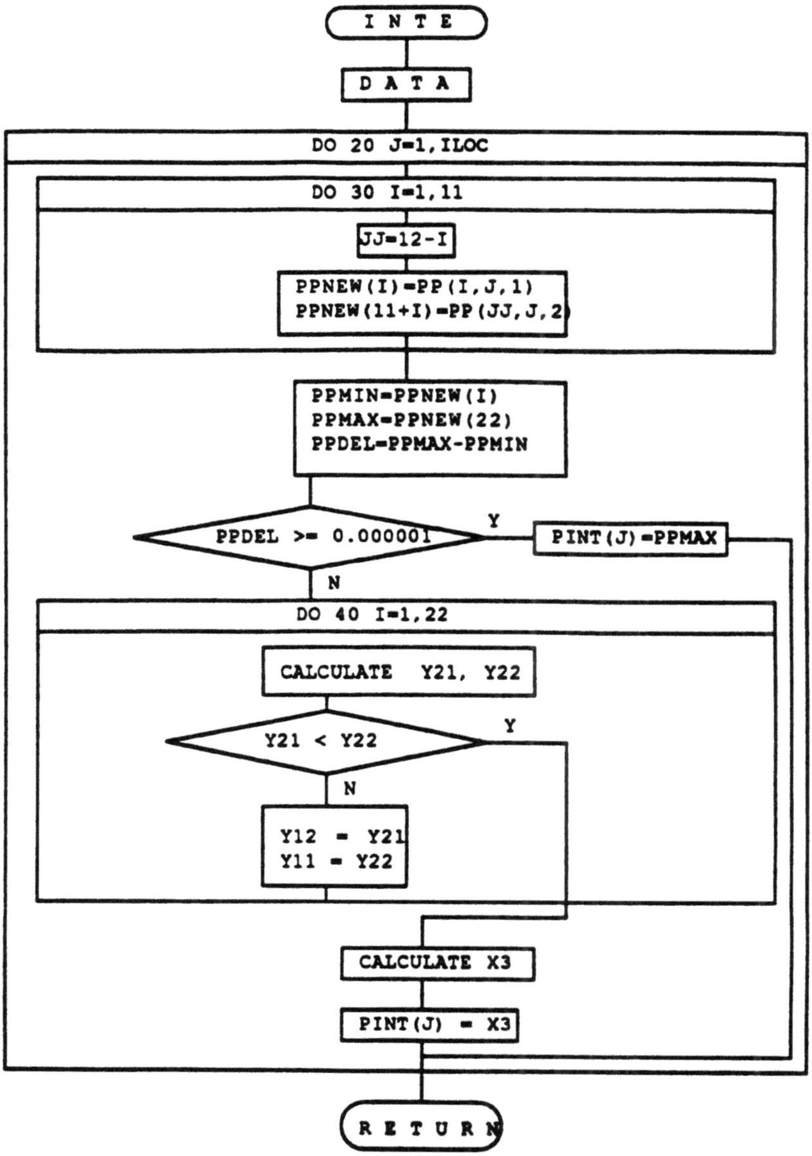

FIGURE 4. Flow-chart of subroutine INTE.

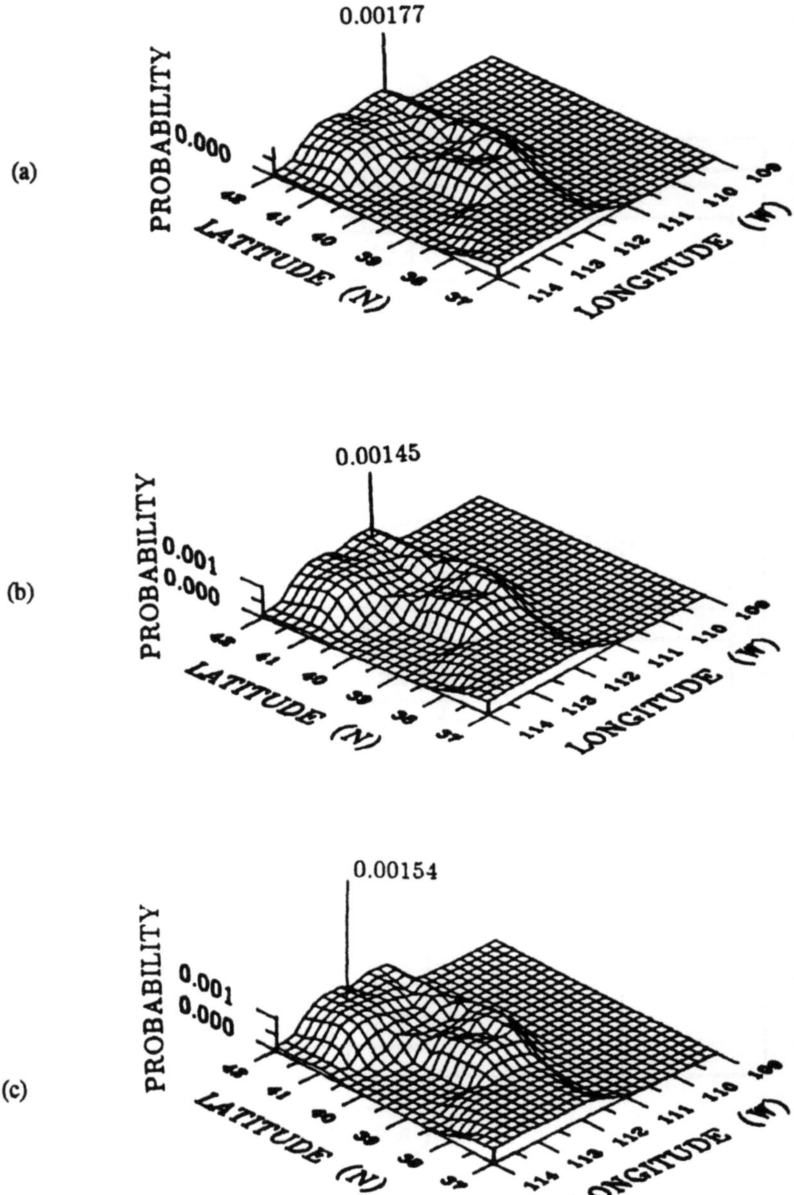

FIGURE 5. 3-D maps of the Utah area for the lower bound of the annual probability of exceedance of 0.2 PGA at membership values of (a) 0.5, (b) 0.8, and (c) 1.0.

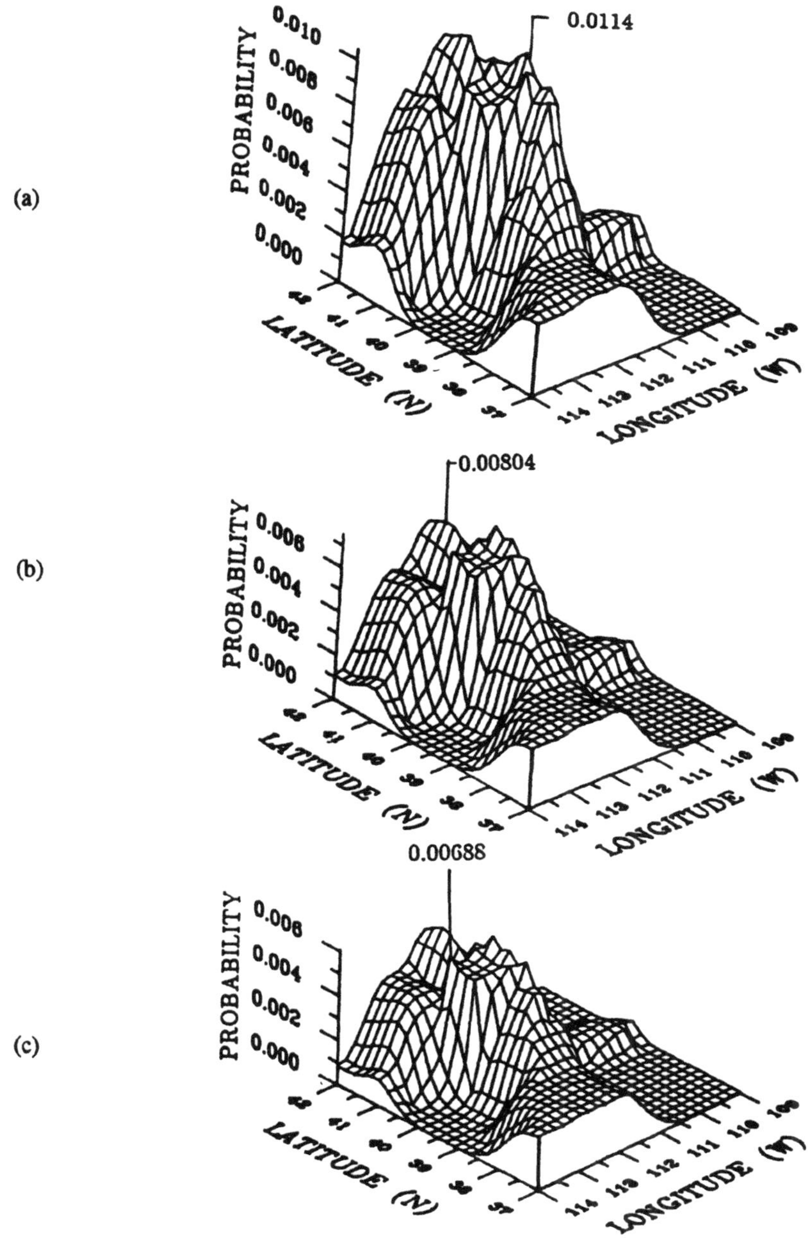

FIGURE 6. 3-D maps of the Utah area for the upper bound of the annual probability of exceedance of 0.2 PGA at membership values of (a) 0.5, (b) 0.8, and (c) 1.0.

(a)

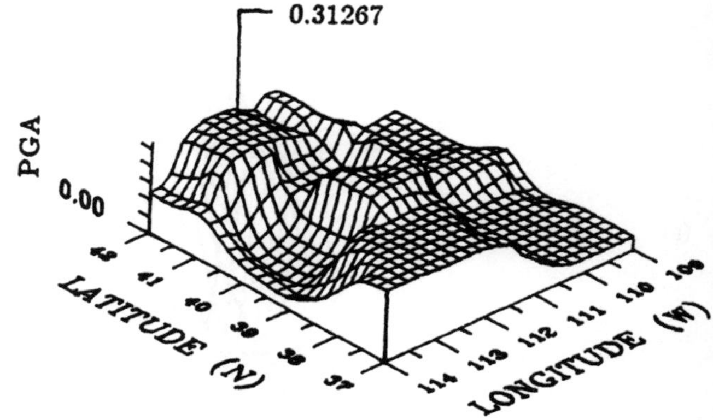

(b)

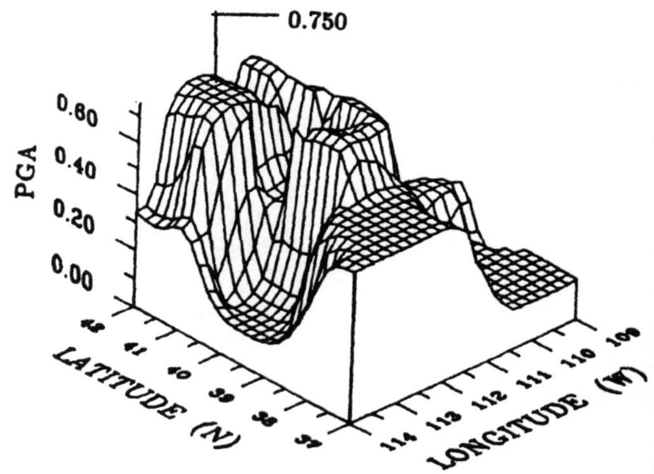

FIGURE 7. 3-D maps of the Utah area for the lower (a) and upper (b) bounds of the PGA at 0.5 membership value for 10% probability of exceedance in 50 years.

(a)

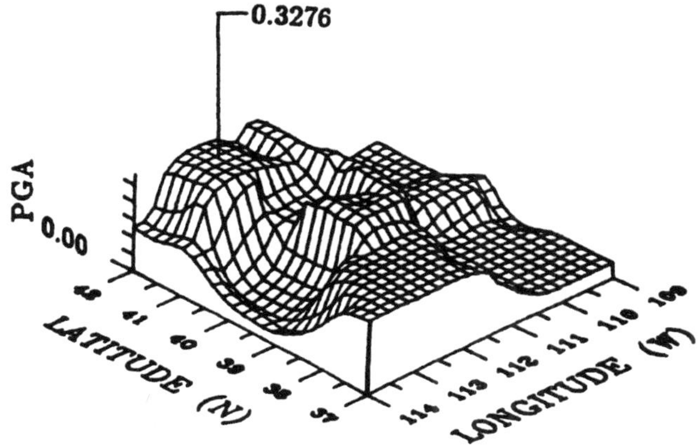

(b)

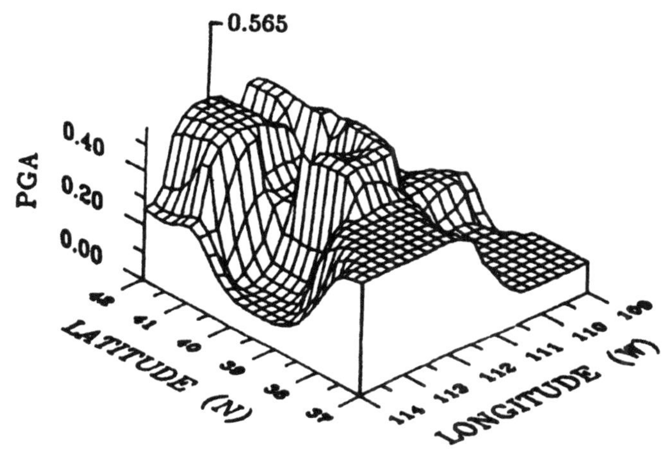

FIGURE 8. 3-D maps of the Utah area for the lower (a) and upper (b) bounds of the PGA at 1.0 membership value for 10% probability of exceedance in 50 years.

(a)

(b)

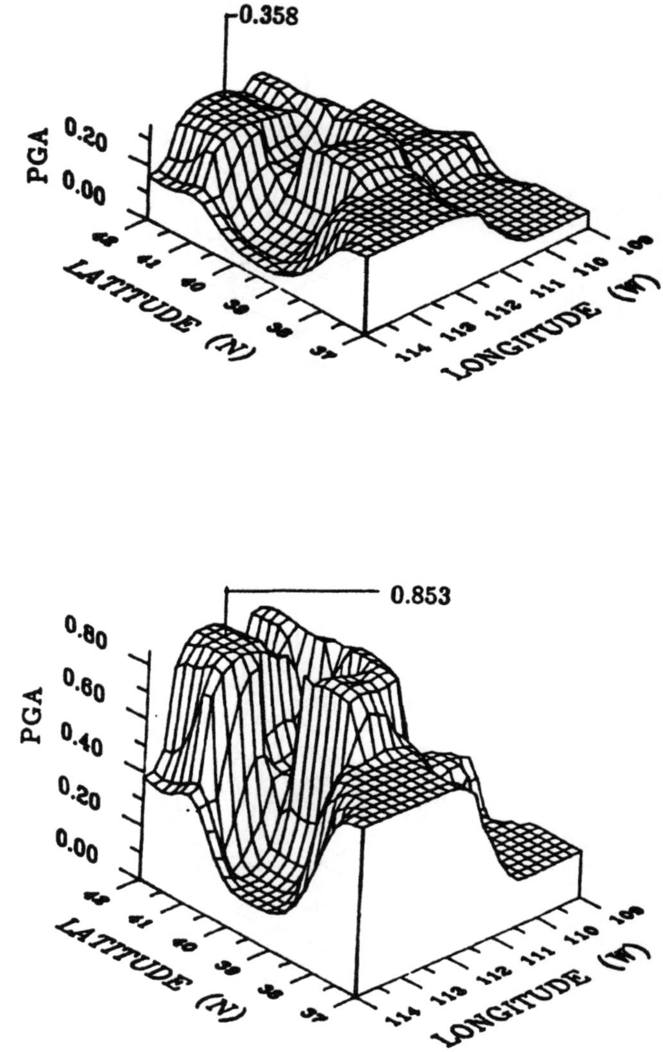

FIGURE 9. 3-D maps of the Utah area for the lower (a) and upper (b) bounds of the PGA at 0.5 membership value for 10% probability of exceedance in 100 years.

(a)

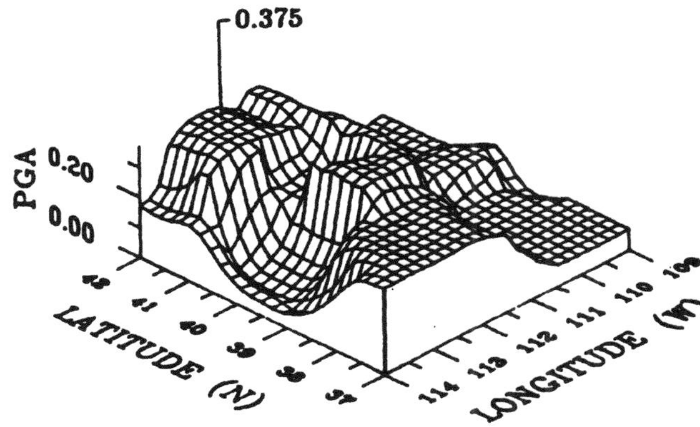

(b)

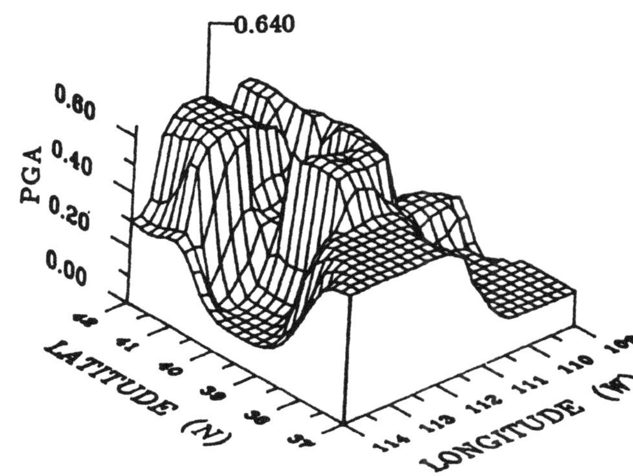

FIGURE 10. 3-D maps of the Utah area for the lower (a) and upper (b) bounds of the PGA at 1.0 membership value for 10% probability of exceedance in 100 years.

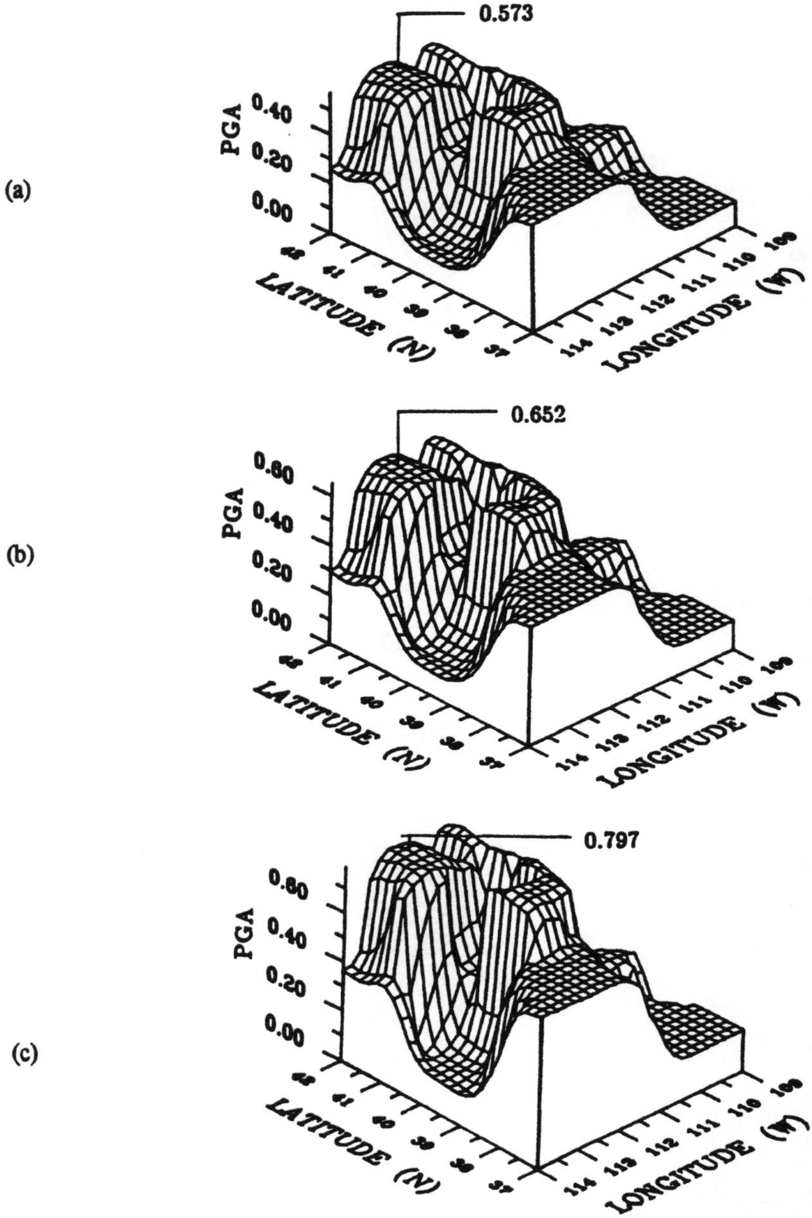

FIGURE 11. Characteristic peak ground acceleration for 10% probability of exceedance in (a) 50 years, (b) 100 years, and (c) 250 years.

tegration are given by Hong (1987). The reader is referred to Zadeh (1965), Brown (1979), Brown and Yao (1983), Dong, Shah and Wong (1985), and Hong (1987), among others, for a comprehensive treatment of fuzzy computation methods.

4. COMPUTER PROGRAM AND RESULTS

SEISRISK II, a computer program for seismic hazard estimation developed by Bender and Perkins (1982), was used as a subroutine to implement the probabilistic-fuzzy model for seismic hazard. The flow-chart of the main program (including SEISRISK II), developed for this study, is shown in Fig. 2. In the subroutine INPUT (see Fig. 2), information about the membership function of each independent fuzzy variable is required. The data required in the subroutine INPUT consists of: historical data on intensities of past earthquakes in each of the fifteen zones considered; membership function of the coefficient β in the attenuation function (see Fig. 1); membership function of the coefficient c_2 in the attenuation function (see Fig. 1); and membership function of maximum magnitude M_{max} for each zone (Eq. 6). In the subroutine REG (see Fig. 2), parameters which will be used several times during the regression analysis to find the membership functions of the coefficients c_1 and c_3 in the attenuation function, and α in the magnitude-frequency relation, are computed.

DO 25 statement in Fig. 2 is performed eleven (number of discretized points of the membership function) times. The subroutine SEIS is a module of the probabilistic seismic hazard analysis program (SEISRISK II) which calculates both the probability of exceedance and the peak ground acceleration for various exposure periods. The subroutine SEIS is performed $2^3 = 8$ times in which 3 is the number of independent fuzzy variables. IC, IM and IB in Fig. 2 are indices to perform seismic hazard analysis (SEIS) for each combination of lower and upper bounds of independent fuzzy variables c_2, M_{max} and β, respectively, at each membership value (i.e., IAL or IALPHA). IDAT is an index to store the calculated results in array. In this study, the value of IDAT varies from 1 to 8. Subroutine SEIS calculates output for each location specified in subroutine INPUT and stores the results in the array PROBG(I,J,K) and ACCEG(I,J,K,L). Annual probabilities of exceedance are stored in the array PROBG(I,J,K) where I is an index of membership values, J is an index of different combinations (i.e., 8 different values), and K is an index of different locations. Ground accelerations are stored in array ACCEG(I,J,K,L) where I,J,K have the same significance as in PROBG and L is an index of different exposure periods.

After the computations for all possible combinations, the lower and upper bounds of the results are calculated by using the subroutine MINMAX as shown in Fig. 3. In this subroutine, the minimum and the maximum values are chosen for PROBG and ACCEG for J=1, IDAT where IDAT is a total number of different combinations. The lower bounds of results are stored in array PR(I,K,1) and AC(I,K,1,L) and the upper bounds are stored in array PR(I,K,2) and AC(I,K,2,L). If requested, these results can be printed for each different membership value as shown in Fig. 2 (output for each ALPHA).

In order to obtain the characteristic values, fuzzy integration is performed at each location of interest (i.e., J=1, ILOC) in the subroutine INTE as shown in Fig. 4.

In the application of the proposed probabilistic-fuzzy approach, membership funcitons of probability of exceedance of $0.2g$ and peak ground accelerations for five different exposure periods (10, 50, 100, 200 and 250 years) are calculated at 500 locations in Utah area. The computer used was VAX/780 with VAX/VMS V3.5 operating system.

For the entire Utah area, Figs. 5 to 11 show examples of maps of probability of exceedance and peak ground acceleration. More examples of such 3-D maps have been produced by Hong (1987) for different probabilities of exceedance, different peak ground accelerations, and different time periods.

5. CONCLUSIONS

Some specific comments and recommendations on seismic hazard assessment, with particular reference to fuzzy sets applications, may be derived from the results presented in this study.

1. The results presented indicate the feasibility of using fuzzy sets theory in assessing the seismic hazard. By using this novel theory, seismic hazard

is predicted in terms of membership values of both ground accelerations and exceedance probabilities. This approach to predict seismic hazard appears more rational than the current prediction models which use conventional probabilistic methods and do not generally include uncertainties associated with incompleteness of information and knowledge.

2. It is seen from the examples presented that an important factor influencing the seismic hazard prediciton is the choice of the input membership functions for the attenuation and magnitude-frequency relations. Another important factor is the choice of the membership function of the maximum magnitude.

3. The probabilistic-fuzzy computation of seismic risk for any membership value and its incorporation in a seismic hazard map, should stimulate studies of both random and fuzzy variables encountered in earthquake prediction to improve their description. In this context, empirical studies are needed to determine reasonable membership values for fuzzy variables associated with earthquake prediction.

4. The area of earthquake engineering applications of fuzzy sets theory is not as well developed to date as that of probability theory. More investigations are needed to extend and implement the probabilistic-fuzzy approach for seismic hazard proposed herein. Further research is needed to extend this model to time-dependent earthquakes and to combine experts' opinions in a consistent manner. Also, for implementation of the proposed model general seismic risk analysis computer programs using both probabilistic and fuzzy information will be needed.

ACKNOWLEDGEMENTS

This study is based upon work supported by the United States National Science Foundation under Grant No. ECE-8610500. The authors are also grateful to Earthquake Engineering Research Institute for permission to reproduce and enhance their contribution to the Fourth U.S. National Conference on Earthquake Engineering (Frangopol and Hong 1990) for publication in this book. Discussions with B. Bender, D.M. Perkins, and J.T.P. Yao are gratefully acknowledged.

REFERENCES

Algermissen, S.T., Perkins, D.M., Thenhaus, P.C., Hansen, S.L., and Bender, B.L. (1982) "Probabilistic estimate of maximum acceleration in rock in the contiguous United States," U.S. Geol. Surv., Golden, Colorado, Open-File Report, 82-1033.

Bender, B. (1985) "Seismic hazard estimation using a finite-fault rupture model," Bull. Seism. Soc. Am., Vol. 74(5), 1899-1923.

Bender, B. and Perkins, D.M. (1982) "SEISRISK II, a computer program for seismic hazard estimation," U.S. Geol. Surv., Open File Report No. 8222-293.

Bonilla, M.G. and Buchanan, J.M. (1970) "Interim report on worldwide historic surface faulting," U.S. Geol. Surv., Open-File Report.

Brown, C.B. (1979) "A fuzzy safety measure," Journal of the Engineering Mechanics Division, Vol. 105 (5), 855-872.

Brown, C.B. and Yao, J.T.P. (1983) "Fuzzy sets in structural engineering," Journal of Structural Engineering, Vol. 109 (5), 1211-1225.

Campbell, K.W. (1987) "Predicting strong motion in Utah," U.S. Geol. Surv., Golden, Colorado, Professional Paper.

Coppersmith, K.J. (1987) "Maximum earthquakes, fault slip rates, and earthquake recurrence," A Two-Day Course on Strong Ground Motion, EERI, San Francisco.

Cornell, C.A. (1968) "Engineering seismic risk analysis," Bull. Seism. Soc. Am., Vol. 58 (5), 1583-1606.

Cornell, C.A. and Vanmarcke, E.H. (1969) "The major influences on seismic risk," Proc. 4th World Conference on Earthquake Engineering, Santiago, Chile, Vol. 1.

Der Kiureghian, A. and Ang, A. H-S. (1977) "A fault-rupture model for seismic risk analysis," Bull. Seism. Soc. Am., Vol. 67 (4), 1173-1194.

Dong, W., Shah, H.C. and Wong, F.S. (1985) "Fuzzy computations in risk and decision analysis," Civil Engineering Systems, Vol. 2, pp. 201-208.

Donovan, N.C. (1977) "A statistical evaluation of strong motion data including the February 9, 1971, San Fernando earthquake," Proc. 5th World Conference on Earthquake Engineering, Rome, Italy, Vol. 1, 1251-1261.

Electric Power Research Institute (1985) "Seismic hazard methodology for nuclear facilities in the eastern United States: Preliminary seismic hazard test computations for parametric analysis and comparative evaluations," EERI Research Project P101-29 (Draft).

Frangopol, D.M. Ikejima, K., and Hong, K. (1988) "Seismic hazard prediction using a probabilistic-fuzzy approach," Structural Safety, Vol. 5 (2) pp. 109-117.

Frangopol, D.M. and Hong, K. (1988) "Fuzzy logic approach in seismic risk prediction," Annual Meeting of the Earthquake Engineering Research Institute, Poster Session, Mesa, Arizona.

Frangopol, D.M. and Hong, K. (1990) "A probabilistic-fuzzy model for seismic hazard applied to the Wasatch Front Range in Utah," Proceedings of Fourth U.S. National Conference on Earthquake Engineering, Palm Springs, California, Vol. 1, 733-742.

Hong, K. (1987) "Assessment of seismic hazard using a probabilistic-fuzzy approach," Thesis presented to the University of Colorado, Boulder, Colorado, in partial fulfillment of the requirements for the degree of Doctor of Philosophy.

Hong, K. and Frangopol, D.M. (1988) "Estimating the seismic hazard using fuzzy sets," Proceedings 7th Annual Meeting of the North American Fuzzy Information Processing Society, San Francisco, California, 91-95.

Kiremidjian, A.S. and Anagnos, T. "Stochastic-slip predictable model for earthquake occurrences," Bull Seism. Soc. Am., Vol. 74, 739-755.

Lamarre, M. and Dong, W.M. (1986) "Evaluation of seismic hazard with fuzzy algorithm," Proc. 3rd U.S. National Conference on Earthquake Engineering, Charleston, South Carolina, Vol. 1, 221-231.

Lawrence Livermore National Laboratory (1981) "Seismic hazard analysis," Vols. 1-5, NUREG/CR-1582.

Yankee Atomic Electric Company (1984) "Review and comments on NUREG /CR-3756 seismic hazared characterization of the eastern United States: Methodology and interim results for ten sites," Report YAEC-1455.

Zadeh, L.A. (1965) "Fuzzy sets," Information and Control, Vol. 8, 338-353.

SEISMIC RELIABILITY ANALYSIS OF EXISTING STRUCTURES BASED ON FUZZY PROBABILITY

Hitoshi Furuta*, Masata Sugito**, Shin-ya Yamamoto***
and Naruhito Shiraishi*

* Department of Civil Engineering, Kyoto University, Kyoto 606-01, Japan

** Department of Transportation Engineering, Kyoto University, Kyoto 606-01, Japan

*** Mitsubishi Heavy Industries, Hiroshima 733, Japan

Abstract: This paper provides a reliability analysis of existing bridge structures for earthquake loads. Using a fuzzy graph and fuzzy relation, an attempt is made to consider the effect of structural deterioration in the evaluation of structural reliability. The analysis method gives the final result in the form of fuzzy probability, which can provide more information to the reliability analysis of the structure in future. Several numerical examples are presented to demonstrate the applicability of the method proposed here.

Keywords: Bridge, Earthquake, Fuzzy Probability, Reliability, Structural Deterioration

1. INTRODUCTION

In Japan, consideration for earthquake is quite important to ensure the reliability of existing structures during their lifetime. In reliability analysis of the existing structures, we can utilize more exact information than that at the design stage. Namely, we can obtain reliable information with respect to the ground condition, the change of traffic loads, the occurrence of earthquakes, and the deterioration of structural resistance, using various data of observation, field test and laboratory test.

In this paper, an attempt is made to develop a method of evaluating the reliability of existing bridge structures for earthquake loads. In general, pier structures are likely to suffer from damage due to earthquakes. Hence, attention is paid to the reliability analysis of bridge piers, especially reinforced concrete (RC) piers. In this reliability analysis, the ultimate limit state is merely considered for possible limit states. The limit state is specified in terms of displacement instead of load effect or stress. It is assumed that failure occurs when the maximum response displacement becomes larger than the allowable displacement prescribed. This formulation is intended to take into account the effect of plastic behavior of RC piers after the yielding of reinforcing steel in the reliability assessment of earthquake resistant structures, without difficulty. The maximum displacement is calculated from the non-stationary power spectrum of earthquake motion[1)]. This response analysis can include the effects of natural frequency of structures, ground condition and seismic zoning in addition to the effect of inelastic behavior of material. Using a fuzzy graph and fuzzy relation, an attempt is made here to consider the effect of structural deterioration in the evaluation of allowable displacement. The analysis method gives the final result(i.e., failure probability) in the form of fuzzy probability, which can provide more information to the reliability assesment of the underlying structure in future. Several numerical examples are presented to demonstrate the applicability of the method proposed here.

2. SEISMIC RESPONSE MODEL USING NON-STATIONARY POWER SPECTRUM

In this paper, the response of structures is related to the intensity of non-stationary power spectrum of input motion. As a representative parameter of structural response, maximum displacement of the top of piers is employed instead of maximum force or stress, because failure of the pier structure can be easily expressed in terms of displacement if the yield displacement and ductility factor are known.

In the past studies[2)], the earthquake load is estimated on the basis of some approximate relation between the maximum acceleration of ground and the maximum response of structures. However, the structural response is greatly influenced by the natural frequency of the structures. Moreover, the duration of earthquake motion affects the failure path that RC structures reach their failures through the successive process of alternative deformations due to the earthquake motion. Considering these facts, it is desirable to take into consideration the non-stationary characteristic of earthquake motion. Especially, when estimating the maximum response deformation, it is preferable to use the non-stationary power spectrum instead of the maximum acceleration of input motion, because the former can account for the spectral characteristic of input ground motion.

An attempt is made here to estimate the structural response which covers both the elastic and inelastic regions, by paying attention to the intensity parameter of non-stationary power spectrum. Considering the non-stationary characteristics[3),4)], the acceleration of earthquake is written as follows:

$$X(t)=\sum_{k=1}^{m}\sqrt{2Gx(t,\omega_k)\Delta\omega}\cos(\omega_k t+\phi_k) \tag{1}$$

where $G_x(t, \omega_k)$ is the non-stationary power spectrum for circular frequency ω and time t, $\Delta\omega$ is the discretization step at ω, ϕ_k is the phase at t=0. The square root of $G_x(t, \omega_k)$ is given as

$$\sqrt{Gx(t,\omega)}=\sqrt{Gx(t,2\pi f)}$$
$$=\begin{cases} 0 & ;\ 0\leq t\leq t_s \\ \alpha_m(f)\dfrac{t-t_s(f)}{t_p(f)}\exp\left\{1-\dfrac{t-t_s(f)}{t_p(f)}\right\} & ;\ t_s<t \end{cases} \tag{2}$$

where $\alpha_m(f)$ is the maximum value of $\sqrt{G_x(t,2\pi f)}$ and its unit is $gal/\sqrt{rad/\sec}$, $t_s(f)$ is the starting time, and $t_p(f)$ is the time from $t_s(f)$ to the appearance time of the maximum value of $\alpha_m(f)$. $\alpha_m(f)$ is given by the following regression equation with respect to the frequency $f(Hz)$, magnitude M, and epicentral distance Δ.

$$\log\alpha_m(f)=B_0(f)+B_1(f)M-B_2(f)\log(\Delta+30) \tag{3}$$

where $B_0(f)$, $B_1(f)$ and $B_2(f)$ are the coefficients which have been given from the regression analysis based on Japanese strong motion data and are functions of frequency f.

Since $\alpha_m(f)$ given by Eq. 3 denotes the intensity at the bedrock, it does not involve the effect of local ground condition overlying bedrock. Using the conversion factor $\beta_a(f)$[5)], it is possible to introduce its effect on nonlinear soil amplification into the estimation of intensity parameter on soil surface. When ground parameter S_n and depth to the bedrock d_p are assumed, $\beta_a(f)$ is given as follows:

$$f>1.0(\mathrm{Hz})$$
$$\log\beta_a(f)=u_0(f)+u_1(f)\alpha_m(f);\quad \alpha_m\geq\alpha'_m$$
$$\log\beta_a(f)=u_0(f)+u_1(f)\alpha'_m(f);\quad \alpha_m<\alpha'_m \tag{4}$$

$$f<1.0(\mathrm{Hz})$$
$$\log\beta_a(f)=u_0(f) \tag{5}$$

where u_0 and u_1 are the functions of frequency f, ground parameter S_n and the depth to the bedrock d_p. S_n is evaluated from blow-counter profile. $\alpha'_m(f)$ denotes the folding point of stress-strain curve. Consequently, the intensity parameter at the ground $\alpha_{ms}(f)$ is calculated as

$$\alpha_{mu}(f)=\beta_{u}(f)\alpha_{m}(f) \tag{6}$$

In this paper, elasto-plastic analysis is performed by use of a bi-linear model. In general, many degrading tri-linear model, i.e., Takeda model, Mutoh model and Fukuda model, have been used for the yielding failure of reinforced concrete members. However, this study aims to show the efficiency of the method that estimates the maximum response displacement. To make the discussion simple, a bi-linear model is employed instead of the tri-linear models.

Table 1 Ground Condition

	Ground 1	Ground 2	Ground 3	Ground 4
S_n	-	-0.2	0.2	0.8
$d_{p(m)}$	-	20.0	50.0	150.0

For the variation of ground motion levels, fifteen combinations of magnitude and epicentral distance are examined. Table 1 presents the ground conditions used in the numerical computation. Ground 1 in Table 1 denotes the bedrock. In the elasto-plastic response analysis, it is necessary to determine the yielding displacement at the top of piers. However, the determination of the yielding displacement is very difficult, because it is affected by the damage state of concrete. Using such a characteristic that velocity response spectrum is constant regardless the natural frequency, it is possible to calculate the yielding displacement σ_y. Namely, σ_y is estimated by dividing the response velocity by the natural circular frequency. A parametric analysis is performed for eight cases with V(velocity)=2.0, 5.0, 8.0, 10.0, 20.0, 30.0, 40.0 and 50.0 cm/sec. For the natural frequency, ten cases are considered as f_0=0.2, 0.5, 0.7, 1.0, 2.0, 3.0, 4.0, 5.0, 7.0 and 10.0 Hz, and also two damping factors h=0.05 and h=0.1 are considered. Table 2 provides the comparison of the proposed method and the method based on the maximum acceleration A_{max}. The values in Table 2 are the correlation coefficients which are obtained by the both methods for the case of V=30.0 cm/sec. As the correlation coefficients are closer to unity, they are better solutions. The values with symbol * shows that the method using the maximum acceleration presents better results than the proposed method. From Table 2, it is seen that the proposed method prefers to the method using the maximum acceleration when the natural frequency is less than 2.0 Hz. Although there are several values with the symbol * in the cases of f_0=0.5 and 0.7 Hz, the differences are neglegibly small. This tendency is due to the fact that when f_0 is greater than 2.0 Hz, the structural response strongly depends on ground motions in high frequency ranges, and hence that the maximum

Table 2 Results of Regression Analysis

f_0		Ground 1		Ground 2		Ground 3		Ground 4	
		0.05	0.1	0.05	0.1	0.05	0.1	0.05	0.1
0.19	α_m	0.989	0.997	0.942	0.997	0.996	0.998	0.998	0.999
	A_{max}	0.929	0.960	0.961	0.976	0.984	0.987	0.961	0.958
0.49	α_m	0.984	0.992	0.983	0.989	0.972	0.987	0.986	0.994
	A_{max}	0.974	0.987	*0.984	*0.988	*0.982	*0.988	0.962	0.974
0.73	α_m	0.967	0.979	0.969	0.974	0.986	0.977	0.982	0.911
	A_{max}	0.942	0.958	*0.974	*0.976	*0.991	*0.985	*0.984	0.976
0.97	α_m	0.993	0.993	0.990	0.993	0.965	0.977	0.938	0.934
	A_{max}	0.991	0.992	0.989	0.989	*0.975	0.971	0.906	0.889
1.99	α_m	0.993	0.993	0.984	0.987	0.981	0.986	0.992	0.989
	A_{max}	0.993	0.993	0.984	0.981	0.963	0.972	0.989	0.985
3.01	α_m	0.899	0.913	0.919	0.923	0.917	0.889	0.877	0.891
	A_{max}	*0.927	*0.939	*0.964	*0.966	*0.975	*0.959	*0.919	*0.925
4.03	α_m	0.921	0.941	0.948	0.948	0.929	0.934	0.935	0.971
	A_{max}	0.952	*0.970	*0.976	*0.982	*0.981	*0.990	*0.967	*0.992
4.99	α_m	0.992	0.995	0.986	0.961	0.977	0.983	0.969	0.967
	A_{max}	0.998	*0.998	0.970	0.977	0.973	0.988	0.990	0.989
7.03	α_m	0.980	0.979	0.956	0.949	0.940	0.922	0.910	0.897
	A_{max}	*0.998	*0.999	*0.998	*0.999	*0.998	*0.996	*0.958	*0.950
10.03	α_m	0.993	0.991	0.975	0.972	0.972	0.970	0.967	0.957
	A_{max}	0.981	0.985	0.963	0.971	0.956	0.963	*0.996	*0.997

acceleration shows a well correspondence to the maximum response displacement. It is also due to that the assumption that the response velocity is constant is valid for the large frequency region. Consequently, it is concluded that the proposed method is suitable for the analysis of structures with long natural period such as towers of suspension bridge and high-rise buildings.

The relation between $\alpha_{ms}(f_0)$ and Y_{max} (maximum response displacement) is influenced by the natural frequency, damping characteristic and yielding displacement. The yielding displacement may be different for structures with the same natural frequency. Hence, the relation between $\alpha_{ms}(f_0)$ and Y_{max} is calculated for various values of natural frequency through the elasto-plastic analysis. Table 3 shows the inclining angles of regression curve and the correlation coefficients obtained here. This table can be used to estimate an approximate value of angle B with the aid of interpolation procedure.

Table 3 Inclining Angle *B* and Correlation Coefficient *r*

h=0.05

fo	*B* / Correlation Coefficient	Ground 1	Ground 2	Ground 3	Ground 4
0.2	*B*	0.4464E+1	0.4658E+1	0.4975E+1	0.5496E+1
	r	0.9895	0.9918	0.9943	0.9891
0.5	*B*	0.9163	0.9219	0.9122	0.1013
	r	0.9832	0.9847	0.9813	0.9740
0.7	*B*	0.7227	0.6491	0.6061	0.4784
	r	0.9852	0.9767	0.9411	0.8909
1.0	*B*	0.4643	0.5023	0.4223	0.2681
	r	0.9852	0.9767	0.9411	0.8909
2.0	*B*	0.1949	0.1362	0.2349	0.1621
	r	0.9919	0.9666	0.9813	0.9930

h=0.1

fo	*B* / Correlation Coefficient	Ground 1	Ground 2	Ground 3	Ground 4
0.2	*B*	0.3664E+1	0.4015E+1	0.4482E+1	0.4870E+1
	r	0.9973	0.9974	0.9985	0.9964
0.5	*B*	0.7749	0.7989	0.7971	0.8622
	r	0.9917	0.9881	0.9935	0.9820
0.7	*B*	0.5920	0.5467	0.5225	0.4490
	r	0.9811	0.9822	0.9897	0.9751
1.0	*B*	0.3245	0.4059	0.3885	0.2678
	r	0.9879	0.9805	0.9461	0.91087
2.0	*B*	0.1748	0.1238	0.1262	0.1418
	r	0.9894	0.9625	0.9827	0.9959

The maximum response displacement of the top of pier structures can be easily calculated using the value of *B* and Eq. 6.

$$Y_{max}=B \cdot \beta_a(f_0)\alpha_m(f_0) \quad (7)$$

3. SEISMIC RELIABILITY ANALYSIS OF EXISTING STRUCTURES

In this paper, the analytical process shown in Fig. 1 is employed to evaluate the seismic reliability of existing structures. The occurrence pattern of earthquake varies according to the location of active fault. Therefore, seismic zoning is carried out by considering the effect of active faults. For two big cities in Japan, i.e.,

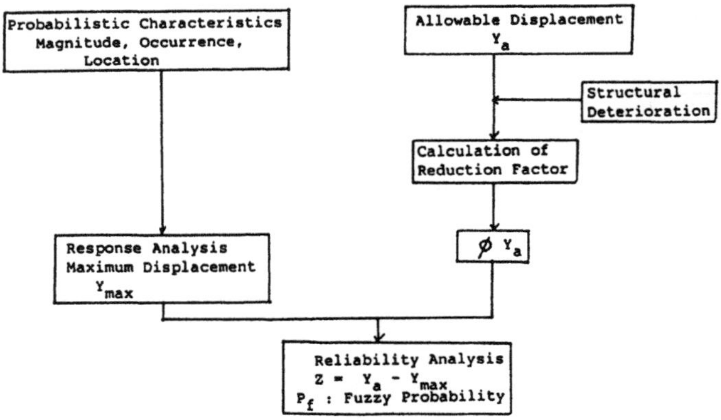

Fig.1 Process of Reliability Analysis

Tokyo and Osaka, seismic zoning is performed using circles with the same center and radial rays. In the zoning, the maximum value of epicentral distance is 300 km, whereas the minimum values depend on each zone.

The cumulative distribution function of earthquake magnitude $F_{Mk}(m)$ is given in the form of the empirical law by Gutenberg-Richter[6].

$$F_{Mk}(m)=\Pr(M<m|M_{Lk}\leq m\leq M_{Uk}) \quad = \frac{1-\exp[-b_k \ln(10)(m-M_{Lk})]}{1-\exp[-b_k \ln(10)(M_{Uk}-M_{Lk})]} \tag{8}$$

where M_{Lk} and M_{Uk} are the maximum and minimum magnitudes in the k-th zone, and b_k is the b-value in the k-th zone. Eq. 8 denotes the cumulative distribution of magnitude of earthquake occurred in the k-th zone. Denoting the area of the k-th zone by A_k, the occurrence number of earthquake within t_r years can be calculated as $\nu_k A_k t_r$. ν_k is the occurrence number of earthquake per year. It is assumed that the magnitude of n earthquakes occurred in the k-th zone M_1, M_2 $\cdots$, M_n are independent each other and follow the same probability distribution $F_{Mk}(m)$. Then, the distribution of the maximum value Z among M_1 to M_n can be obtained as

$$F_Z(z)=P_r(Z\leq z)=P_r(\bigcap_{i=1}^{n} M_i\leq z)=\{F_Z(z)\}^n \tag{9}$$

Replacing n by $\nu_k A_k t_r$, the probability density function of the maximum magnitude within t_r years $f_{Mek}(m)$can be derived as

$$f_{Mek}(m)=\nu_k A_k t_r \{F_{Mk}(m)\}^{\nu_k A_k t_r - 1} f_{Mk}(m) \tag{10}$$

where $f_{Mk}(m)$ is calculated through the differentiation of $F_{Mk}(m)$. Obviously, Eq. 10 is calculated by differentiating Eq. 8.

The occurrence location of earthquake in each zone is considered to be independent of the year and have the linear distribution such as

$$f_{Rk}(\Delta)=\begin{cases}\frac{2\Delta}{\Delta_{Uk}^2-\Delta_{Lk}^2} & ; \quad \Delta_{Lk}\leq\Delta\leq\Delta_{Uk}\\ 0 & ; \quad \Delta<\Delta_{Lk},\ \Delta>\Delta_{Uk}\end{cases} \tag{11}$$

where Δ_{Lk} and Δ_{Uk} are the lower and upper bounds of epicentral distance, respectively. Using the probability density functions given by Eqs. 10 and 11, the maximum response displacement Y_{max} can be calculated based on the relation expressed by Eq. 7. In the calculation, the allowable displacement Y_a, earthquake magnitude M and epicentral distance Δ are treated as random variables, while others are treated as deterministic quantities. Using the probability density functions given by Eqs. 10 and 11, the maximum response displacement Y_{max} can be calculated as Eq.7.

Here, failure probability is calculated based on AFOSM(Advanced First Order Second-Moment Method)[7]. The reason is that AFOSM can include any kind of random variable and deal with non-normal variables as well as normal variables through a transformation from non-normal to normal distributions. The limit state function used here is as follows:

$$Z=g(Y_a,M,\Delta)=Y_a-B\cdot\beta_a(f_0)\alpha_m(f_0) \tag{12}$$

where $\beta_a(f_0)$ can be given by Eqs. 4 and 5, and $\alpha_m(f_0)$ is given by Eq. 3. The allwabale displacement Y_a is determined on the basis of some design specifications and experimental results.

4. EVALUATION OF STRUCTURAL DETERIORATION

Using a reduction factor ϕ, the effect of structural deterioration is introduced into the determination of allowable displacement[8].Then, the limit state function Z is obtained as

$$Z=\phi Y_a-Y_{max} \tag{13}$$

The reduction factor ϕ is calculated by use of an evaluation graph shown in Fig. 2.

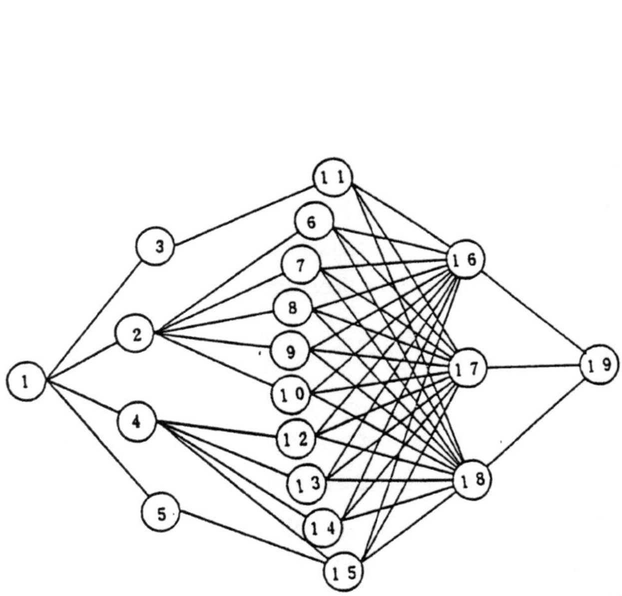

1:dummy
2:visual inspection
3:measurement of displacement
4:non-destructive test
5:strength test for concrete
6:horizontal cracks
7:diagonal cracks
8:spalling
9:excessive deformation of reinforcing steel
10:failure of reinforcing steel
11:movement, sink, leaning
12:cracks of concrete
13:corrosion of reinforcing steel
14:allocation of reinforcing steel
15:stress and strain
16:damage due to deformation
17:damage due to displacement
18:damage sue to cracks
19:evaluation of reduction factor

Fig.2 Evaluation Graph

The evaluation graph is constructed based on the report by Public Works Research Institute(PWRI)of Japanese Government[9]. The damage pattern is classified into three categories, 1)pattern 1: collapse at the bottom of pier due to bending action, 2)pattern 2: collapse at the middle part of pier due to bending action, and 3)pattern 3: collapse due to shearing force.

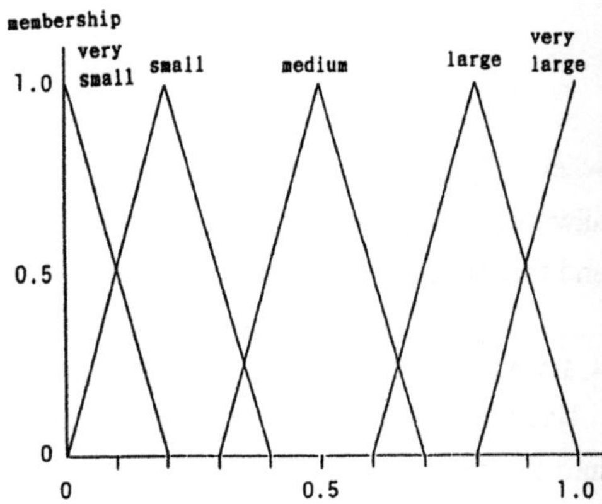

Fig.3 Membership Functions

In general, the assessment of damage degrees performed mostly on the basis of visual inspection. This implies that it is unavoidable to include some uncertainty or ambiguity in the damage assessment procedure. In order to account for such uncertainty, the reduction factor is defined here by a fuzzy set.

$$\tilde{Z}=\tilde{\phi}Y_a-Y_{max} \tag{14}$$

where the symbol $\sim$ denotes a fuzzy quantity. The calculation of Eq. 14 should be performed by use of the extension principle, because the reduction factor is defined by a fuzzy set. The damage degree is expressed in terms of "very large(Vl)", "large(L)", "medium(M)", "small(S)"and "very small(Vs)". Their membership functions are given in Fig.3, in which the abscissa means that u=1 indicates the collapse and u=0 indicates no damage.

Based on the method of fuzzy graph developed by Furuta and Shiraishi[10)], the reduction factor can be calculated according to the evaluation graph. In this method, the concepts of fuzzy relation, fuzzy composition and pattern recognition are used. Table 4 shows the fuzzy relation for the damage pattern 1, and Table 5 presents the weights between nodes 6 through 15 and 16 through 18.

Table 4 Fuzzy Relations

Input		Output node			Input		Output node			Input		Output node
Node	Inf.	16	17	18	Node	Inf.	16	17	18	Node	Inf.	19
6	Vs	Vs	Vs	Vs	11	Vs	Vs	S	Vs	16	Vs	Vs
	S	Vs	Vs	S		S	S	S	S		S	S
	M	Vs	Vs	M		M	L	L	L		M	M
	L	S	Vs	L		L	Vl	Vl	Vl		L	Vl
	Vl	M	S	L		Vl	Vl	Vl	Vl		Vl	Vl
7	Vs	Vs	Vs	S	12	Vs	Vs	Vs	Vs	17	Vs	Vs
	S	Vs	Vs	M		S	Vs	Vs	S		S	S
	M	S	Vs	L		M	S	Vs	M		M	L
	L	M	S	L		L	M	S	L		L	Vl
	Vl	M	S	Vl		Vl	M	S	Vl		Vl	Vl
8	Vs	Vs	Vs	M	13	Vs	Vs	Vs	Vs	18	Vs	Vs
	S	S	Vs	L		S	Vs	Vs	S		S	Vs
	M	M	S	L		M	S	S	M		M	S
	L	L	M	Vl		L	M	S	L		L	L
	Vl	L	L	Vl		Vl	L	M	Vl		Vl	Vl
9	Vs	S	Vs	L	14	Vs	Vs	Vs	Vs			
	S	S	Vs	Vl		S	S	Vs	S			
	M	M	S	L		M	M	S	M			
	L	L	M	Vl		L	L	S	L			
	Vl	Vl	L	Vl		Vl	Vl	L	Vl			
10	Vs	S	Vs	L	15	Vs	Vs	Vs	Vs			
	S	M	S	Vl		S	S	S	S			
	M	M	M	Vl		M	M	M	M			
	L	L	L	Vl		L	L	L	L			
	Vl	Vl	L	Vl		Vl	Vl	Vl	Vl			

Vs : Very Small
S : Small
M : Medium
L : Large
Vl : Very Large

VG : Very Good
G : Good
M : Medium
B : Bad
VB : Very Bad

5.NUMERICAL EXAMPLE

A few numerical examples are presented to describe the method developed here. In these examples, seismic reliability is evaluated paying attention to the response displacement of the top of piers, in which the ductility factor and the coefficient of variation of allowable displacement are assumed to be 4 and 0.1, respectively.

Table 5 Weights

	16	17	18
6	0.4	0.3	0.5
7	0.5	0.3	0.6
8	0.7	0.3	0.7
9	0.9	0.3	0.9
10	1.0	0.3	1.0
11	1.0	1.0	0.5
12	0.5	0.3	0.6
13	0.5	0.3	0.5
14	0.5	0.3	0.3
15	0.6	0.4	0.4

	19
16	1.0
17	0.9
18	0.8

1)Example 1

Consider a pier whose natural frequency period is 0.626 sec and yielding displacement is 0.618 cm. Table 6 presents the failure probabilities obtained for two big cities in Japan, i.e., Tokyo and Osaka. These results are calculated by AFOSM and FOSM(First Order Second-Moment Method). The results by FOSM are compared with those by AFOSM. AFOSM cannot provide appropriate solutions for some cases; zone 3 through 5 in Tokyo,, where the design points needed in AFOSM can not exist beyond the lower bound of epicentral distance. This phenomenon may be due to the fact that when the magnitude is rather small, the design point for the variable of epicentral distance is smaller than the lower bound prescribed by the past earthquake records. This leads to the underestimation of failure probability in the underlying zone. For the reason, two failure probabilities are calculated, where the former provides the upper bound and

Table 6 Comparism of Failure Probabilities fpr Tokyo and Osak

AFOSM

Years	Tokyo		Osaka
10	0.85739E-1	0.859703E-1	0.393431E-1
20	0.119889	0.120344	0.689941E-1
30	0.142386	0.143138	0.945736E-1
40	0.159350	0.160407	0.117471
50	0.173082	0.174408	0.138367
60	0.184705	0.186245	0.157669
70	0.194840	0.196536	0.175645
80	0.203867	0.205663	0.192993
90	0.212037	0.213889	0.208340
100	0.219521	0.221385	0.223359

FOSM

Years	Tokyo	Osaka
10	0.183105E-3	0
20	0.189668E-2	0
30	0.486881E-2	0.119209E-6
40	0.825227E-2	0.447035E-5
50	0.116341E-1	0.463720E-4
60	0.148569E-1	0.230193E-3
70	0.178819E-1	0.728964E-3
80	0.207113E-1	0.122874E-2
90	0.233752E-1	0.335276E-2
100	0.258992E-1	0.567657E-2

the latter provides the lower bound of failure probabilities. Namely, true failure probabilities lie between these two probabilities. Although AFOSM may not provide a unique solutionm, it can give more accurate solution than FOSM.

In general, earthquakes occur in Tokyo more frequently than in Osaka. However, in Table 6 AFOSM shows that the failure probability for Tokyo becomes larger than that of Osaka after 100 years. One reason of this result is that only several zones have higher occurrence rates of earthquake in Tokyo, while all zones have uniform occurrence rates in Osaka. Comparing the occurrence rates of some zones in Tokyo and Osaka, it is evident that the zones in Tokyo have much more occurrence rates than those of the zones in Osaka. Therefore, the failure probabilities within about 90 years in Tokyo is smaller. However, after 100 years, the total failure probability over all Osaka region proceeds to that of Tokyo. Another reason may be the approximation error with regards to the transformation from non-normal variables to normal variables.

Comparing both the results by AFOSM and FOSM, there are significant differences for the failure probabilities within 10 and 20 years. For these two cases, FOSM provides zero probabilities whereas AFOSM provides 0.0393 and 0.0690. Naturally, the solutions by AFOSM are more reliable than FOSM. This discrepancy may be due to such approximation employed in FOSM as Taylor expansion for the linearization of the limit sate function and the ignorance of the information regarding the probability distribution function.

2)Example 2

Consider a pier whose natural frequency period is 1.03 sec and yielding displacement is 2.14 cm. Assuming that the reduction factors as 0.8, 0.6, 0.4, and 0.2, the variation of failure probability is examined. Fig.4 and Fig.5 present the change of failure probability due to the change of damage degree and reference time(year). In the both figures, some points have double values, whose large value indicates the upper bound and smaller value indicates the lower bound. Namely, true failure probabilities lie between these two probabilities.

Using the fuzzy graph method, the reduction factor of allowable displacement is calculated for the damage pattern 1. The calculation is done for four cases; case 1: all damage degrees are "very small(Vs)", case 2: all damage degrees are "very large(Vl)", case 3 and case 4: damage degrees are given according to the past representative collapse failures. In case 3, the damage degree input to nodes 6 through 15 in the evaluation graph shown in Fig.2 are S, Vs, 0, 0, 0, S, S, M, Vl, Vl, respectively, and in case 4 they are M, S, Vs, 0, 0, L, M, M, Vl, Vl,

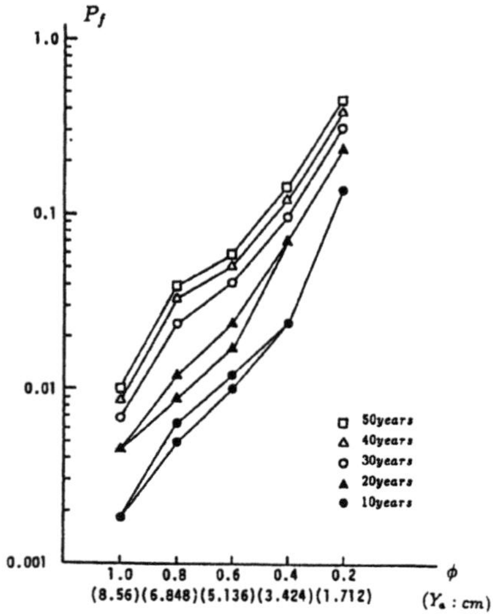

Fig.4 Relation between Reduction Factor and Failure Probability

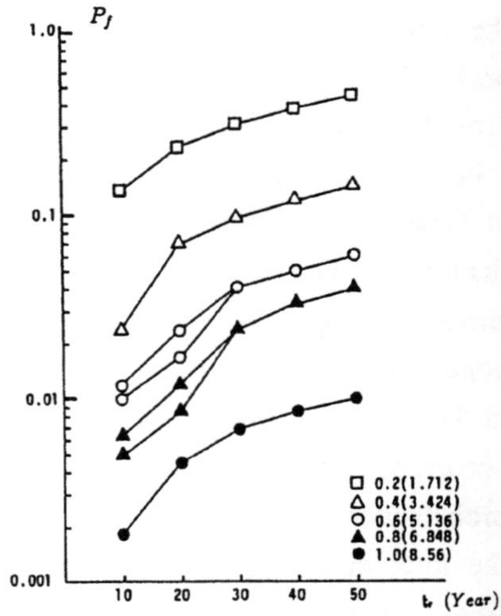

Fig.5 Relation between Remaining Life and Failure Probability

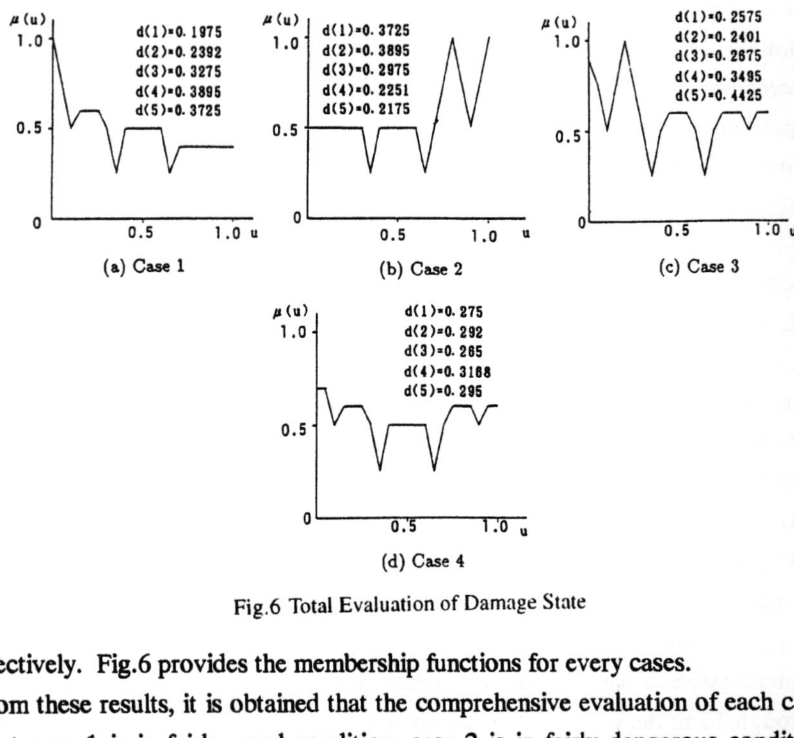

(a) Case 1 (b) Case 2 (c) Case 3

(d) Case 4

Fig.6 Total Evaluation of Damage State

respectively. Fig.6 provides the membership functions for every cases.

From these results, it is obtained that the comprehensive evaluation of each case is that case 1 is in fairly good condition, case 2 is in fairly dangerous condition,

case 3 is in very good state, the failure probabilities after 30 years are calculated as:
for very good,

$$P_f=1/0.0068+0.8/0.0239+0.6/0.0403+0.4/0.0953+0.2/0.3097 \quad (15)$$

for good,

$$P_f=1/0.0068+0.8944/0.0239+0.7746/0.0403+0.6325/0.0953+0.4472/0.3097 \quad (16)$$

for fairly good,

$$P_f=0.5/0.0068+0.7/0.0239+0.9/0.0403+0.9/0.0953+0.7/0.3097 \quad (17)$$

for very dangerous,

$$P_f=0.2/0.0068+0.4/0.0403+0.6/0.0953+0.8/0.3097 \quad (18)$$

6.CONCLUSIONS

In this paper, an attempt was made to develop a simple method of evaluating the seismic reliability of existing structures. To introduce the spectral characteristic of earthquake, the intensity parameter of non-stationary spectrum was utilized for estimating the maximum response displacement in a simple manner. The reliability analysis was performed paying attention to the response displacement of the top of pier. This enables to make the calculation of elasto-plastic analysis simpler. Needless to say, it is noted that this simplification must be checked by performing a sufficient number of numerical calculation. Moreover, it is necessary to classify the structural behavior into the elastic behavior and inelastic behavior. Using the classification, it is possible to obtain a more reliable regression curve for the estimation of maximum response displacement.

Using the method of fuzzy graph, it is possible to introduce the effect of structural deterioration into the calculation of failure probability. The numerical examples confirm the applicability of the method proposed here.

REFERENCES

1) Kameda, H.: Evolutionary Spectra of Seismogram by Multifilter, Jour. of Eng., Mech., Div., ASCE, Vol.101, pp.787-801, 1975.

2) Kanda, J.: Safety Evaluation of Inelastic Building Structures in a Seismic Region, Proc. of Korea-Japan Joint Seminar on Emerging Technologies in Structural Engineering and Mechanics, pp.216-225, 1988.

3) Goto, H., Sugito, M., Kameda, H., Saito, H. and Ohtaki, K.:Prediction of Nonstationary Earthquake Motions for Moderate and Great Earthquakes on Rock Surface, Annuals, Disaster Prevention Research Institute, Kyoto Univ., No.27, B-2, pp.19-48, 1984.

4) Sugito, M. and Kameda, H.: Prediction of Nonsationary Earthquake Motions on Rock Surface, Proc. of JSCE Structural Eng./Earthquake, Vol.2, No.2, pp.149-159, 1985.
5)Sugito, M., Goto, H. and Takeyama, S.: Conversion Factor between Earthquake Motion on Soil Surface and Rock Surface with Nonlinear Soil Amplification Effect, Proc. of 7th Japan Earthquake Engineering Symposium, pp.571-576, 1986.
6) Kameda, H. and Takagi, H.: Seismic Hazard Estimation Based on Non- Poisson Earthquake Occurrence, The Memoirs of the Faculty of Eng., Kyoto Univ., Vol.13, pp.397-433, 1981.
7) Toft-Christensen, P. and Baker, M. J.: Structural Reliability Theory and Its Applications, Springer-Verlag, 1982.
8) Shiraishi, N. and Furuta ,H. : System Reliability Analysis of Damaged Structures, Proc. of International Workshop on Structural System Reliability, Boulder, pp. 253-264, 1988
9) PWRI: Manuals for Repair Method for Civil Engineering Structures Damaged by Earthquake, 1986.
10) Furuta, H. and Shiraishi, N. : Integrity Assessment of Existing Structures, Proc. of Symposium on Reliability Based Design in Civil Engineering, Lausanne, pp. 185-192, 1987.

COMBINED PROBABILITY-POSSIBILITY EVALUATION THEORY FOR STRUCTURAL RELIABILITY

Hiroshi Kawamura and Yasuhiko Kuwamoto

Department of Architecture and Civil Engineering

Faculty of Engineering, Kobe University

Rokkodai, Nada, Kobe, 657, JAPAN

ABSTRACT: In structural identification and planning, as pointed by Yao and Natke (1991), it is very important to evaluate the uncertainties of future loads and structural responses. In this paper, probability functions in which additive law holds are combined with possibility functions in which maxitive law holds, so that their evaluation measures belong to a kind of fuzzy measures instead of probability measures. By using their joint evaluation distributions and measures, one can evaluate the uncertain strength of parallel or serial structural systems and the survival or fracture possibility of structures. This paper presents some examples of case studies and shows that the proposed combined probability-possibility evaluation theory is effective to evaluate the combined objective and subjective reliability of structures from practical and theoretical points of view.

KEYWORDS: Structural Reliability, Possibility Measure, Probability Measure, Evaluation Measure, Fuzzy Measure, Parallel System, Serial System, Survival Possibility, Fracture Possibility

1. INTRODUCTION

The correct probability of a certain future event can be obtained, only when the sufficient data of its past events are given and their repetition is assured in future. For the evaluation of the probabilities of uncertain future events which do not satisfy the above conditions, subjective probability was described for example in Ang (1984) and probability of a fuzzy event was proposed by Zadeh (1968) instead of the usual probability theory. In the case that the additive law is not valid in probability measures, Dempster-Shafer's theory was proposed by Shafer (1976) and possibility theory was proposed by Zadeh (1978).

In this paper, for the subjective evaluation of the safety of future structures under planning, a combined probability-possibility theory is proposed. This new theory is based on the relative frequency histograms derived from the numbers of "yes" answerers to questionnaires asking about structural strength and loading distributed to structural experts. Therefore, probability-possibility measure can be defined as the total number of "yes" answerers about concerned items, and it belongs to a kind of fuzzy measures proposed by Sugeno (1972).

Consider a future, unique and complex structure under planning. No one can estimate the objective reliability of the structure. This structure has to be designed and planned based on subjective judgments of many kinds of experts (strictly saying, based on many kinds of pseudo-objective theories of concerned many experts). The purpose of this paper is to present a theoretical and practical method for the evaluation of the safety of such a future structure.

2. THEORY

2.1 Definitions of Evaluation Distribution and Measure

Suppose that experts are asked to answer "yes" or "no" to a question (e.g., is it safe?) about a given structure under given conditions x_i ($i=1,\ldots,n$). The number of the experts is assumed to be m, and it is assumed that they are allowed to give one or more answers "yes" within the total conditions.

Let the number of experts who answer "yes" to a question under the condition

x_i be l_i, and a relative frequency μ_i *can* be given by

$$\mu_i = l_i / m. \tag{1}$$

μ is plotted as shown in Fig. 1 which is here named evaluation mass function. X is the set of x_i. Generally, $\Sigma\mu_i$ is larger than the unity, so a new theoretical treatment is needed.

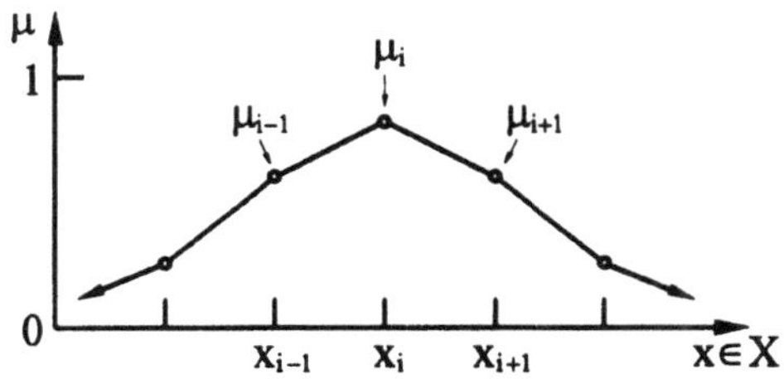

Fig. 1 Evaluation Distribution as Normalized Frequencies in which Multiple Events are Countable

Now let's define evaluation measure M as the total number of the experts who answer "yes" under a subset of conditions x_i. Then the following equations can be derived:

$$M(\phi) = 0, \quad M(X) = 1, \tag{2}$$

$$X_1 \subset X_2 \rightarrow M(X_1) \leq M(X_2), \tag{3}$$

where ϕ is empty set, X_1 and X_2 are subsets of X. Eqs. 2 and 3 show that M belongs to a kind of fuzzy measure proposed by Sugeno (1972).

Here, every set is assumed to be a finite discrete one.

2.2 Compositional Assumption

In this paper, it is assumed that evaluation mass function μ_i (i=1,...,n) is expressed by the sum of possibility mass function π_i and probability mass function f_i as follows:

$$\mu_i = \pi_i + f_i. \tag{4}$$

This assumption is based on a fundamental idea that if evaluations are performed by means of questionnaires, an evaluation of some items has two extreme types, i.e., a possibility evaluation (the maximal evaluation of the items) and a probability evaluation (the sum of the evaluations of the items). So, in ordinary cases, it exists between possibility and probability evaluations.

Some additional discussions on the classification of μ_i into π_i and f_i are described in Appendix A.

A distribution of π_i can be drawn as shown in Fig. 2 where a horizontal bar corresponds to an expert who answers "yes" regarding the subset x_i included within the bar. So the number of bars are the number of experts who answer "yes". A distribution of f_i can be drawn as shown in Fig. 3 where a point corresponds to an expert who answers "yes" only one time, so the number of points is the number of experts who answer "yes".

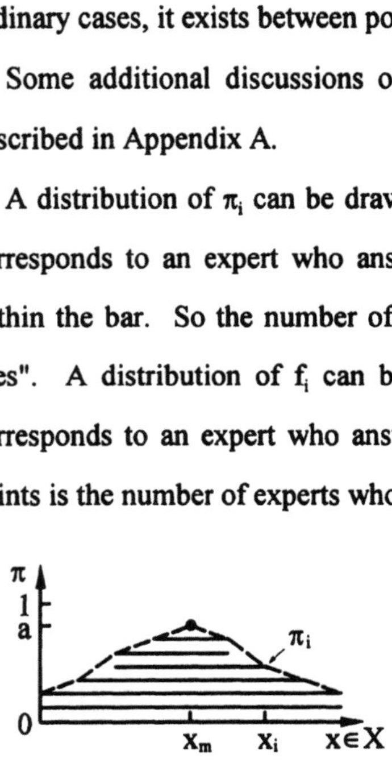

Fig. 2 Possibility Mass Function π_i

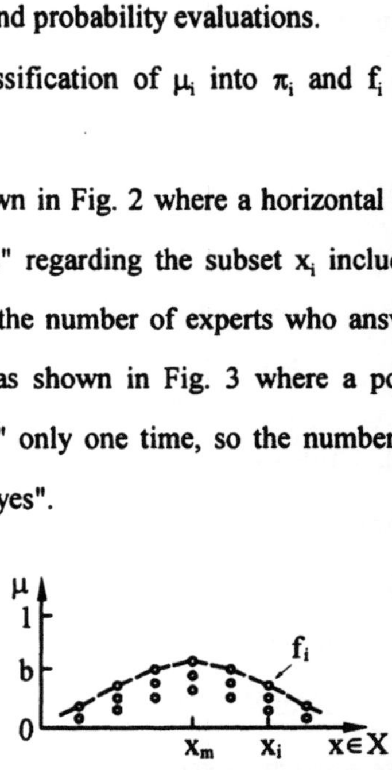

Fig. 3 Probability Mass Function f_i

According to the above assumption, evaluation measure $M(x_i \in X_0)$ can be reduced to

$$M(x_i \in X_0) = \max_{x_i \in X_0}(\pi_i) + \sum_{x_i \in X_0} f_i, \qquad (5)$$

which also satisfies the conditions of fuzzy measure, i.e., Eqs. 2 and 3 in which X_0 is replaced by ϕ, X, X_1 and X_2. When a, α, b and β are given by

$$\max_{x_i \in X}(\pi_i) = a, \quad \sum_{x_i \in X} \pi_i = \alpha, \qquad (6)$$

$$\max_{x_i \in X}(f_i) = b, \quad \sum_{x_i \in X} f_i = \beta, \tag{7}$$

the following relations can be derived:

$$a + \beta = 1, \quad a + b \leq 1, \tag{8}$$

the former of which can be given by the substitution of X for X_0 in Eq. 5.

2.3 Joint Evaluation Mass Function

As for the above evaluation distribution, joint evaluation mass function is needed as well as joint probability mass function. Fig. 4 shows that μ_{3ij} under the condition x_{3ij} $(=g(x_{1i},x_{2j}))$ is joint evaluation of μ_{1i} under the condition x_{1i} and μ_{2j} under the condition x_{2j}. According to Eq. 4, μ_{1i}, μ_{2j} and μ_{3ij} are expressed by

$$\begin{aligned} \mu_{1i} &= \pi_{1i} + f_{1i}, \\ \mu_{2j} &= \pi_{2j} + f_{2j}, \\ \mu_{3ij} &= \pi_{3ij} + f_{3ij}, \end{aligned} \tag{9}$$

where i=1,...,n, and j=1,...,o.

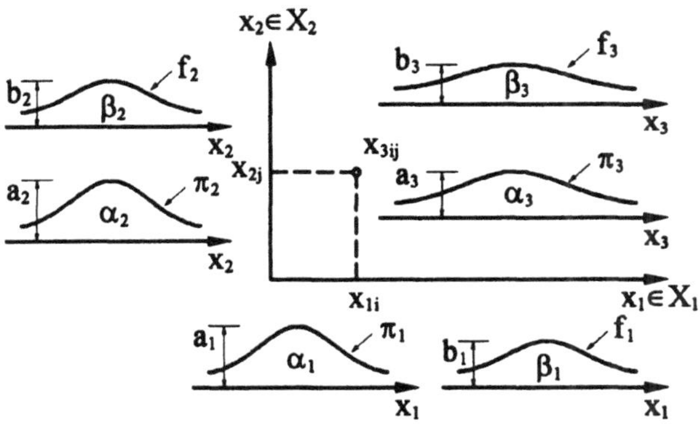

Fig. 4 Joint Evaluation Mass Function

In this paper, it is assumed that μ_3 is given by

$$\mu_3 = \mu_1 \cdot \mu_2, \tag{10}$$

so that π_3 and f_3 are given by

$$\pi_3 = \pi_1 \cdot \pi_2 + \gamma(\pi_1 \cdot f_2 + \pi_2 \cdot f_1),$$
$$f_3 = f_1 \cdot f_2 + (1-\gamma)(\pi_1 \cdot f_2 + \pi_2 \cdot f_1), \tag{11}$$

where an interpolation coefficient γ is introduced to satisfy Eq. 9.

This assumption ($\mu_3 = \mu_1 \cdot \mu_2$) can be justified in the case where μ_1 and μ_2 are given by the same expert group. When the experts are identified one by one, $\mu_3 = \mu_1 \wedge \mu_2$ may be effective, because the common "yes" experts play an important role in decision-making. However, if the number of experts is very large, it is difficult to identify each expert. On the other hand, $\mu_3 = (\mu_1 + \mu_2)/2$ may be effective in the case where μ_1 and μ_2 are derived from different expert groups (where numbers are equal to each other).

As shown by Eqs. 6, 7 and 8, a_3, β_3 are given by

$$a_3 = \max_{x_3 \in X_3}(\pi_3), \quad \beta_3 = \sum_{x_3 \in X} f_3, \tag{12}$$

and have the following relation:

$$a_3 + \beta_3 = 1. \tag{13}$$

If each of the pair of π_1 and f_1, and f_2 and π_3 and f_3 have single extreme values with respect to the same x_{m1}, x_{m2} and x_{m3}, respectively, as shown in Figs. 2 and 3, according to Eq. 12, a_3 and β_3 are given by

$$a_3 = a_1 \cdot a_2 + \gamma(a_1 \cdot b_2 + a_2 \cdot b_1),$$
$$\beta_3 = \beta_1 \cdot \beta_2 + (1-\gamma)(\alpha_1 \cdot \beta_2 + \alpha_2 \cdot \beta_1), \tag{14}$$

where a_1, α_1, b_1, β_1 and a_2, α_2, b_2, β_2 are given regarding π_1, f_1 and π_2, f_2, respectively, by Eqs. 6 and 7.

Using Eqs. 13 and 14, γ is given by

$$\gamma = \frac{1 - a_1 \cdot a_2 - \alpha_1 \cdot \beta_2 - \alpha_2 \cdot \beta_1 - \beta_1 \cdot \beta_2}{a_1 \cdot b_2 + a_2 \cdot b_1 - \alpha_1 \cdot \beta_2 - \alpha_2 \cdot \beta_1}. \tag{15}$$

When the functions of π and f have single peaks at the same x_m, it can be proved that γ exists between 0 and 1 as shown in Appendix B.

An interesting characteristic in an idealized joint evaluation is described in Appendix C.

3. APPLICATION

3.1 Objective Structural Systems

In this paper, as a case study, the whole strengths x_s and x_p of serial and parallel systems composed of two elements with strengths x_1 and x_2 as shown in Figs. 5 and 6 are supposed to be evaluated by using μ_s and μ_p. As another example, let's consider the survival and fracture evaluations μ_s and μ_f of a structure with strength x_1 subjected to a load x_2 as shown in Fig. 7. These evaluations can be performed as evaluation measures which belong to fuzzy measures.

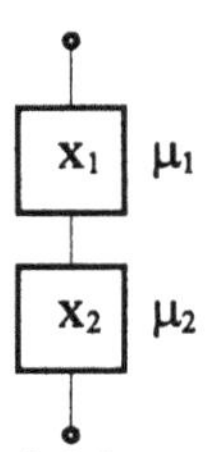

Fig. 5 Serial Structural System

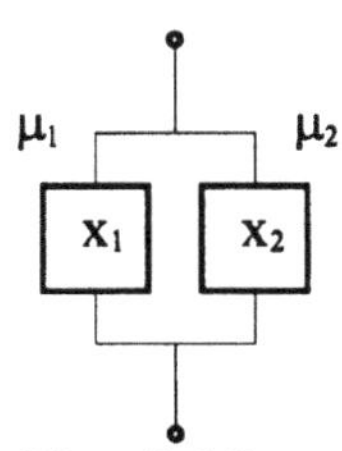

Fig. 6 Parallel Structural System

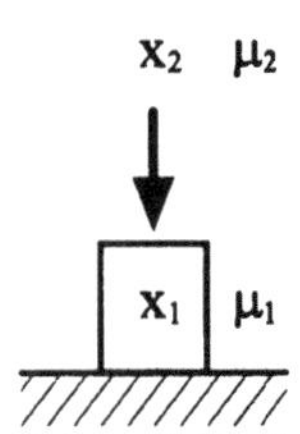

Fig. 7 Load-Resistance Structural System

3.2 Joint Evaluation Mass Functions

Firstly, let's assume that the evaluation mass functions, μ_1 and μ_2, are given as shown in Figs. 8 and 9, where μ_1 and μ_2 mean that x_1 is about 50 kgf and that x_2 is about 30 kgf, respectively. The values μ_1 and μ_2 may change according to the change of the pitch 10 kgf of the discrete values x_1 and x_2. Judging from an engineering point of view, it is desirable to make the pitch less so long as it does not lose practical meanings. Of course, μ_1 and μ_2 satisfy the above equations 4 to 8, then the joint evaluation mass function μ_3 can be calculated according to the equations 9 to 14, where γ is given as 0.680 by using Eq. 15, and a_3 and β_3 are reduced to 0.3760 and 0.6240, respectively, by using the above γ and Eq. 14, which satisfies Eq. 13, i.e., $a_3+\beta_3=0.3760+0.6240=1$. Another condition which π and f should satisfy is discussed in Appendix D.

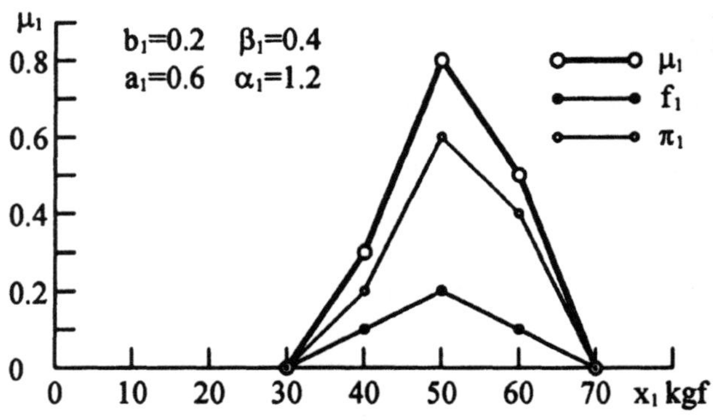

Fig. 8 Assumed Evaluation Mass Function μ_1 of x_1

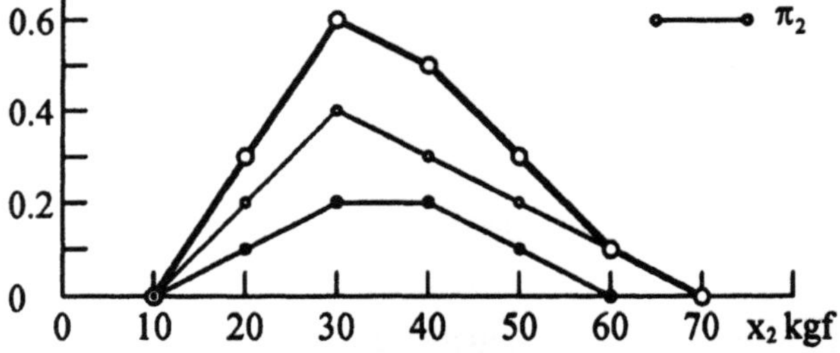

Fig. 9 Assumed Evaluation Mass Function μ_2 of x_2

The relation among x_1, μ_1, f_1,π_1, x_2, μ_2, f_2, π_2, x_3, μ_3, f_3 and π_3 are shown in Table 1 in which $g(x_{1i},x_{2j})$ means x_{3ij}.

Table 1 Joint Evaluation Mass Function

0 0 0	70	0 0 0 g(30,70)	0 0 0 g(40,70)	0 0 0 g(50,70)	0 0 0 g(60,70)	0 0 0 g(70,70)
0.1 0 0.1	60	0 0 0 g(30,60)	0.03 0.0032 0.0268 g(40,60)	0.08 0.0064 0.0736 g(50,60)	0.05 0.0032 0.0468 g(60,60)	0 0 0 g(70,60)
0.3 0.1 0.2	50	0 0 0 g(30,50)	0.09 0.0228 0.0672 g(40,50)	0.24 0.0520 0.1880 g(50,50)	0.15 0.0292 0.1208 g(60,50)	0 0 0 g(70,50)
0.5 0.2 0.3	40	0 0 0 g(30,40)	0.15 0.0424 0.1076 g(40,40)	0.40 0.0976 0.3026 g(50,40)	0.25 0.0552 0.1948 g(60,40)	0 0 0 g(70,40)
0.6 0.2 0.4	30	0 0 0 g(30,30)	0.18 0.0456 0.1344 g(40,30)	0.48 0.1040 0.3760 g(50,30)	0.30 0.0584 0.2416 g(60,30)	0 0 0 g(70,30)
0.3 0.1 0.2	20	0 0 0 g(30,20)	0.09 0.0228 0.0672 g(40,20)	0.24 0.0520 0.1880 g(50,20)	0.15 0.0292 0.1208 g(60,20)	0 0 0 g(70,20)
0 (μ_2) 0 (f_2) 0 (π_2)	10	0 (μ_3) 0 (f_3) 0 (π_3) g(30,10)	0 0 0 g(40,10)	0 0 0 g(50,10)	0 0 0 g(60,10)	0 0 0 g(70,10) (= x_3)
	x_2 kgf / x_1 kgf	30	40	50	60	70
		0 (μ_1) 0 (f_1) 0 (π_1)	0.3 0.1 0.2	0.8 0.2 0.6	0.5 0.1 0.4	0 0 0

3.3 Evaluation Measures

In the above case, the evaluation measure given by Eq. 5 is expressed by $M_k(x_{3ij} \in X_{0k})$ in which X_{0k} is subset of x_3 and $k=1,2,\ldots\ldots,i \times j$.

As for the serial structural system shown in Fig. 5, the whole strength x_s is given by

$$x_{sk}=x_{3ij}=g_s(x_{1i},x_{2j})=\min(x_{1i},x_{2j}), \tag{16}$$

and considering

$$x_{sk}=X_{0k},\ f_{sk}=f_{3ij},\ \pi_{sk}=\pi_{3ij},\ \text{and}\ \mu_{sk}=M_k(x_{3ij}=x_{sk}), \tag{17}$$

f_s, π_s and μ_s can be plotted with respect to x_s as shown in Table 2 and Fig. 10, where f_s, π_s and μ_s present the probability, possibility and evaluation distributions of x_s, respectively, which shows that x_s is about 30 kgf.

Table 2 Calculated Evaluation Mass Function μ_s of x_s as Measures (Strength of Serial Structural System)

x_s kgf	$\sum f_3=f_s$	$\max(\pi_3)=\pi_s$	$M=\mu_s$
10	0	0	0
20	0.1040	0.1880	0.2920
30	0.2080	0.3760	0.5840
40	0.2212	0.3024	0.5236
50	0.0876	0.1880	0.2756
60	0.0032	0.0468	0.0500
70	0	0	0

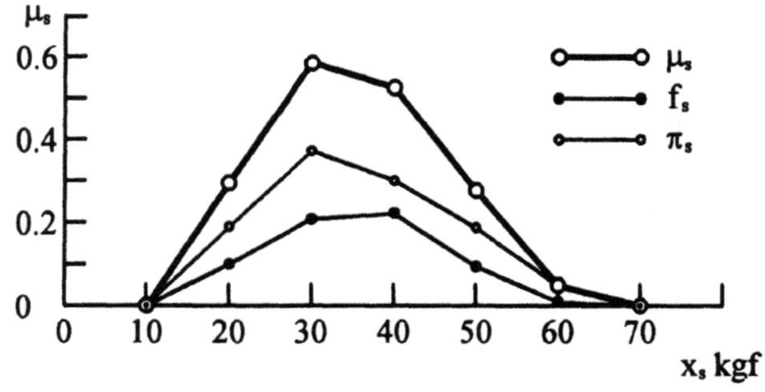

Fig. 10 Calculated Evaluation Mass Function μ_s of x_s as Measures (Strength of Serial Structural System)

As for the parallel structural system in Fig. 6, if its elements have sufficient ductility, the whole strength x_p is given by

$$x_{pk}=x_{3ij}=g_p(x_{1i},x_{2j})=x_{1i}+x_{2j}, \tag{18}$$

the probability, possibility and evaluation distributions, f_p, π_p and μ_p of x_p can be shown with respect to x_p as shown in Table 3 and Fig. 11 in the same manner as

x_s, which means x_s is about 80 kgf.

Table 3 Calculated Evaluation Mass Function μ_p of x_p as Measures (Strength of Parallel Structural System)

x_p kgf	$\sum f_3 = f_p$	$\max(\pi_3) = \pi_p$	$M = \mu_p$
50	0	0	0
60	0.0220	0.0670	0.0900
70	0.0976	0.1880	0.2856
80	0.1756	0.3760	0.5516
90	0.1778	0.3024	0.4812
100	0.1104	0.1948	0.1564
110	0.0356	0.1208	0.0500
120	0.0032	0.0468	0.0500
130	0	0	0

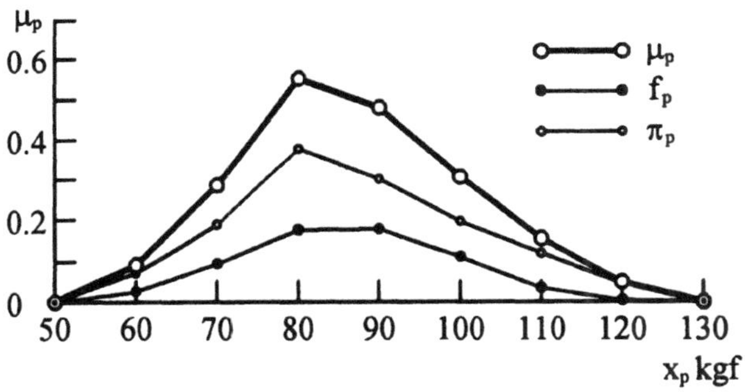

Fig. 11 Calculated Evaluation Mass Function μ_p of x_p as Measures (Strength of Parallel Structural System)

Regarding the structure shown in Fig. 7, suppose that strength x_1 and the load x_2 are evaluated by means of Figs. 8 and 9, respectively, and the survival and fracture evaluations μ_s and μ_f can be calculated as shown in Table 4, where

$$\mu_s = M(x_{3ij} \in X_{0s}), \tag{19}$$

in which

$$X_{0s} = \{x_{3ij} | x_{3ij} = g_s(x_{1i}, x_{2j}) = s | x_{1i} \geq x_{2j}\}, \tag{20}$$

and

$$\mu_f = M(x_{3ij} \in X_{0f}), \tag{21}$$

in which

$$X_{0f} = \{x_{3ij} \mid x_{3ij} = g_f(x_{1i}, x_{2j}) = f \mid x_{1i} \leq x_{2j}\}, \tag{22}$$

where in case of $x_{1i}=x_{2j}$, M defined as Eq. 5 has to be replaced by

$$M(x_i \in X_0) = \max_{x_i \in X_0}(\pi_i) + \sum_{x_i \in X_0} f_i / 2. \tag{23}$$

Table 4 Survival and Fracture Evaluations μ_s , μ_f as Measures

S or F	$\sum f_3$	$\max(\pi_3)$	μ_s or μ_f
S	0.5428	0.3760	0.9188
F	0.0812	0.1880	0.2692

4. CONCLUSIONS

By introducing the compositional assumption that evaluation distribution is given by the sum of possibility and probability distributions, a very flexible evaluation theory can be presented. This theory called a combined probability-possibility evaluation theory has the following features.

(1) The numerical foundation is based on normalized frequencies in which multiple events are countable under the same conditions.

(2) The sum compositional assumption holds not only in evaluation distributions but also in their measures.

(3) Therefore, by using joint evaluation distributions, one can analyze the reliabilities not only of a unit structure but also of serial and parallel structural systems.

However, there remain the following some problems which should be solved in future pointed by Kawamura (1992):

(1) Development into infinite continuous sets of variables.

(2) Extension into joint evaluation distributions of three or more marginal evaluation distributions.

ACKNOWLEDGEMENT

The authors are very grateful to Prof. Alberto Bernardini at Universita Degli Studi di Padova for his kind comments on this paper.

REFERENCES

Ang, A.H-S. and Tang, W.H. (1984) Probability Concepts in Engineering Planning and Design, Vol. II-Decision, Risk, and Reliability: John Wiley & Sons.

Kawamura, H. (1992) Definition and Application of Fuzzy Sets and Fuzzy Measures for Structural Reliability Evaluation, Proc. of 8th Fuzzy System Symposium, Hiroshima, Japan, 17-20. (in Japanese)

Shafer, G. (1976) Mathematical Theory of Evidence: Princeton University Press.

Sugeno, M. (1972) Fuzzy Measure and Fuzzy Integral, Trans. of the Society of Instrument and Control Engineers, Vol. 8, No. 2, 94-102. (in Japanese)

Yao, J.T.P. and Natke, H.S. (1991) Uncertainties in Structural Identification and Control, Proc. of International Fuzzy Engineering Symposium on Fuzzy Engineering toward Human Friendly Systems, 2, Japan, 844-849.

Zadeh, L.A. (1968) Probability Measures of Fuzzy Events, J.Math.Analysis and Appl., 10, 421-427.

Zadeh, L.A. (1978) Fuzzy Sets as a Basis for a Theory of Possibility, Fuzzy Sets and Systems, 1, 3-28.

APPENDIX A

Suppose that a distribution of fuzzy measure M and an evaluation mass function μ are given by means of questionnaire and that π, f and μ have single peaks at the same x_m as shown in Fig. A-1, where M_j means relative frequencies who answer "yes" under the condition $x \leq x_i$ and μ_i means relative frequencies who answer "yes" under the condition $x = x_i$.

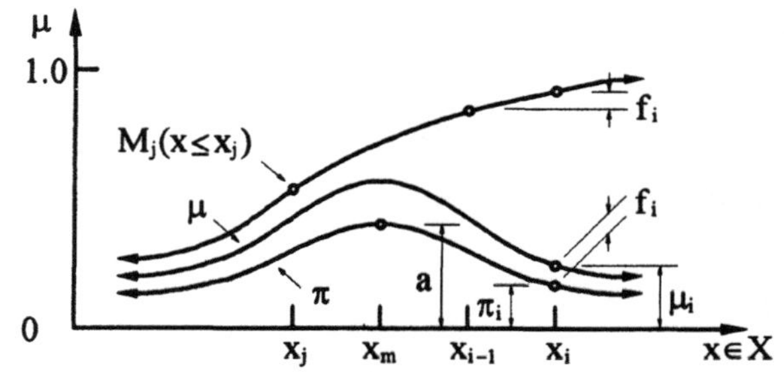

Fig. A-1 Discrimination between π and f by Using M

In case that $x \geq x_m$: As for M, M_i and M_{i-1} are given by Eq. 24

$$M_i = a + \sum_{k=0}^{i} f_k, \quad M_{i-1} = a + \sum_{k=0}^{i-1} f_k, \tag{24}$$

so the difference $M_i - M_{i-1}$ is reduced to

$$M_i - M_{i-1} = f_i, \tag{25}$$

which represents possibility mass function f_i. Remembering Eq. 4, π_i is given by

$$\pi_i = \mu_i - f_i = \mu_i - (M_i - M_{i-1}). \tag{26}$$

In case that $x \leq x_m$: Suppose that M_j means relative frequencies who answer "yes" under the condition $x \geq x_j$, and M_i, M_{i-1} are given by

$$M_{i-1} = a + \sum_{k=i-1}^{\infty} f_k, \quad M_i = a + \sum_{k=i}^{\infty} f_k, \tag{27}$$

so the difference $M_{i-1} - M_i$ is reduced to

$$M_{i-1} - M_i = f_{i-1}, \tag{28}$$

and π_{i-1} is given by

$$\pi_{i-1} = \mu_{i-1} - f_{i-1} = \mu_{i-1} - (M_{i-1} - M_i). \tag{29}$$

When M is not given and it is able to be assumed that π and f have distributions proportional to μ as follows:

$$\pi_i = r \cdot \mu_i, \quad f_i = (1 - r)\mu_i. \tag{30}$$

By using Eqs. 6, 7, 8, r is given by

$$r = \left(\sum \mu_i - 1\right) / \left(\sum \mu_i - \max \mu_i\right), \tag{31}$$

where Σ and max are operative for all the range of x_i.

In the joint evaluation mass function shown in Fig. 4, the proportionality of Eq. 30 is reserved as shown in Appendix C.

APPENDIX B

The numerator of γ is given by

$$A = a_1 \cdot b_2 + a_2 \cdot b_1 - \alpha_1 \cdot \beta_2 - \alpha_2 \cdot \beta_1, \tag{32}$$

and

$$a_1 \le \alpha_1, \quad a_2 \le \alpha_2, \quad b_1 \le \beta_1, \quad b_2 \le \beta_2, \tag{33}$$

then

$$A \le \alpha_1 \cdot \beta_2 + \alpha_2 \cdot \beta_1 - \alpha_1 \cdot \beta_2 - \alpha_2 \cdot \beta_1 = 0. \tag{34}$$

The denominator of γ is given by

$$B = 1 - a_1 \cdot a_2 - \alpha_1 \cdot \beta_2 - \alpha_2 \cdot \beta_1 - \beta_1 \cdot \beta_2, \tag{35}$$

and

$$a_1 + \beta_1 = 1, \quad a_2 + \beta_2 = 1, \tag{36}$$

then

$$\begin{aligned} B &= 1 - (1 - \beta_1)(1 - \beta_2) - \alpha_1 \cdot \beta_2 - \alpha_2 \cdot \beta_1 \\ &= \beta_1 (a_2 - \alpha_2) + \beta_2 (a_1 - \alpha_1) \le 0. \end{aligned} \tag{37}$$

On the other hand,

$$\begin{aligned} A - B &= 1 - a_1 \cdot a_2 - \beta_1 \cdot \beta_2 - a_1 \cdot b_2 - a_2 \cdot b_1 \\ &\le 1 - a_1 \cdot a_2 - \beta_1 \cdot \beta_2 - a_1 \cdot \beta_2 - a_2 \cdot \beta_1 \\ &= 1 - a_1(\beta_2 + a_2) + \beta_1(\beta_2 + a_2) \\ &= 1 - a_1 + \beta_1 = 2\beta_1 \ge 0, \end{aligned} \tag{38}$$

then

$$-A \le -B. \tag{39}$$

Therefore,

$$0 \leq \gamma\,(=A/B) \leq 1. \tag{40}$$

APPENDIX C

Consider an idealized joint evaluation of μ_1 and μ_2 which are composed of π_1, f_1, and μ_2, f_2, respectively, and are expressed by

$$\pi_1 = r_1 \cdot \mu_1, \quad r_1 = (1-r_1)\mu_1, \tag{41}$$

$$\pi_2 = r_2 \cdot \mu_2, \quad f_2 = (1-r_2)\mu_2. \tag{42}$$

Suppose that μ_3 is composed of π_3, f_3 which are shown by

$$\pi_3 = r_3 \cdot \mu_3, \quad f_3 = (1-r_3)\mu_3. \tag{43}$$

By using the upper Eq. 11, (as for π_3), r_3 is reduced to

$$r_3 = r_1 \cdot r_2 - \gamma(r_1 + r_2 - 2r_1 \cdot r_2). \tag{44}$$

On the other hand, by using the lower Eq. 11 (as for f_3), r_3 is reduced to

$$1 - r_3 = (1-r_1)(1-r_2) + (1-\gamma)(r_1 + r_2 - 2r_1 \cdot r_2). \tag{45}$$

Any way, r_3 can be calculated according to Eq. 31.

The most interesting thing is that γ is given by

$$\gamma = (r_3 - r_1 \cdot r_2)/(r_1 + r_2 - 2r_1 \cdot r_2), \tag{46}$$

which is effective to both Eqs. 44 and 45, which means the effectiveness of Eq. 43. It is proved that the proportionality is reserved in joint evaluation.

APPENDIX D

The distributions of π and f should satisfy the following equations:

$$\sum_x \pi \geq 2\left(1 - \sum_x f\right), \tag{47}$$

$$\max_x(\pi) = 1 - \sum_x f. \tag{48}$$

Eq. 47 means that the total number of answers given by experts who answer "yes" more than once ($\Sigma\pi$) is more than two times the number of experts who answer more than once ($1-\Sigma f$). Eq. 48 means that the maximal number of answers about an item of experts who answer "yes" more than once ($\max(\pi)$) is equal to the number of experts who answer more than once ($1-\Sigma f$), However, these equations are necessary conditions (not sufficient conditions) to satisfy Fig. 2 and Fig. 3. In reality, we can not get such idealized Fig. 2 and Fig. 3 conditions but such approximate ones as shown in Fig. 8 and Fig. 9.

RELIABILITY ASSESSMENT USING VARIANCE REDUCTION TECHNIQUES BASED ON FUZZY SYSTEM PERFORMANCE

Kwan-Ling Lai[1], and Bilal M. Ayyub[2]
[1]China Consultants, Inc., Taipei, Taiwan.
[2]Department of Civil Engineering, University of Maryland, College Park, MD 20878, USA.

ABSTRACT

Classical reliability methods require a crisp definition of system failure according to some performance measure that is of interest to an analyst. Therefore, they do not consider any form of uncertainty in the system failure. This limitation might result in reliability assessments that are not realistic. The uncertainty associated with qualifying the performance of the system as inadequate, i.e., failed, should be considered in reliability assessment. The objective of this contribution is to generalize currently used reliability assessment methods by treating failure as a fuzzy event. Therefore, the reliability of the system can be provided over a non-performance spectrum. Monte Carlo simulation with variance reduction techniques will be used in the computation of the reliability.

KEYWORDS

Failure, Fuzzy set, Structural reliability, Variance, Simulation

1. INTRODUCTION

Uncertainties in engineering systems can be classified into two broad categories, i.e., ambiguity and vagueness. The different types of uncertainty in structural engineering are discussed by Ayyub (1992). Classical structural reliability assessment methods determine the probability of failure of a structural component or system according to a crisp failure definition. They successfully model the ambiguity type of uncertainty, using the theory of probability and statistics. However, the vagueness in the perception of damage and the imprecision of the definition of some parameters are not properly considered in the classical methods. The objective of this contribution is to develop a method for structural reliability assessment based on a generalized treatment of uncertainty. A fuzzy-based assessment model is suggested to provide the reliability of a structure over a damage spectrum.

2. DEFINITION OF FAILURE

The model for crisp failure consists of two basic, mutually exclusive events, i.e., complete survival and complete failure. The transition from one to another is abrupt rather than continuous. This model is illustrated in Figure 1, where R_f is the structural response at the limiting state for a selected design criterion. If the structural response R is smaller than R_f, i.e., $R < R_f$, the complete survival state exists and it is mapped to the zero failure level ($\alpha = 0$). If the structural response R is larger than R_f, i.e., $R > R_f$, the complete failure state occurs and it is mapped to $\alpha = 1$. The limit state is defined where $R = R_f$.

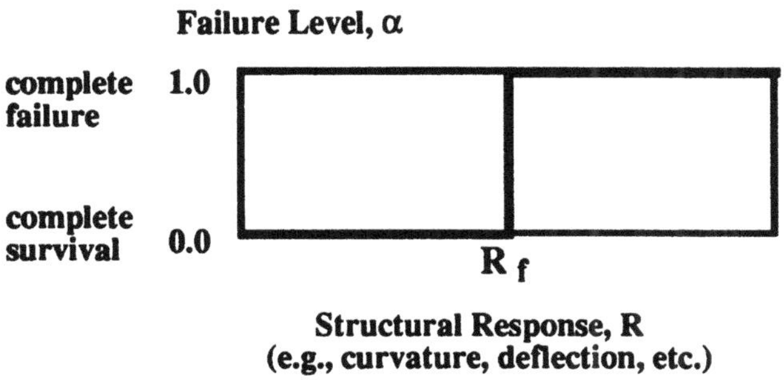

Figure 1. Crisp failure

The fuzzy failure model is illustrated by introducing a subjective failure level index α as shown in Figure 2, where R_l and R_u are the lower bound and upper bound of structural response for the region of transitional or partial failure, respectively. The complete survival state is defined where $R \leq R_l$, $R_l < R < R_u$ is the transitional state, and $R \geq R_u$ is the complete failure state. In Figure 2, the structural response is mapped to the failure-level scale to model a certain performance event as follows:

$$\text{Performance event A:} \quad R \rightarrow A = \{ \alpha : \alpha \in [0,1] \} \tag{1}$$

where 0 = the failure level for complete survival, 1 = the failure level for complete failure, and [0,1] = all real values in the range of 0 to 1 for all failure levels.

The index α can also be interpreted as a measure of degree of belief in the occurrence of some performance condition. In this case, $\alpha = 0$ is interpreted as no belief of the occurrence of an event, $\alpha = 1$ means absolute belief of the occurrence of an event. Figure 3 illustrates this interpretation of the index α. According to this interpretation, a performance event A can be modeled as

$$A = \{ r_1 | \alpha_A(r_1) , r_2 | \alpha_A(r_2) , \dots , r_n | \alpha_A(r_n) \} \tag{2}$$

in which "=" should be interpreted as "is defined to be;" and " | " = a delimiter. Equation 2 shows the structural response to take discrete values. Generally, structural response can take any value in a continuous range, and the membership function of A is expressed as

$$A = \{\, r|\ \alpha_A(r) : \text{ for all } r \in R \,\} \tag{3}$$

where $\alpha_A(r) \in [0,1]$. The use of discrete values is common in fuzzy set modeling for easy computation and representation. However, this does not constitute any limitation to the approach, since computations can also be performed using continuous performance values.

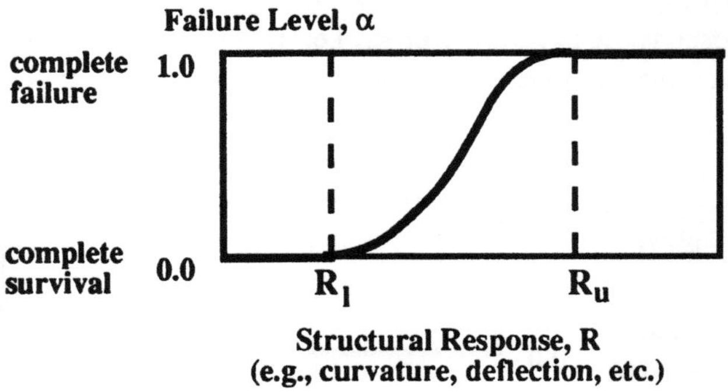

Figure 2. Fuzzy failure interpreted as failure level

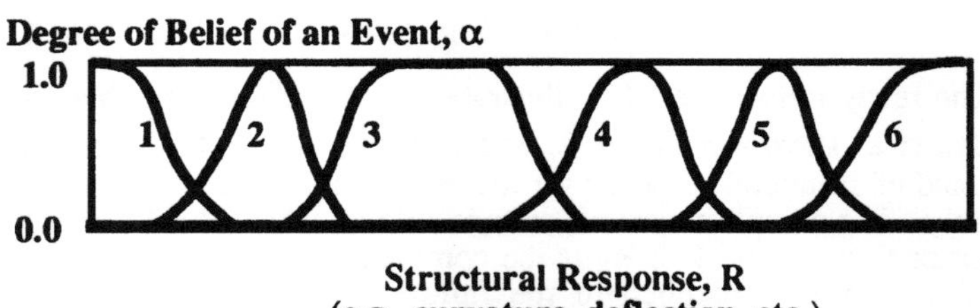

Event Number	Definition
1	complete survival
2	low serviceability failure
3	serviceability failure
4	high serviceability failure
5	partial collapse
6	complete collapse

Figure 3. Fuzzy failure interpreted as degree of belief

3. STRUCTURAL RELIABILITY ASSESSMENT

3.1 Classical Structural Reliability Assessment

Classical structural reliability assessment techniques are based on precise and crisp definitions of failure and survival of a structure in meeting a set of strength, function and serviceability criteria. Consider the following performance function:

$$Z = g(X_1, X_2, ..., X_n) \qquad (4)$$

where $X_1, X_2, ..., X_n$ = basic random variables; and Z = performance measure or safety margin. Equation 4 is defined such that failure occurs where $g(.) < 0$, survival occurs where $g(.) > 0$, and the limit state equation is defined as $g(.) = 0$. The probability of failure can then be determined by solving the following integral:

$$P_f = \int\int...\int f_{\underline{X}}(X_1, X_2, ..., X_n)\, dx_1\, dx_2 ... dx_n \qquad (5)$$

where $f_{\underline{X}}$ is the joint probability density function of $\underline{X} = \{X_1, X_2, ..., X_n\}$ and the integration is performed over the region where $g(.) < 0$. For example, consider the crisp failure model for a stochastic moment-curvature relationship as shown in Figure 4, in which ϕ_f = the curvature at failure, f_L = probability density function of external load L, and f_M = conditional probability density function of the resisting (strength) moment at $\phi = \phi_f$. In this case, the performance function is simply expressed as follows:

$$Z = M - L = M(\phi) - L \qquad (6)$$

where $M(\phi)$ = the resisting moment as a function of the curvature ϕ. The probability of failure P_f can then be determined by the following:

$$\begin{aligned} P_f &= \text{Prob}\{\text{ failure level} = 1 \} = \text{Prob}\{ L > (M \text{ at } \phi = \phi_f) \} \\ &= \int_0^{\infty} \text{Prob}\{ L > (m \text{ at } \phi = \phi_f) \}\, f_M(m)\, dm \\ &= \int_0^{\infty} [1 - F_L(m)]\, f_M(m)\, dm \qquad (7) \end{aligned}$$

where F_L = cumulative distribution function of L. The probability of failure can be estimated by moment methods, or simulation methods with or without variance reduction techniques.

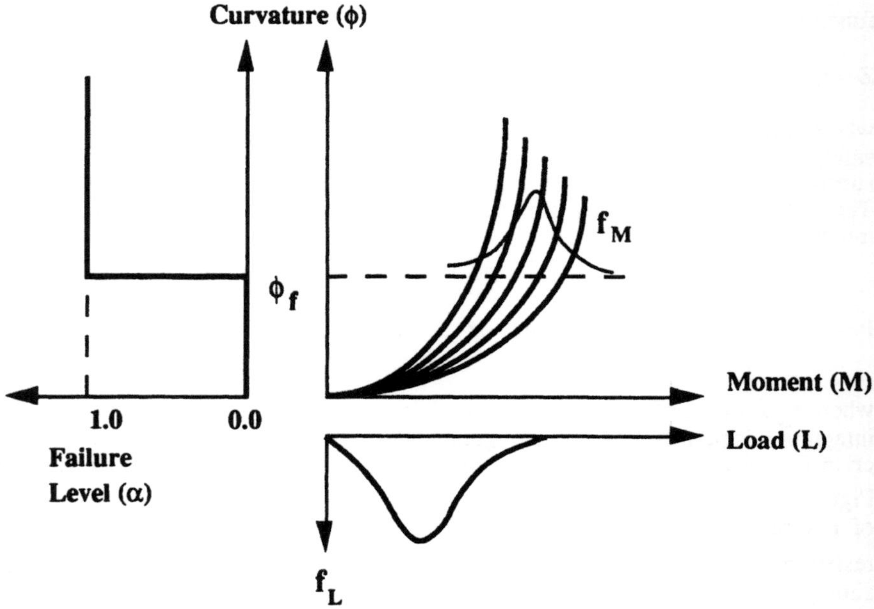

Figure 4. Structural reliability assessment based on crisp failure definition

3.2 Structural Reliability Assessment Based on Fuzzy Definition of Failure

A mathematical model for structural reliability assessment that includes both ambiguity and vagueness types of uncertainty was first suggested by Alvi and Ayyub (1990). The model was suggested to result in the likelihood of failure over a damage spectrum. Since the structural reliability assessment is based on the fuzzy failure model, the probability of failure, in this case, is a function of α. For example, consider the fuzzy failure model for a stochastic moment-curvature relationship as shown in Figure 5, in which ϕ_f = curvature at some failure level α_f; and ϕ_l and ϕ_u = lower bound and upper bound of curvature for partial failure. The performance function is the same as give by Equation 6. The probability of failure at failure level α_f is then determined by the following:

$$P_f(\alpha_f) = \text{Prob}\{ L > (M \text{ at } \phi = \phi_f) \} = \int_0^{\infty} \text{Prob}\{ L > (m \text{ at } \phi = \phi_f) \} f_M(m)\, dm$$

$$= \int_0^{\infty} [1 - F_L(m)]\, f_M(m)\, dm \qquad (8)$$

It can be estimated using moment methods, or simulation methods with or without variance reduction techniques.

The failure probability content over the performance event can be mathematically defined by an arithmetic average P_{fa} or geometric average P_{fg} as shown in the following:

$$\text{arithmetic average, } P_{fa} = \frac{\int_0^1 P_f(\alpha)\, d\alpha}{\int_0^1 d\alpha} \tag{9a}$$

$$\text{geometric average, } \log_{10}(P_{fg}) = \frac{\int_0^1 \log_{10}(P_f(\alpha))\, d\alpha}{\int_0^1 d\alpha} \tag{9b}$$

where the geometric average can be considered as the weighted power order average of the probability of failure. These probabilities are numerically evaluated by the following
respective equations:

$$P_{fa} = \sum_{i=1}^{k} \frac{P_f(\alpha_i) + P_f(\alpha_{i-1})}{2} \Delta\alpha_i \tag{10a}$$

$$\log_{10}(P_{fg}) = \sum_{i=1}^{k} \frac{\log_{10}(P_f(\alpha_i)) + \log_{10}(P_f(\alpha_{i-1}))}{2} \Delta\alpha_i \tag{10b}$$

where $\Delta\alpha_i = \alpha_i - \alpha_{i-1}$; and k = number of increments for α.

4. VARIANCE REDUCTION TECHNIQUES

In the use of direct simulation method, the probability of failure is estimated as the ratio of the number of failures to the total number of simulation cycles. Therefore, large numbers of simulation cycles are needed for small probabilities of failure in order to estimate the probability of failure within an acceptable level of statistical error. The fundamentals of this method is available in several references by Harbitz (1986), Law and Kelton (1982), Melchers (1987), Rubinstein (1981), etc.

The efficiency of simulation can be largely improved by using variance reduction techniques (VRT). One of the commonly used methods is conditional expectation combined with antithetic variates variance reduction techniques. The method is highly efficient, and converges to the correct probability of failure in a relatively small number of simulation cycles. The method is mathematically simple, and does not require large spaces of computer memory. Therefore, it can be programmed on small micro- and personal computers.

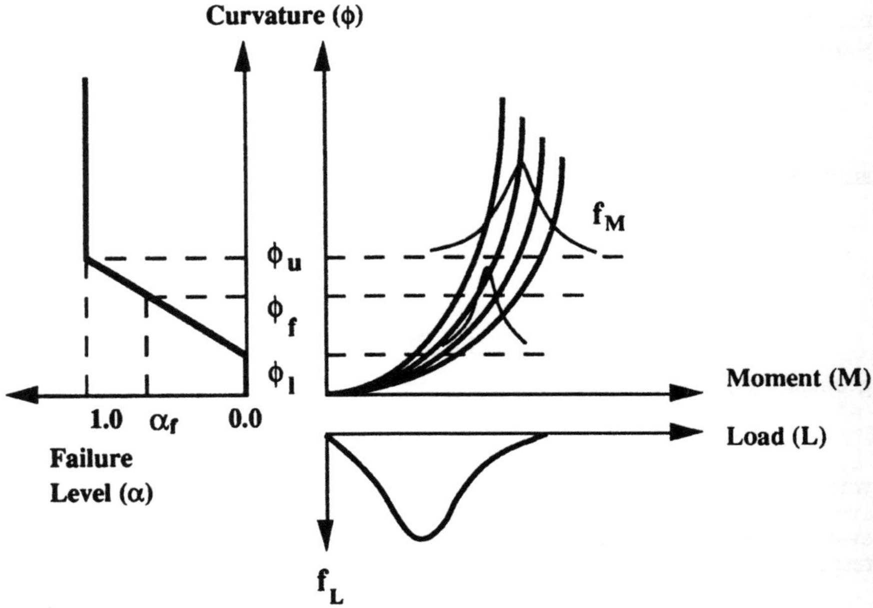

Figure 5. Structural reliability assessment based on fuzzy failure definition

The common random numbers variance reduction technique was used for structural reliability assessment by Haldar and Ayyub (1984). It was used to compare the structural reliability levels of alternative designs. The method is based on the fact that the variance of the estimated difference in the probability of failure of two or more alternative designs can be reduced by using the same streams of random numbers for common random variables among the designs.

Importance sampling procedures with design points were suggested for structural reliability assessment by Melchers (1987), Bourgund and Bucher (1986), and Schueller and Stix (1986). Based on their studies, a drastic reduction in computational effort and time can be accomplished. The method was used for the reliability assessment of structural systems that have failure modes and/or components in series.

Latin hypercube sampling was used for structural reliability assessment by Ayyub and Lai (1989). The technique provides a constrained sampling scheme instead of random sampling. In Latin hypercube sampling, N different values in N non-overlapping intervals, which are selected to be of the same probability of occurrence, are generated for each random variables. The selected values for different variables are then grouped by associating a random permutation of the N integers. The method results in stabilizing the estimation process of the failure probability. In addition, the generated random values are affected to a lesser extent by the choice of the seed in the random number generator.

Other methods of sampling and variance reduction techniques are also available. They include, for example, adaptive importance sampling, correlation methods, control variates, directional simulation, biased estimation, weighted integration, linear programming, etc. In the following section, the conditional expectation (CE) VRT is discussed in detail.

4.1 Conditional Expectation VRT

The performance function of a fundamental structural reliability assessment case is given by

$$Z = R - L \tag{11}$$

where R = function of structural strength or resistance; and L = function of the corresponding load effect. Therefore, the probability of failure, P_f, is given by

$$P_f = P(Z < 0) = P(R < L) \tag{12}$$

For a randomly generated value of L (or R), say l_i (or r_i), the probability of failure is given by

$$P_{f_i} = \text{Prob}(R < l_i) = F_R(l_i) \tag{13a}$$

or

$$P_{f_i} = \text{Prob}(L > r_i) = 1 - F_L(r_i) \tag{13b}$$

where F_R and F_L are cumulative distribution functions of R and L, respectively. In this formulation R and L are assumed to be statistically uncorrected random variables. Thus, for N simulation cycles, the mean value of the probability of failure is given by the following equation:

$$\bar{P}_f = \frac{\sum_{i=1}^{N} P_{f_i}}{N} \tag{14}$$

The variance (Var) and the coefficient of variation (COV) of the estimated probability of failure are given by

$$\text{Var}(\bar{P}_f) = \frac{\sum_{i=1}^{N} (P_{f_i} - \bar{P}_f)^2}{N(N-1)} \tag{15a}$$

$$\text{COV}(\bar{P}_f) = \frac{[\text{Var}(\bar{P}_f)]^{1/2}}{\bar{P}_f} \tag{15b}$$

The randomly generated variables should be selected as the ones of least variabilities, and the resulting conditional expectation can be evaluated by some known expression, e.g. the cumulative distribution function of the random variable that was not randomly generated. This variable is called the control variable. This method can be used for any performance function with any probability distributions for the random variables. The only limitation is that the control variable, X_k, must be statistically uncorrected to the $g_k(.)$ as shown in the following equation:

$$P_f = E_{X_{i:i=1,2,\ldots,n;i\neq k}} [F_{X_k} (g_k (X_{i:i=1,2,\ldots,n;i\neq k}))] \tag{16}$$

where $g_k (X_{i:i=1,2,\ldots,n;i\neq k})$ is defined as follows:

$$P_f = \text{Prob} [g(X_1, X_2, \ldots, X_n) < 0] \tag{17a}$$

or

$$P_f = \text{Prob} [X_k < g_k ((X_{i:i=1,2,\ldots,n;i\neq k})] \tag{17b}$$

The concept and the steps involved in this method are further explained by Ayyub & Haldar (1984) and White & Ayyub (1985).

5. EXAMPLES

A typical structural performance curve of a ship hull, in the form of a moment-curvature relationship, due to hogging and sagging moments is shown in Figure 6. Two examples considering the hull structure subjected to hogging moment only are discussed. The first example involves the crisp and three different fuzzy failure definitions. The fuzzy failure definitions are interpreted as failure levels. In the second example, two cases of failure definitions which are interpreted as measures of degree of belief are discussed.

5.1 Example I

Considering the resisting-moment versus curvature relationship of the hull structure of a ship subjected to a hogging moment only. The performance function for this case is expressed as

$$Z = M - L = M(\phi) - L \tag{18}$$

where M = resisting moment, L = external loading in the units of ft-tons, and ϕ = curvature. In this example, both the crisp and fuzzy failure definitions were examined using the same external loading. The statistical characteristics of the external loading are given in Table 1.

The transition from survival to failure in the crisp case was assumed to be attained at curvature level of $\phi_f = 0.3\text{x}10^{-5}$. The statistical characteristics of the corresponding resisting moment M_f are also shown in Table 1. The probability of failure for the flexural member can be expressed as

$$P_f = \text{Prob} (Z < 0) = \text{Prob} (M_f - L < 0) \tag{19}$$

Since M_f and L were assumed to follow normal distribution and extreme value-Type I probability distribution, respectively, a close-form solution to this structural reliability problem is not available. Therefore, Monte Carlo simulation with conditional expectation variance reduction technique was used to estimate the probabilities of failure. The results for the crisp failure definition based on 1000 simulation cycles are shown in Figure 7 using a logarithmic scale for the probability of failure.

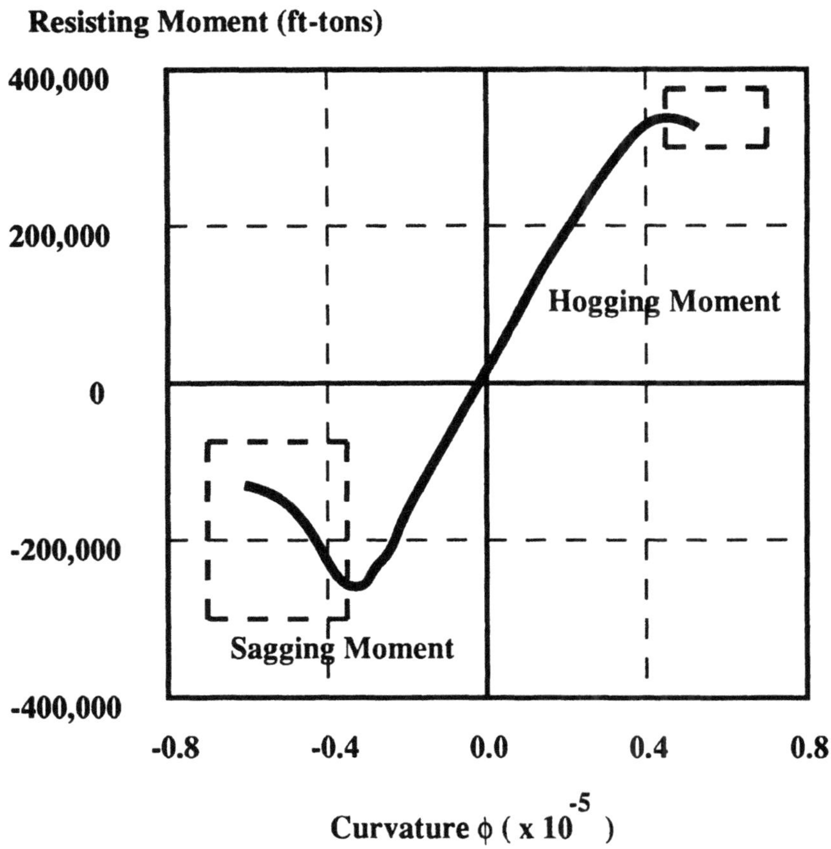

Figure 6. Structural performance curve for hogging and sagging moments

Table 1. Statistical characteristics of loading and strength for the example

Random Variable	Mean Value	Coefficient of Variation (COV)	Probability Distribution Type
L	100×10^3 ft-tons	0.30	Extreme Value Type I
M_f	244×10^3 ft-tons	0.10	Normal

Three failure definitions were used as shown in Figure 8. These failure definitions were carefully selected to study the convergence of the resulting probability of failure. Figure 7 shows the resulting probabilities of failure P_f, which are based on 1000 simulation cycles, for the three failure definitions using logarithmic scale. From this figure, it is evident that the curve of P_f versus α based on a fuzzy failure definition converges to the one for the crisp failure definition as the failure definition approaches the crisp case.

For the purpose of studying the convergence and sensitivity analysis of results to the number of simulation cycles, Monte Carlo simulation with conditional expectation variance reduction technique was performed for 100, 200, 500, and 1000 simulation cycles for failure definition 2. Direct simulation method was also applied for comparison using different simulation cycles. The expected values and the corresponding coefficients of variation (COV) of the probabilities of failure for the two methods are shown in Figures 9 and 10 for failure levels of 0.2, 0.4, 0.6, 0.8, and 1.0. According to the results shown in Figures 9 and 10, it is significant to note that the simulation method with conditional expectation variance reduction technique converges to the true value of the probability of failure in a relatively small number of simulation cycles. It was also concluded that 1000 simulation cycles for the simulation method with conditional expectation variance reduction technique are sufficient for the purpose of obtaining acceptable estimates of the probabilities of failure.

The average probabilities of failure for the three failure definitions were calculated using Equation 10. The results are shown in Table 2. According to this table, the arithmetic averages of P_f based on fuzzy failure are larger than the P_f based on the crisp failure. But the geometric averages of P_f are not necessarily larger than the P_f for the crisp case. This is because the probability of failure versus failure level is plotted using a logarithmic scale. It was also observed that the arithmetic average of P_f based on a fuzzy failure increased as the vagueness, i.e., uncertainty, in the failure definition was increased. However, this characteristic does not apply to the geometric average of P_f. The geometric average of P_f fluctuates above and below the crisp-case P_f with the increased degree of vagueness.

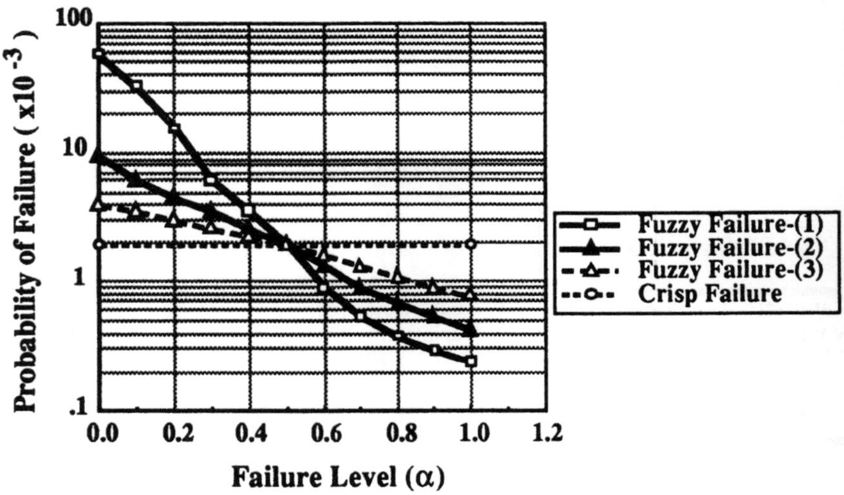

Figure 7. Probabilities of failure for Example I

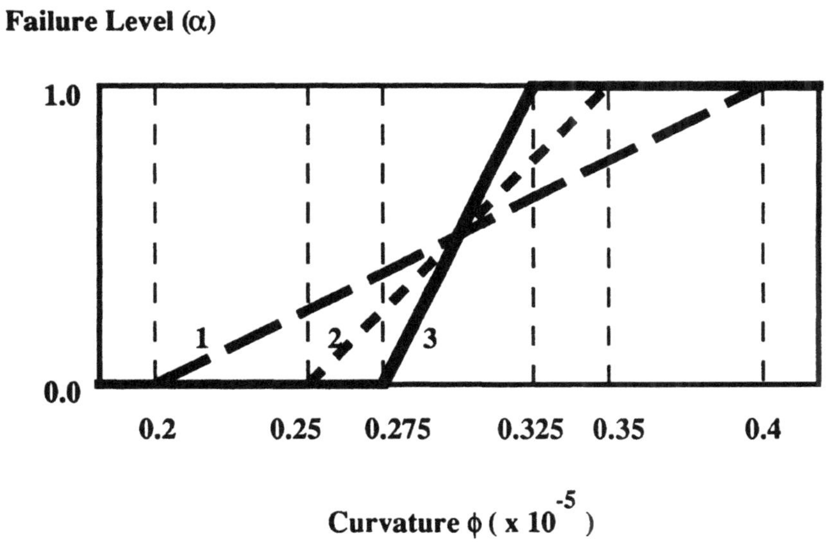

Figure 8. Three fuzzy failure definitions for Example I

Table 2. Average probabilities for different failure models in Example I

Failure Model	Curvature at Failure, ϕ_f	Arithmetic Average of Probability of Failure	Geometric Average of Probability of Failure
Crisp	$0.3\text{x}10^{-5}$	$1.973\text{x}10^{-3}$	$1.973\text{x}10^{-3}$
Fuzzy 1	$0.2\text{x}10^{-5}$ to $0.4\text{x}10^{-5}$	$9.137\text{x}10^{-3}$	$2.320\text{x}10^{-3}$
Fuzzy 2	$0.25\text{x}10^{-5}$ to $0.35\text{x}10^{-5}$	$2.752\text{x}10^{-3}$	$1.851\text{x}10^{-3}$
Fuzzy 3	$0.275\text{x}10^{-5}$ to $0.325\text{x}10^{-5}$	$2.086\text{x}10^{-3}$	$1.854\text{x}10^{-3}$

5.2 Example II

Consider the same performance curve of the shiphull structure as described in Example I. Monte Carlo simulation method with conditional expectation variance reduction technique was used to estimate the probability of occurrence of the performance conditions.

Example II - Case A. Six events of structural performance were assumed in this case as shown in Figure 11. These six events were selected for the purpose of illustrating a damage spectrum. Since the concept of the degree of belief for an event is used here rather than failure level, the resulting probabilities can reasonably be interpreted as the probabilities of occurrence of the events. The estimated values of the probabilities of occurrence at the associated degrees of belief were calculated using 1000 simulation cycles. The resulting probabilities of occurrence were plotted on a logarithmic scale versus the degree of belief on an arithmetic scale in Figure 12. It can be clearly seen from Figures 11 and 12 that these curves are similar to the curves that were used to define the events.

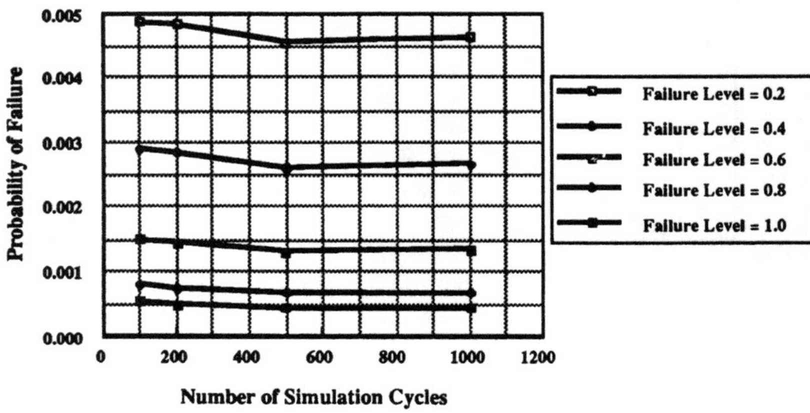

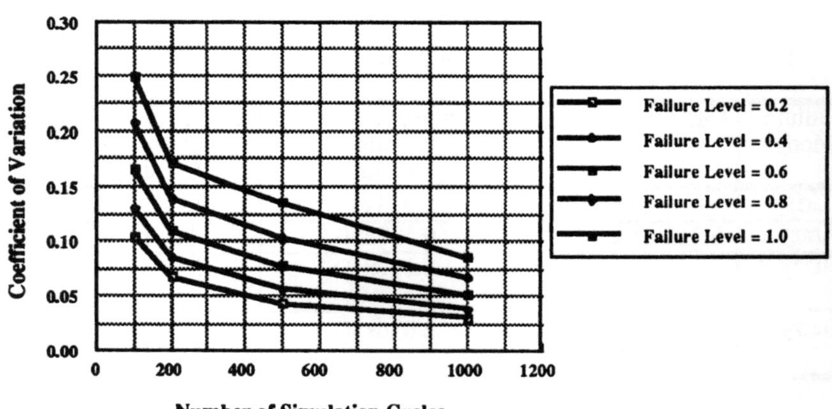

Figure 9. Probabilities of failure and the corresponding coefficients of variation for the simulation method with conditional expectation VRT in Example I

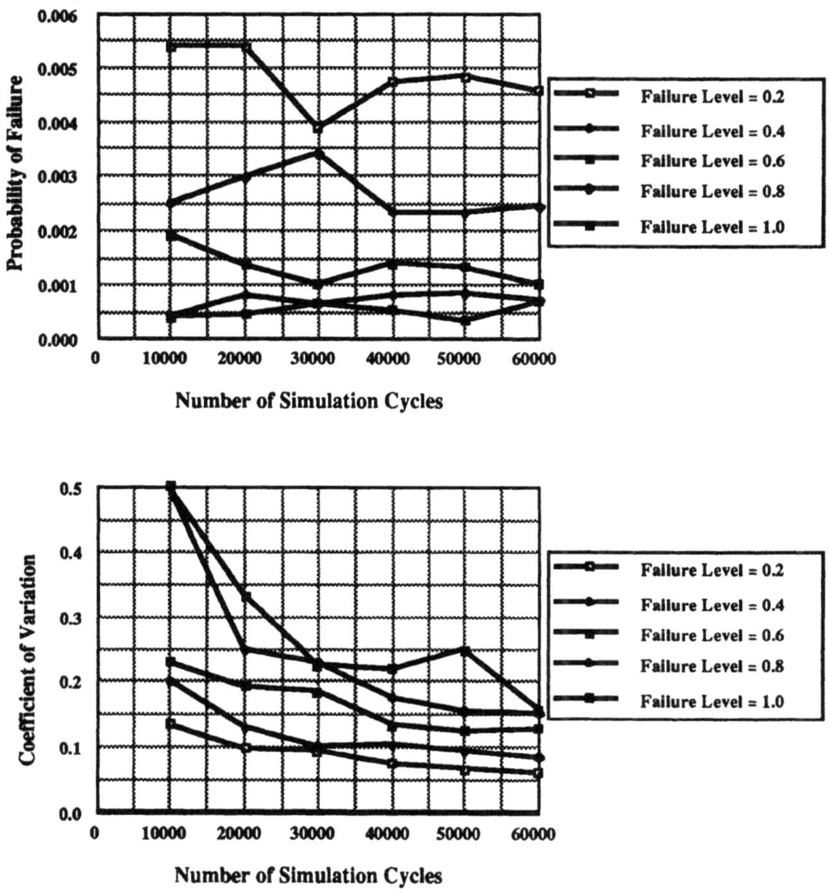

Figure 10. Probabilities of failure and the corresponding coefficients of variation for direct simulation method in Example I

The average probabilities of occurrence for the six events were calculated using Equations 9 and 10 resulting in the values shown in Table 3. One might expect the probability of occurrence to decrease as its underlying event is varied from 1 to 6. However, the probability of occurrence increases as the definition of failure is changed from event 5 to event 6. This discrepancy can be attributed to the definition of failure as given by Figure 11. The failure definitions in this case were not consistent with the intuitive interpretation of these linguistic descriptions of failure levels. Intuitively, engineers tend to interpret a certain failure level as "at least" the specified failure level, rather than a failure level that is constrained within upper and lower limits. Therefore, the failure definitions were revised according to this intuitive interpretation as discussed in Case B of this example.

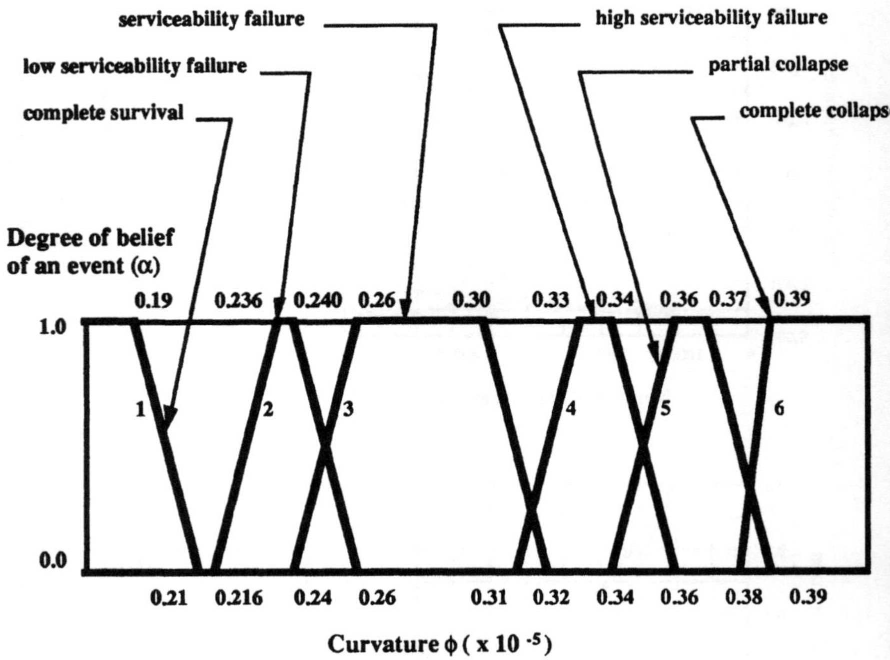

Figure 11. Six events of structural response for Example II - Case A

Example II - Case B. The definitions of the six events that were used in Case A were modified as shown in Figure 13. They are interpreted in the context of "at least," e.g., the event *serviceability failure* is commonly interpreted as *at least serviceability failure*. In this case, the functional relationship between the structural response and degree of belief α is a one-to-one mapping. The estimated values of the probabilities of occurrence were calculated using 1000 simulation cycles. Figure 14 shows the resulting probabilities of occurrence on a logarithmic scale versus the degree of belief on an arithmetic scale.

The average probabilities of occurrence for the six events were then calculated using Equations 9 and 10. The results are shown in Table 4. In this case, the results are in agreement of our intuitive expectation, i.e., the average probability of occurrence decreases as its underlying event is varied from 1 to 6.

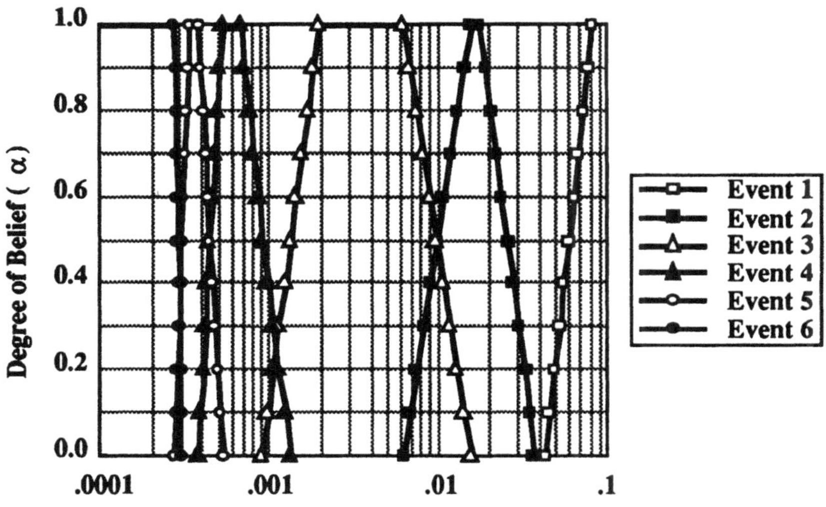

Figure 12. Probability of occurrence versus belief degree for Example II - Case A

Table 3. Average probabilities of occurrence for the six events of Example II - Case A

Event No.	Definition	Arithmetic Average of Probability of Occurrence	Geometric Average of Probability of Occurrence
1	complete survival	0.940	0.940
2	low serviceability failure	1.611×10^{-2}	1.583×10^{-2}
3	serviceability failure	8.718×10^{-3}	8.396×10^{-3}
4	high serviceability failure	4.944×10^{-4}	4.782×10^{-4}
5	partial collapse	1.439×10^{-4}	1.424×10^{-4}
6	complete collapse	2.847×10^{-4}	2.846×10^{-4}

6. CONCLUSIONS

A mathematical model for structural reliability assessment that includes both ambiguity and vagueness types of uncertainty was developed. Based on the developed methods and the results of the examples, the following main conclusions can be drawn:

1. The developed method for structural reliability assessment results in the probabilities of failure over a damage spectrum and satisfied the conditions of stability, convergence, consistency and limiting behavior.
2. The probability of failure corresponding to a failure level decreased as the

failure level was increased using the same random loading conditions. This is in agreement of the intuitive understanding of structural behavior, since larger loads that have smaller probabilities of occurrence are needed to cause the increased failure level.

3. The curve of probability of failure versus failure level based on a fuzzy failure definition converged to the one for the crisp failure definition as the failure definition approached the crisp case.
4. The resulting arithmetic average probability of failure was larger than the probability of failure based on the crisp failure definition. This increase was attributed to the vagueness uncertainty in the failure definition.
5. The examples illustrated that reliability levels of structures were highly dependent on failure definitions. Therefore, care should be exercised in establishing these definitions in order to obtain realistic assessments of structural reliability.

The proposed approach is suitable for the non-decreasing portions of this performance curve. For relatively large curvatures due to either hogging or sagging moments, as shown in the dashed boxes in Figure 6 where the resisting moments start to decrease, the moment-curvature relationship cannot be used as the basis for the performance functions. In such cases, the strain energy concept should be used. Since an energy-curvature relationship is non-decreasing, it is suitable for performing this type of assessment instead of the moment-curvature relationship. However, the stochastic distribution of the external load should also be expressed in an energy form. This type of analysis is not performed herein.

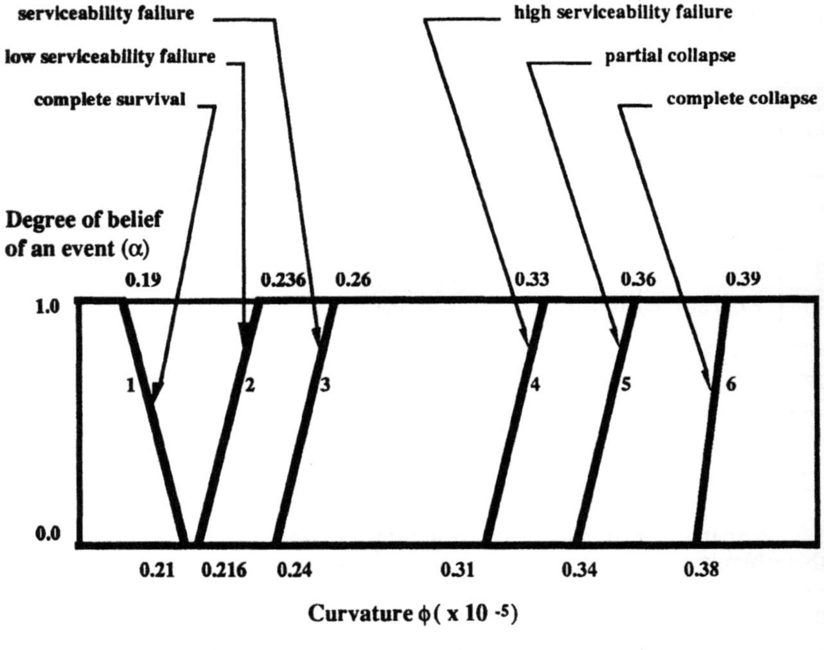

Figure 13. Six modified events for Example II - Case B

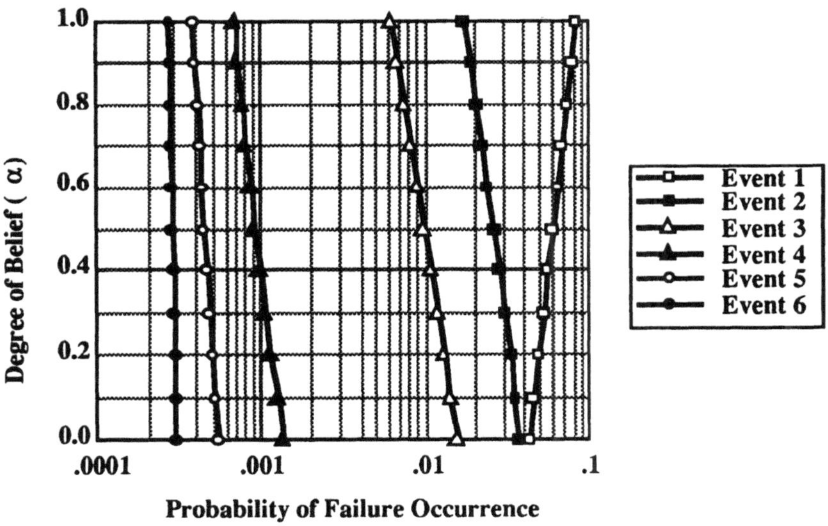

Figure 14. Probability of occurrence versus belief degree for Example II - Case B

Table 4. Average probabilities of occurrence for the six events of Example II - Case B

Event No.	Definition	Arithmetic Average of Probability of Occurrence	Geometric Average of Probability of Occurrence
1	complete survival	0.940	0.940
2	low serviceability failure	2.619×10^{-2}	2.555×10^{-2}
3	serviceability failure	1.008×10^{-2}	9.722×10^{-3}
4	high serviceability failure	9.388×10^{-4}	9.206×10^{-4}
5	partial collapse	4.444×10^{-4}	4.424×10^{-4}
6	complete collapse	2.847×10^{-4}	2.846×10^{-4}

References

Alvi I.A.and Ayyub B.M. (1990) "Analysis of System Performance Under Stochastic and Fuzzy Uncertainty," North American Fuzzy Information Processing Society, *NAFIPS 1990*, University of Toronto.

Ayyub B.M., Gupta M.M. and Kanal L.N. (1992) *Analysis and Management of Uncertainty: Theory and Applications*, Elsevier Science Publishing Company, Inc., New York.

Ayyub B.M. and Haldar A. (1984) "Practical Structural Reliability Techniques," *Journal of Structural Engineering*, ASCE, Vol. 110, No. 8, August 1984, pp. 1707-1724.

Ayyub B.M. and Lai K-L. (1989) "Structural Reliability Assessment Using Latin

Hypercube Sampling," *the proceedings of the Fifth International Conference on Structural Safety and Reliability*, ICOSSAR, Volume 2, published by ASCE, Edited by A. H-S. Ang, M. Shinozuka and G.I. Schueller, San Francisco, pp. 1177-1184.

Bourgund U. and Bucher C.G. (1986) *Importance Sampling Procedures Using Design Points (ISUDP) - A User Manual*, Report No. 8-86, Institute of Engineering Mechanics, University of Innsbruck, Austria.

Haldar A. and Ayyub B.M. (1984) "Practical Variance Reduction Techniques in Simulation," *Publication PVP-Vol. 93*, ASME, June 1984, pp. 63-74.

Harbitz A. (1986) "An Efficient Sampling Method for Probability of Failure Calculation," *Structural Safety*, Vol. 3, No. 2, pp. 100-115.

Law A.M. and Kelton W.D. (1982) *Simulation Modeling and Analysis*, McGraw Hill Book Co., Now York.

Melchers R.E. (1987) *Structural Reliability Analysis and Prediction*, Ellis Horwood Limited, UK.

Rubinstein R.Y. (1981) *Simulation and Monte Carlo Method*, John Wiley and Sons, New York.

Schueller G.I. and Stix R. (1986) "A Critical Appraisal of Methods to Determine Failure Probabilities," *Bericht/ Report 4-86*, Institute of Engineering Mechanics and Institute of Mathematics, University of Innsbruck, Austria.

White G.J. and Ayyub B.M. (1985) "Reliability Methods for Ship Structures," *Naval Engineers Journal*, ASNE, Vol. 97, No. 4, May 1985, pp. 86-96.

GPSR Compliance
The European Union's (EU) General Product Safety Regulation (GPSR) is a set of rules that requires consumer products to be safe and our obligations to ensure this.

If you have any concerns about our products, you can contact us on

ProductSafety@springernature.com

In case Publisher is established outside the EU, the EU authorized representative is:

Springer Nature Customer Service Center GmbH
Europaplatz 3
69115 Heidelberg, Germany

www.ingramcontent.com/pod-product-compliance
Ingram Content Group UK Ltd.
Pitfield, Milton Keynes, MK11 3LW, UK
UKHW020132290926
15471UKWH00003B/441